SILKWORM TO SILK

SILKWORM TO SILK
TRADITION, INNOVATION AND FUTURE APPLICATIONS

Henry Mitchell
Lillian Cross

Kruger Brentt
Publishers
2026

Kruger Brentt Publishers UK. LTD.
Company Number 9728962

Regd. Office: 68 St Margarets Road, Edgware, Middlesex HA8 9UU

ISBN: 978-1-78715-647-0

For information on all our publications visit our website at http://krugerbrentt.com/

PREFACE

Sericulture, the art and science of silk production, has a rich heritage spanning thousands of years. From the domestication of the silkworm to the creation of exquisite silk textiles, this industry embodies a blend of tradition, craftsmanship, and scientific innovation. In the contemporary era, sericulture continues to evolve, integrating modern biotechnological techniques, sustainable practices, and advanced processing methods to meet global demands.

Silkworm to Silk: Tradition, Innovation, and Future Applications provides a comprehensive exploration of the entire silk production process, from silkworm biology and breeding to cocoon harvesting, silk reeling, weaving, and value-added product development. The book also highlights recent innovations in silk biotechnology, genetic improvement of silkworms, and applications of silk in medical, industrial, and technological fields.

The text is intended for students, researchers, practitioners, and entrepreneurs interested in sericulture, textile engineering, and biotechnology. By combining traditional knowledge with modern scientific advances, the book offers a holistic understanding of silk production and its diverse applications, emphasizing sustainability, innovation, and future prospects.

Henry Mitchell

Lillian Cross

CONTENTS

1 Chapter

SILKWORM AND SILK: TRADITIONAL AND INNOVATIVE APPLICATIONS

1. INTRODUCTION: AN OVERVIEW

The diverse topics addressed in the current Special Issue "Silkworm and Silk: Traditional and Innovative Applications" illustrate how sericulture, a practice deeply embedded in human history, can serve as a conduit for uniting a remarkably broad spectrum of scientific and technical knowledge in both traditional subjects and advanced technologies. Similar to numerous other species, Bombyx mori was derived from a wild progenitor by anthropogenic selection; but, unlike other animals, the genetic lineages of silkworm strains have been conserved in public research institutions for decades. Consequently, in countries with a tradition of sericulture, extensive collections of diverse genotypes must be preserved through established practices. Moreover, most sericultural locations globally persist in the agricultural production of cocoons by traditional silkworm rearing methods. As the mulberry leaf serves as the sole sustenance for the larvae of this insect (or an artificial diet derived from it), there has consistently been a demand for moriculture. Consequently, organizations focused on preserving B. mori genetic resources also maintain collections of mulberry germplasm. A considerable amount of research is dedicated to elucidating the characteristics of these varieties and their relationship to the silkworm. This issue reports a multi-level study on the impact of regional mulberry genotypes on silkworm health; the authors illustrated that selecting mulberry varieties with specific qualitative and quantitative traits can significantly influence larval development, cocoon production, and raw silk characteristics [1]. Researchers have also discovered methods to utilize mulberry kinds for feeding other animals or as a source of raw materials and active compounds; additionally, they have implemented new technology to get sophisticated materials from silk. This document features a bibliometric study by Giora et al. [2], which examines contemporary research trends in sericulture and moriculture, offering a comprehensive review of the last two decades of research in this increasingly significant field. Two more evaluations [3,4] sought to link advancements in or critiques of sericulture to cultural and economic transformations across diverse global areas, emphasizing that this practice may be significantly influenced not just by technological progress but also by several external influences.

One of the newest trends in sericulture is to promote the development of circular economy supply chains to employ silk by-products for feed and food, a topic that is com- prehensively discussed and documented in the article by Tassoni et al. [5]. Nowadays, silk by-products are easily exploitable via melting silk with the right solvents to obtain solutions of silk proteins, opening the door to entirely new applications in industry including in the textile sector, which is still of the utmost importance for an economy dependent on traditional sericultural agroindustry. Furthermore, this issue contains numerous articles that address these novel applications and cover both the general role of silk fibroin in biomedicine [6] and methods for creating hybrid materials, for instance, by adding nanoparticles [7]. The paper by Bassani et al. [8] examined particular devices, such as silk nerve conduits, in a study case illustrating how research findings can be translated into medical products for use in clinical settings.

Advances in methodology and knowledge are not limited to the new biotechnological uses of silk proteins. A significant part of the Special Issue is devoted to papers using the most recent techniques to study silkworm physiology to enhance silk production. This can improve the general knowledge of silkworm biology and hasten the potential use of this organism in various applied fields. Specifically, Yokoi et al. [9] undertook an RNA-seq investigation of several silkworm larvae tissues, including the midgut, fat body, testis, ovary, and the different subregions of the silk gland, generating reference transcriptome data for a *B. mori* strain. Masuoka et al. [10] identified novel silk protein regulation factors, using a proprietary software, to conduct co-expression networks and time-course expression analyses on transcriptomic data from various tissues, including different subregions of the silk glands, in reference silkworm strains. Zhang et al. [11] performed a transcriptomic analysis of fat body derived from silkworms treated with hydrogen sulfide, a molecule that, at specific doses, exerts beneficial effects on the developmental and economical traits of *B. mori*. All these results provide data and methods with which to develop strategies for enhancing silk production in *B. mori*. Moreover, de la Peña et al. [12] compared RNA-seq transcriptional profiles of silk glands belonging to various silkworm strains commonly reared in the Philippines and maintained under different temperature conditions, produc- ing a set of information useful for the genetic improvement of *B. mori* strains and to increase their silk productivity in this geographical area. Finally, the article by Baci et al. [13] reports a comprehensive overview of the most advanced genome-editing techniques performed in *B. mori*, using the CRISPR/Cas9 methodology. In their paper, the authors illustrated strategies used to obtain both knock-out and knock-in *B. mori* strains. They also focused on the possible applications of CRISPR/Cas9 to generate silkworm strains characterized by a higher resistance to viral infections, highlighting the importance of this technology for both basic research and applied purposes in *B. mori*. Since *B. mori* is considered a model species among Lepidoptera, it has been extensively employed in studies of physiology, genetics, molecular biology, and pathology. Zhang et al. [14] made use of this lepidopteron to investigate how hemolymph ecdysteroid titer affects maternal genes during oogenesis, demonstrating that maternal mRNAs begin accumulating in the ovary during the larval period, at the wandering stage. In the article by Chang et al. [15], the role of polyamines in regulating cell cycle progression and DNA replication was explored as well as the expression of genes coding for enzymes involved in the polyamine pathway in different tissues. Since it is known that the administration of polyamines to *B. mori* larvae increases the ex- pression of the gene coding for fibroin heavy chain and improves the quantity and quality of silk, this study opens the door to manipulating the expression of key genes involved in the polyamine pathway in silk glands to improve silk production. The use of silkworms and insects in general as model organisms began in response to growing concerns about animal health and welfare, with the goal of preventing, or at least reducing, the usage of higher animals in all sectors of the life sciences. In fact, the principles of the 3Rs (Replacement, Reduction, and Refinement), developed in the 1950s, offer a foundation for conducting more compassionate animal research. Since then, they have become an integral part of national and international laws and regulations governing the use of animals in research and are progressively leading to the development of suitable alternative models, among invertebrates, for experimental aims. Holometabolous insects represent a promising option due to their lower costs, convenience in handling, lack of ethical restrictions, and features that can successfully reproduce different biological mechanisms occurring in mammals. In this Special Issue, three articles pave the way for the development of the silkworm as an alternative model for screening antimicrobial compounds against nosocomial pathogens, such as *Staphylococcus epidermidis* [16], and bacteria that cause skin diseases and systemic infection, such as *Cutibacterium acnes* [17], as well as for studying the effects of natural compounds with hypoglycemic activity [18]. Although it is still up for debate whether these insect-based models are applicable in clinical settings, the findings of these three studies are undoubtedly encouraging. Another article focused on the health status of the silkworm and its pathologies. In particular, the study dealt with nuclear polyhedrosis, a major viral disease in sericulture that causes high larval lethality and great economic losses. By using a genomic approach that considered different BmNPV strains, the authors demonstrated that mutation and rearrangement of the genome can lead to differential pathogenicity of this virus [19]. The holobiont concept, which has recently emerged as a theoretical and experimental context in which to study the interactions between hosts and their associated microbial communities, is worth mentioning while discussing the silkworm as a laboratory model. It is becoming increasingly clear that the development, growth, and health of macro-organisms are influenced by the complex microbial communities they host. The article by Dee Tan and Bautista [20] deals with this topic; the authors studied the gut microbial composition of four *B. mori* strains reared in the Philippines to construct a database of useful information fostering an improved understanding of whether and how the gut microbiota affects silk production and to develop strategies for future improvement of the strains. Another topic that has attracted increasing attention in recent years is the administration of probiotics to provide health benefits, generally by improving or restoring the gut microbial community. Unban et al. [21] isolated from the midgut of *Samia ricini*, an- other lepidopteran species used for silk production, a xylose-utilizing lactic acid bacterium, *Enterococcus hirae* SX2, which acts as a probiotic in this insect due to its tannin tolerance and antimicrobial activity against insect pathogens. The findings in this article indicate that the oral administration of E. hirae SX2 to Eri silkworm enhances growth and decreases mortality, thereby favorably impacting economic traits. Additionally, it serves as a foundation for devising strategies to identify new probiotics that can beneficially influence B. mori larvae and advance sericulture research..

REFERENCES

1. AUrbanek Krajnc, A.; Bakonyi, T.; Ando, I.; Kurucz, E.; Solymosi, N.; Pongrac, P.; Bercˇicˇ, R.L. The Effect of Feeding with Central European Local Mulberry Genotypes on the Development and Health Status of Silkworms and Quality Parameters of Raw Silk. *Insects* 2022, 13, 836. [CrossRef] [PubMed]
2. Giora, D.; Marchetti, G.; Cappellozza, S.; Assirelli, A.; Saviane, A.; Sartori, L.; Marinello, F. Bibliometric Analysis of Trends in Mulberry and Silkworm Research on the Production of Silk and Its By-Products. *Insects* **2022**, *13*, 568. [CrossRef]
3. Tzenov, P.; Cappellozza, S.; Saviane, A. Black, Caspian Seas and Central Asia Silk Association (BACSA) for the Future of Sericulture in Europe and Central Asia. *Insects* **2022**, *13*, 44. [CrossRef]
4. Andadari, L.; Yuniati, D.; Supriyanto, B.; Murniati, M.; Suharti, S.; Widarti, A.; Steven, E.; Sadapotto, A.; Winarno, B.; Minarningsih, M.; et al. Lens on Tropical Sericulture Development in Indonesia: Recent Status and Future Directions for Industry and Social Forestry. *Insects* **2022**, *13*, 913. [CrossRef]
5. Tassoni, L.; Cappellozza, S.; Dalle Zotte, A.; Belluco, S.; Antonelli, P.; Marzoli, F.; Saviane, A. Nutritional Composition of *Bombyx mori* Pupae: A Systematic Review. *Insects* **2022**, *13*, 644. [CrossRef]
6. Lujerdean, C.; Baci, G.M.; Cucu, A.A.; Dezmirean, D.S. The Contribution of Silk Fibroin in Biomedical Engineering. *Insects* **2022**, *13*, 286. [CrossRef]
7. Singh, M.; Dey, E.S.; Bhand, S.; Dicko, C. Supercritical Carbon Dioxide Im-

pregnation of Gold Nanoparticles Demonstrates a New Route for the Fabrication of Hybrid Silk Materials. *Insects* **2022**, *13*, 18. [CrossRef]

8. Bassani, G.A.; Vincoli, V.; Biagiotti, M.; Valsecchi, E.; Zucca, M.V.; Clavelli, C.; Alessandrino, A.; Freddi, G. A Route to Translate a Silk-Based Medical Device from Lab to Clinic: The Silk Biomaterials Srl Experience. *Insects* **2022**, *13*, 212. [CrossRef]
9. Yokoi, K.; Tsubota, T.; Jouraku, A.; Sezutsu, H.; Bono, H. Reference Transcriptome Data in Silkworm *Bombyx mori*. *Insects* **2021**, *12*, 519. [CrossRef]
10. Masuoka, Y.; Cao, W.; Jouraku, A.; Sakai, H.; Sezutsu, H.; Yokoi, K. Co-Expression Network and Time-Course Expression Analyses to Identify Silk Protein Regulatory Factors in *Bombyx mori*. *Insects* **2022**, *13*, 131. [CrossRef]
11. Zhang, R.; Cao, Y.Y.; Du, J.; Thakur, K.; Tang, S.M.; Hu, F.; Wei, Z.J. Transcriptome Analysis Reveals the Gene Expression Changes in the Silkworm (*Bombyx mori*) in Response to Hydrogen Sulfide Exposure. *Insects* **2021**, *12*, 1110. [CrossRef] [PubMed]
12. de la Peña, P.N.O.; Lao, A.G.D.; Bautista, M.A.M. Global Profiling of Genes Expressed in the Silk Glands of Philippine-Reared Mulberry Silkworms (*Bombyx mori*). *Insects* **2022**, *13*, 669. [CrossRef] [PubMed]
13. Baci, G.M.; Cucu, A.A.; Giurgiu, A.I.; Muscă, A.S.; Bagameri, L.; Moise, A.R.; Bobis, O.; Rat iu, A.C.; Dezmirean, D.S. Advances in Editing Silkworms (*Bombyx mori*) Genome by Using the Crispr-Cas System. *Insects* **2022**, *13*, 28. [CrossRef] [PubMed]
14. Zhang, M.; Xu, P.; Chen, T. Hemolymph Ecdysteroid Titer Affects Maternal Mrnas during *Bombyx mori* Oogenesis. *Insects* **2021**, *12*, 969. [CrossRef]
15. Chang, L.; Li, Z.; Guo, H.; Zhang, W.; Lan, W.; Wang, J.; Shen, G.; Xia, Q.; Zhao, P. Function of Polyamines in Regulating Cell Cycle Progression of Cultured Silkworm Cells. *Insects* **2021**, *12*, 624. [CrossRef]
16. Montali, A.; Berini, F.; Saviane, A.; Cappellozza, S.; Marinelli, F.; Tettamanti, G. A *Bombyx mori* Infection Model for Screening Antibiotics against *Staphylococcus epidermidis*. *Insects* **2022**, *13*, 748. [CrossRef]
17. Matsumoto, Y.; Tateyama, Y.; Sugita, T. Evaluation of Antibacterial Drugs Using Silkworms Infected by *Cutibacterium acnes*. *Insects* **2021**, *12*, 619. [CrossRef]
18. Aznar-Cervantes, S.D.; Santesteban, B.M.; Cenis, J.L. Products of Sericulture and Their Hypoglycemic Action Evaluated by Using the Silkworm, *Bombyx mori* (Lepidoptera: Bombycidae), as a Model. *Insects* **2021**, *12*, 1059. [CrossRef]
19. Guo, H.; Zhang, B.; Zheng, X.; Sun, J.; Guo, H.; Li, G.; Zhao, G.; Xu, A.; Qian, H. Pathogenicity Detection and Genome Analysis of Two Different Geographic Strains of Bmnpv. *Insects* **2021**, *12*, 890. [CrossRef]

20. Dee Tan, I.Y.; Bautista, M.A.M. Bacterial Survey in the Guts of Domestic Silkworms, *Bombyx mori* L. *Insects* **2022**, *13*, 100. [CrossRef]
21. Unban, K.; Klongklaew, A.; Kodchasee, P.; Pamueangmun, P.; Shetty, K.; Khanongnuch, C. Enterococci as Dominant Xylose Utilizing Lactic Acid Bacteria in Eri Silkworm Midgut and the Potential Use of *Enterococcus hirae* as Probiotic for Eri Culture. Insects 2022, 13, 136. [CrossRef] [PubMed]

2 Chapter

EXAMINATION OF TROPICAL SERICULTURE ADVANCEMENT IN INDONESIA: CURRENT STATUS AND PROSPECTIVE DIRECTIONS FOR INDUSTRY AND SOCIAL FORESTRY

1. INTRODUCTION: AN OVERVIEW

TThe COVID-19 pandemic has exerted pressure on the global agenda, including the Sustainable Development Goals (SDGs) for 2030. This crisis has resulted in widespread unemployment and severe health issues, leading to poverty for millions of individuals. The pandemic's impact has been particularly severe in developing nations, such as Indonesia, especially in regions with a strong reliance on forest resources due to the scarcity of both agricultural and non-agricultural employment opportunities. The situation has deteriorated due to diminished availability to inputs, labor, and agricultural land, leading to a decline in productivity, household income, and nutrition. [7].

Many attempts have been made by the government to prevent the most severe nega- tive impacts and to boost the local economy by developing several alternative forest-based agribusinesses [4,8]. In Indonesia, an economic recovery program, including social forestry, has been implemented, with the double objective of enhancing consumption and economic productivity [9]. Forests, as natural resource systems, have the potential to provide multiple benefits. In addition to wood products, forests can provide non-timber forest outputs and environmental benefits. NTPs are biological products other than wood of high value, gen- erally obtained from wild biodiversity in natural or human-modified environments [10,11]. Many studies have revealed that non-timber forest products, including ecotourism, small- scale timber enterprises, and environmental services, could play a significant role in social and economic recovery during and after a pandemic [8,12,13]. Non-timber forest products contribute 80% of the social forestry business model in Indonesia [14].

One of the non-timber forest products with the potential to resolve economic challenges during and after a crisis is the natural silk produced by sericulture agribusiness. Sericulture is a prospective and potential activity that can regularly generate higher in- come [15,16]. Apart from its ability to provide gainful employment and economic improvement for people in rural areas due to the high selling price of its products [15,16], sericulture also plays an important role in preventing the migration of rural people to urban areas in search of employment [17,18]. The silk industry is labor-intensive [19] and so provides employment for 7.9 million people in India and 20,000 weaver families in Thailand [20]. In the context of Indonesia, there are 1200 silk farmers [21] and 4900 weavers [22] involved in silk agro-industry activities. Sericulture agribusiness has been determined to be one of the five NTFP priorities, having the potential to contribute to the economy of the country and tackle poverty [23–25].

Sericulture was first introduced to Indonesia through trade in the 10th century; since 1953 [26], it has developed rapidly as it is suitable to the agroclimatic conditions as well as the local culture. Silk yarn production reached its peak in 1971 with a production of 140 tons but then tended to decrease due to several factors [27], such as the low quality of silkworm seeds, pebrine disease attacks, and mulberry plants as feed have begun to be shifted to other horticultural commodities, low prices for silk products, and a lack of support from silkworm production programs and organization.

The silk agro-industry chain is a sequence of activities starting with mulberry cultiva- tion (moriculture) and silkworm rearing. These stages are in the upstream segment and are commonly known as sericulture. The next phase is reeling raw silk from the cocoons and producing the yarn and its derivative products. These products then deliver to the weaving industry for further processing and marketing. Yarn processing, referred to as the manufacturing sector, is the downstream sector of the agro silk industry [21,24].

Several factors support sericulture and make it become one of the NTFP priorities. Economically, the silk industry is an important way of fulfilling domestic as well as export needs, either in the form of cocoons, yarns, or finished goods [19,28]. It is a potential commodity that can significantly contribute to foreign exchange earnings, raise living standards, and reduce poverty, hunger, gender discrimination, and disease risk. Sericulture is mostly developed in rural and suburban areas and hence provides job opportunities for rural communities [29]. In addition, the sericulture and silk industry in Indonesia represents an economically viable rural enterprise for inclusive development, which can effectively have fall-outs in various industrial sectors [20].

Ecologically, the silk industry involves an environmentally friendly production process. The mulberry tree ensures green land covering, soil conservation, and erosion protection and allows for the use of land that is not suitable for other crop cultivation. The silkworm rearing does not cause pollution, CO_2 emissions are very low, and waste can be easily degraded [30], so it can coexist well with already inhabited areas [31,32]. Silk produced from sericulture is a non-synthetic, renewable, and biodegradable fiber [15]. Furthermore, sericulture also contributes to the improvement of the microclimate of the area of devel- opment, which will eventually improve the surrounding environmental conditions [25]. In addition, mulberry cultivation can only be developed in tropical countries, including Indonesia, which has agroclimatic conditions that are suitable for mulberry and silkworm cultivation [33]. Because it has moderate dry and rainy seasons, this country allows mul- berry plants to serve as the main food for silkworms to grow and be cultivated throughout the year [34–36].

In terms of market prospects, there is still a gap between the annual world demand for natural silk (92,743 tons per year) and the supply of yarn (83,393 tons) [24,25]. Although it is only ranked twelfth among silk yarn-producing countries in the world, Indonesia has a comparative advantage as it produces better quality yarn than other counties [37] and so has the potential to fill the gap. However, filling the market gap is not easy since there are high-quality standards for the international market. At a domestic level, the demand for silk has not yet been met, and hence must be supplied from imported products. The annual need for silk yarn is 500–800 tons, while the national production in 2015 was only about 12.13 tons, 8.95 tons of which were from South Sulawesi [21]. In the current pandemic situation, the need for cocoons as a raw material for domestically produced silk, which was previously imported from other silk-producing countries, such as China, cannot be fulfilled. The above-mentioned characteristics of the silk industry indicate that sericulture has several social, economic, and ecological advantages [25].

However, until now, the competitive advantage of sericulture has not encouraged the establishment of a strong and independent national sericulture industry in Indonesia. During the period 2011–2015, Indonesia continued to experience a decline in the production of both cocoons and silk yarn, correlated with a decline in the natural silk fabric industry in various regions in Indonesia. Some factors causing the decline of the silk agro-industry in Indonesia include: the limited availability of quality silkworm eggs and feed (mulberry leaves), limited skilled labor, the uncertainty of the product selling price [19,38,39], and pest and disease attacks [21]. Apart from those challenges, Sadapotto [27] stated that ineffective policy formulation for sericulture is actually the main factor behind the failure of sericulture in Indonesia due to neglecting institutional factors and focusing more on physical and technical factors in the analysis of problems regarding sericulture. Sericulture industry development involves lots of stakeholders from upstream to downstream. At the upstream level, the development of silk is the responsibility of the Ministry of Environment and Forestry [40]. At the downstream level, activities are under the authority of several relevant stakeholders, including the Ministry of Industry and Trade and the Ministry of Small–Medium Enterprise and Cooperatives. Strong coordination and collaboration among all stakeholders involved in natural silk production are essential for the development of the sericulture industry in Indonesia [21,27].

At the farm level, although it is not a capital-intensive business, sericulture requires sufficient capital to start by buying silkworm eggs, preparing the land to plant mulberries as silkworm feed, and setting up rearing rooms; not all farmers can afford the costs related to these investments. These challenges have led natural silk agribusiness activities to be increasingly abandoned by farmers who tend to choose other farm activities [41].

Various efforts have been made to overcome this situation by developing good-quality silkworm eggs and mulberry seeds and providing technical assistance to silk farmers; unfortunately, these have not provided significant results yet. The key to the successful development of natural silk lies mainly in the support from government policies, the availability of good quality silkworm eggs and feed, the availability of land resources and easy access to adequate funding, as well as training of workers who have skills and expertise in silk cultivation [42].

Numerous research have been undertaken concerning the advancement of the silk industry in Indonesia; nevertheless, the most remain incomplete. An assessment of the disparity between current practices and necessary solutions is required to optimize sericulture agriculture through appropriate upstream and downstream laws and governance tailored to Indonesian circumstances. This research seeks to elucidate the diverse developmental potentials and challenges that must be addressed to maximally advance sericulture agribusiness.

2. EXAMINING SERICULTURE DEVELOPMENT IN INDONESIA

Indonesia's sericulture production centers are in several provinces, including West Java, Central Java, and South Sulawesi. South Sulawesi contributes 70–80% of national silk yarn production [21,43,44]. Silk production has been developed in Indonesia since 1953 [26], and mulberry trees began to be cultivated in South Sulawesi in the early 1960s [19].

The sericulture in Indonesia has decreased significantly due to the pebrine disease epidemic [21]. Before 2012, the development areas for sericulture were located on the islands of Java, Sulawesi, Kalimantan, and Sumatra, regions that cover 11 provinces. After 2012, the sericultural activity continued in two provinces only, South Sulawesi and West Java (Figure 1).

Various internal and external factors have influenced the decline in the distribution of sericulture development areas; therefore, it is crucial to study the dynamics of sericulture in Indonesia. This section will review the dynamics of sericulture in Indonesia, including the cultivation of mulberry plants (moriculture), rearing of silkworms, products, and social, economic, cultural, and institutional factors.

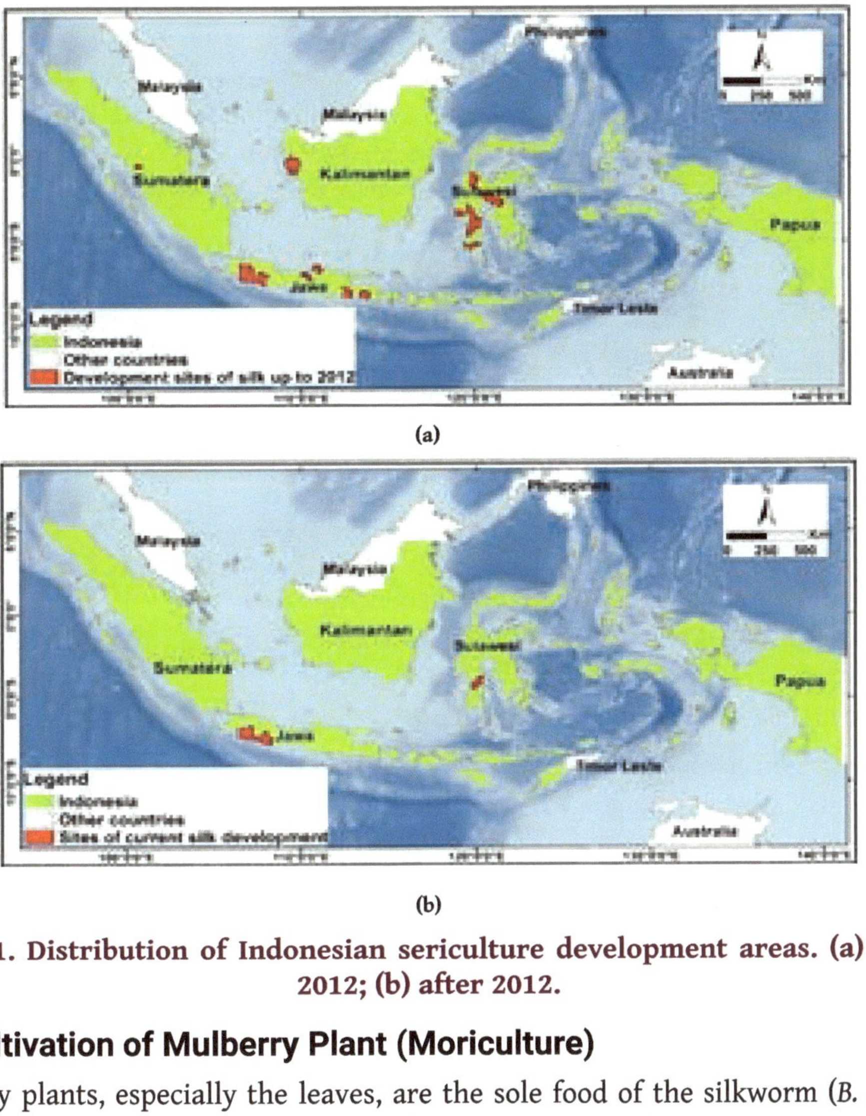

(a)

(b)

Figure 1. Distribution of Indonesian sericulture development areas. (a) before 2012; (b) after 2012.

2.1 Cultivation of Mulberry Plant (Moriculture)

Mulberry plants, especially the leaves, are the sole food of the silkworm (*B. Mori* L.) [45,46]. Recently, more attention has been paid to mulberry plant production improvement (quality or quantity)—more than 60% of the total cost of cocoon production goes to mulberry cultivation [45].

Mulberry cultivation in Indonesia uses *Morus cathayana* Hemsl. [47–50], *M.s alba* L. [47–50], *M. nigra* L., *M. multicaulis* (Perr.) Loudon, *M. australis* Poir, and *M. bombycis* Koidz [47–49]. The mulberry species most widely cultivated to date is *M. cathayana* [50]. This species is preferred because it has comparative advantages such as being relatively easy to grow, adaptive to any location, and producing a higher number of leaves compared with other species of *Morus* developed in Indonesia [50]. The scientific name reference of the mulberry species or variety refers to https://powo.science.kew.org/.

In Indonesia, productivity is relatively low, ranging from 8 to 23 tons/ha/year [51,52] compared with in India (the majority of which are *M. alba* var. *multicaulis Loud* and *M. alba* var. *arthoputpurea* (Shahtut)), which reaches 10–30 tons/ha/year in tropical conditions and 25–30 tons/ha/year in temperate regions [53]. China, a major global silk producer, has mulberry productivity ranging from 15 to 46.5 tons/ha/year, depending on the species of mulberry and the conditions of the area of development [54], while production in India ranges from 18.40–23.06 tons/ha/year [55,56].

Mulberry leaves determine the productivity and quality of silkworm products, so we must consider their nutritional quality. The main factors affecting the quality and quantity of mulberry leaves are the genetic factors of each mulberry species, the environment, the cultivation methods, and pest and disease control [57].

Techniques to increase the productivity and quality of mulberry leaves via genetic factors are through type/variety selection and hybridization [47,58,59]. Several studies conducted from 1999 to 2015 in Indonesia have created several mulberry hybrids with 40–53 tons of leaves/ha/year that are resistant to drought, pests, and diseases and are easy to cultivate [60,61]. The crosses came from several sources, including *M. cathayana* Hemsl.,

M. australis Poir, *M. indica* L., *M. nigra* L., *M. amakusaguwa (Hybrid M. bombycis* Koidz *x M. acidosa* Griff.*), M. acidosa* Griff. and *M. latifolia* Poir. [62,63]. Information on the productivity of mulberry hybridization in Indonesia can be seen in Table 1.

Table 1. The productivity of mulberry hybridization in Indonesia.

The Origin of the Cross	Leaf Production (ton/ha/year)
M. australis Poirx *M. indica* L.	23.2 *
M. nigra L.x *M. indica* L.	23 *
M. cathayana Hemsl. x*M. amakusaguwa (Hybrid M. bombycis* Koidzx *M.s acidosa* Griff.*)*	52.35 **
M. acidosa Griff.x *M. latifolia* Poir.	35.42 **
M. australis Poirx *M. indica* L.	29.56 **

Source: * [62], ** [63].

The result of hybridization between *M. cathayana* Hemsl. x *M. amakusaguwa (Hybrid M. bombycis* Koidz x *M. acidosa* Griff.*)*, having the highest leaf productivity, was launched by the Ministry of Forestry in 2013 and named SULI 01 [16]. The SULI 01 mulberry hybrid is suitable to be planted in lowlands and highlands. This accession can produce better quality cocoons than other mulberry varieties and has greater resistance to pests and diseases [64]. Unfortunately, this type of SULI 01 is not well known by the public, so its use as a silkworm feed is still limited. So far, silk farmers have not used hybrid types that have higher productivity than conventional ones.

The environment that affects the productivity of mulberry plants must also be considered. Some of the requirements for mulberry growth are soil conditions that are not loamy, not too sandy, and not acidic; optimal and balanced fertilization; and maintenance according to Standard Operational Procedures (SOPs) at each growth stage and water availability [65]. In Indonesia, the mulberry is found from the lowlands to the highlands (over 3600 m above sea level—m asl). It can grow well even if the leaf production is lower than that of mulberry plants cultivated in areas with an optimum altitude of 400–700 m asl and an optimum temperature of 23.9–26.6°C [61]. Problems arise because, in extreme areas, silkworm cultivation experiences pose many challenges.

Although many areas of Indonesia are suitable for the growing of mulberry plants, the tropical climate in Indonesia causes the productivity and supply of mulberry leaves to be discontinuous [28,66]. This is related to the differences in the dry and rainy seasons. Mulberry plants require a lot of water in their growth process, so that in the rainy season, mulberry plants generally grow faster and produce more leaves. Apart from this, the decrease in leaf production in the dry season also depends on the genetic characteristics of the cultivated mulberry varieties [63]. Most of the mulberry plants in Indonesia are propagated vegetatively using stem cuttings, although they can be reproduced by gamic propagation (through seeds). Some mulberry plants in Indonesia can be seen in Figure 2.

Figure 2. Cultivation of mulberry plants in Indonesia. (a) South Sulawesi; (b) West Java; (c) Hybrid SULI 01.

Another challenge related to mulberry cultivation is pest and disease attacks. Mulberry plants have some diseases and pests that affect the quality and productivity of the leaves and interfere with the growth and nutrient content in the leaves [67]. Eight types of pests and four types of diseases are commonly found in Indonesia, especially in the central part of South Sulawesi [68], such as grasshoppers (*Valanga nigricornis* (Burmeister), leaf webber (*Glyphodes pulverulentalis* Hampson), snails (*Achatina fulica*), mealybug (*Maconellicoccus hirsutus* Green), white peach scale (*Pseudaulacapsis pentagona* Targioni-Tozzetti), stem borers (*Epepeotes plorator* Newman), powdery mildew (*Phyllactinia corylea* (Pers.) P.Karst), red rust fungus (*Aecidium mori* (Barclay) Barclay), and leaf spot (*Cercospora moricola* Cooke and *Sirosporium mori* (Sydow et P. Sydow) M.B. Ellis). It was stated that these pests and diseases affect the rate of leaf loss up to 15.32% and are influenced by factors related to where the mulberry plants grow. Whitefly (*Bemisia tabaci*) is the most common pest on mulberry plants in Indonesia and can cause damage of up to 80% or crop failure [69]. In India, whitefly attacks cause a leaf yield loss of 4500 kg/ha/year [70].

The use of insecticides and fungicides needs to be controlled because mulberry leaves are used as feed for silkworms in addition to other uses for humans. Due to the application of pesticides over the years, sericulture faces various problems, such as silkworm sensitivity and death. The use of pesticides can increase the mortality of silkworms in the 5th instar. *B. mori* L. silkworms are sensitive to the smell of pesticides, even though they are used for other plants close to mulberry fields or where silkworms are kept [40,71,72]. Pesticides containing dimethoate in mulberry leaf at rearing facilities silkworms will affect cocoon quality, the weight of instar fifth silkworms, and are contained as chemical residues; for this reason, harvest losses are recorded [73].

The area of the mulberry plantation will greatly affect cocoon production. This can be seen in the trend of cocoon production in South Sulawesi Province. Currently, the area of mulberry production in South Sulawesi is 65.2 ha, with cocoon production in two districts (Soppeng and Wajo) where are produced 14,075 kg [40]; in 2014, in the same two districts, there was an area of 238.35 ha of mulberries with cocoon production of 29,249 kg [74]. However, on average, cocoon productivity per hectare of mulberry plants is currently better than in 2014. This is possible because the use of superior eggs of silkworms and mulberry is more intensive and because SOPs for silkworm rearing with supervision extension workers in the area. Other challenges need to be anticipated. This relates to the trend of farmers converting their mulberry plantations into other commodity crops [40,41]. However, they still grow mulberries on part of their land and continue silkworm rearing.

2.2 Silkworm Rearing (Sericulture)

Nowadays, the domestic silkworm or mulberry silk moth, *B. mori* L., belonging to the Bombycidae family, provides more than 99% of the silk in the world. Therefore, we can divide silkworm accessions into Chinese, Japanese, European, and Tropical races [75]. However, there are only three accessions of silkworms in Indonesia, Chinese, Japanese and Tropical races.

The Chinese race has greenish eggs, yellow eggshells, plain silkworm patterns, blue belly color between segments, shorter silkworm life, oval/oval cocoon larvae shape, and high silk percentage. The Japanese race has gray eggs, white eggshells, heavier eggs, skin spots, pink color between segments of the abdomen, longer silkworm life, stronger silkworm, and peanut-shaped cocoons. Tropical races have white eggshell color, plain silkworm pattern, strong/disease resistant silkworm, short silkworm life, small cocoons, difficult cocoon shells to spin, poor productivity and fiber quality, but high adaptability in Indonesia. The differences in eggs, silkworms, and cocoons of each race in Indonesia can be seen in Figure 3.

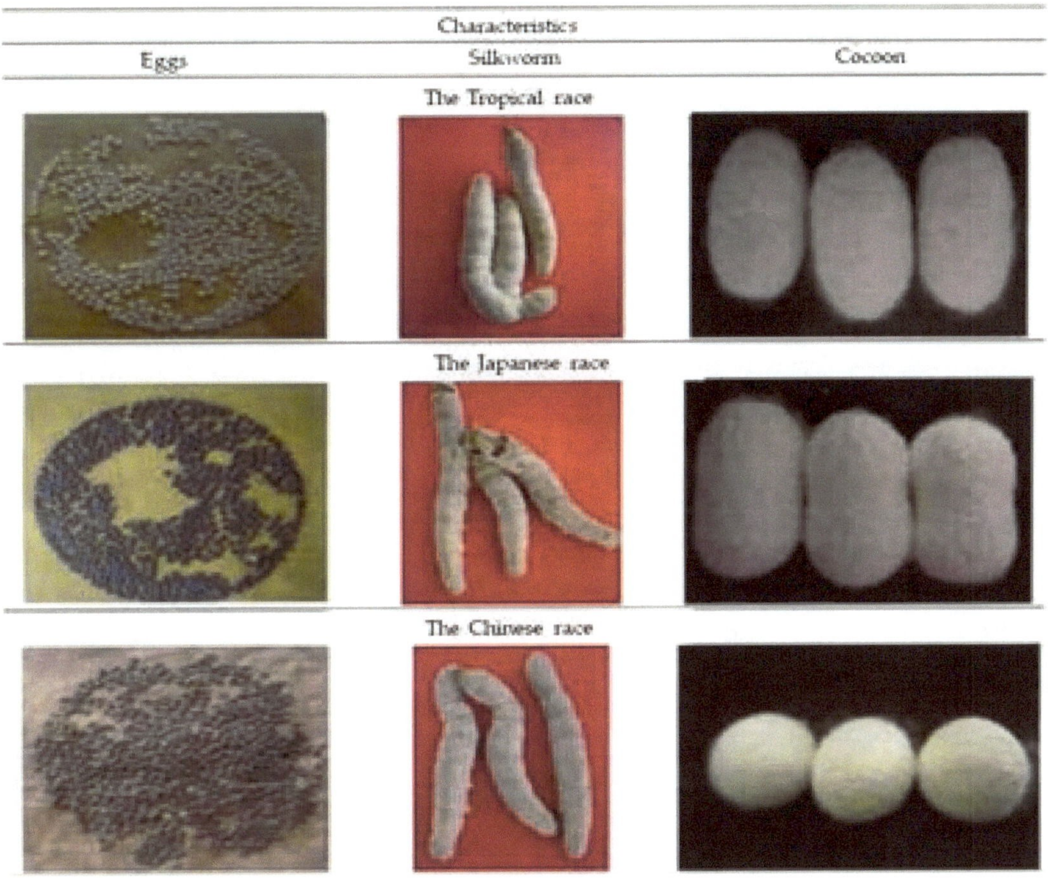

Figure 3. The characteristics of each group of races.

In Indonesia, what has been developed are the Chinese race silkworms, the Japanese race, and the result of crossing between the Japanese strains with the Chinese strains [26]. The quality of hybrid silkworm cocoons was better than that of the Chinese and Japanese silkworms [76].

Problems in the rearing of silkworms in Indonesia include: (1) the low quality of silk- worm eggs [19]; (2) limited number and supply of silkworm eggs [77]; (3) incompatibility of the type of silkworm with the location of cultivation [78]; (4) inadequate conditions of facilities and infrastructure [78]; (5) lack of farmers' compliance with good silkworm- rearing standards (SOPs) [79]; and (6) silkworm pests and disease attacks [21]. The limited supply and low quality of silkworm eggs have caused farmers to use the same silkworm eggs for various geographic conditions [78], resulting in varying productivity [77]. Infras- tructure and compliance with silkworm rearing standards (SOPs) are important points in silkworm rearing. Silkworm diseases are the most significant factor that impacts cocoon production [21,80,81].

2.2.1 Silkworm Breeding

In general, the silkworm eggs used are of three accessions, coming from three subtropical regions (China, Japan, and Europe), and are usually kept under optimum conditions. The silkworm eggs currently circulating are the result of crossing parents from subtropical areas (bivoltine), apparently not able to adapt well to tropical conditions such as Indone- sia [82,83]. For tropical conditions such as those in Indonesia, where the agroclimate fluctuates, the mulberry leaf quality is low, and the silkworm rearing ability is limited, stronger silkworms are needed.

The success of silkworm rearing is highly dependent on the genetic quality of the eggs and their suitability for the environment. Therefore, it is recommended to obtain and use eggs suitable to the conditions of the rearing environment and with the desired properties as targets [84]. Hybridization in Indonesia is mostly carried out by the Sericulture Laboratory under the Ministry of Environment and Forestry and Perum Perhutani (a state forest enterprise). There are five silkworm hybridization products that are developed and used in Indonesia: Hybrid C-301 [52,75,77,85], BS 08 [86], BS 09 [86], PS 01 [87] and SINAR [88]. Information about silkworm hybridization products can be found in Supplementary Table S1.

2.2.2 Silkworm Nursery

After obtaining high-quality silkworm eggs, nursery development is needed so they can be commercialized. Until 2012, Indonesia had only two producers of silkworm eggs, namely Pusat Produksi Ulat Sutera/PPUS (Center for Silkworm Production) Candiroto and Kesatuan Pengusahaan Sutera Alam/KPSA (the Natural Silk Concession Unit) Soppeng [35,86,89,90]. Both institutions are under the management of Perum Perhutani Unit I Central Java (PPUS Candiroto) and Perum Pehutani Unit II East Java (KSPA Soppeng). The production of these two nurseries results from the silkworm hybridization that has been carried out, namely C301 [52,75,77,85] and BS 09 [86]. The following conveys the productivity of silkworm eggs in Indonesia and the institutions that produce them (Table 2).

Table 2. The productivity of silkworm eggs in Indonesia.

Egg Type	Producer	Hatching Percent-age (%)	Characteristics	Cocoon Quality	Filament Quality
C301 *	K S P A Soppeng and PPUS Candiroto	81.65–81.95	- egg color: gray - eggshell color: white - silkworm pattern: spots silkworm segment color: gray - silkworm life: 24–26 days - silkworms cannot stand being kept in standard conditions - The cocoon is white and oval shape	- Normal cocoons: 86.75–93.25% - Weight of cocoon: 1.7–2.04 g - Weight of cocoon shell: 0.41–0.46 g - Cocoon shell ratio: 22.3– 23.97%	- Filament length: 1026–1127 m, - Filament percent-age: 21% - Filament denier: 2.71–3.63 - Decompo-sition: 76–89%
BS 09 **	KSPA Soppeng dan PPUS Candiroto	>90	- egg color: gray - eggshell color: white - silkworm pattern: spots - color between segments: blue - silkworms are somewhat sensitive to minimum conditions - cocoon shape: oval - cocoon color: white	- Normal cocoons: 90.00–96.00% - Cocoon shell ratio: 21.28–23.49%	- Filament length: 1060–1216 m, - Filament percentage: 16.64–19.09% - Filament denier: 2.3–3.26
PS 01 ***	Bina Mandiri Forest Farmers Group, Sukabumi	90.74–96.99	- egg color: gray - eggshell color: white - silkworm pattern: spots - silkworm color between segments: slightly bluish gray - silkworm growth is more uniform - resistant to disease - the life of the silkworm is 25–27 days - cocoon color: white - stable cocoon shape	- Normal cocoons: 97.5–100% - Weight of cocoon: 1.7–1.9 g - Weight of cocoon shell: 0.38–0.44 g - Cocoon shell ratio: 22.29–24.82%	Filament length: 808–1003 m, - Rolling power: 95–100% - Filament denier: 2.3–2.43 - Spun yield: 13.29–15.31

Source: * [77], ** [91], *** [87].

Hybrid C301 is a silkworm strain widely used in various environmental conditions [39]. Silkworm eggs from two different nurseries reared in the same place sometimes resulted in different productivity [35,86]. Rahmathulla [92] confirmed that seasonal differences in the environmental components significantly affect the genotypic expression in the form of the phenotypic output of silkworm crops, such as cocoon weight, shell weight, and cocoon shell ratio.

The average rearing carried out by each silk farmer is one box containing 25,000 eggs. The average cocoon production in Indonesia from the rearing of one box is 25.03 kg, while the cocoon production in China, the highest producer in the world, can reach 39.97 kg/box [19]. The productivity of the C301 egg hybrid is 25–35 kg/box [74,84], that of the BS 09 hybrid is 27–30 kg/box [74], and that of the PS 01 hybrid is 35–40 kg/box [74,84]. The productivity of the PS 01 hybrid silkworm was better than that of the BS-09 hybrid, which is reared at an altitude of 250 m above sea level; as can be observed according to the body features, mortality of the instar first–four instar silkworms, and the quality of the silkworm cocoons [93].

The number and quality of eggs have experienced ups and downs, thus encouraging importing [94]. In general, silkworm eggs are imported from China. This can be seen from the source of silkworm eggs used by silkworm farmers in South Sulawesi Province, which come from the production of Perum Perhutani (KPSA Soppeng) and are also imported egg seeds from China [19,95].

The price of eggs from Perhutani is more affordable for silk farmers with relatively good productivity, although the production quality is unstable due to changes in weather conditions. The production of silkworm eggs from China is higher than that of *Perhutani* eggs [94,96], but it is only available in certain months [19]. Chinese eggs have higher adaptability and steadiness in yield than hybrid C301, but hybrid C301 showed better cocoon and yarn quality in the observed parameters. Importing silkworm eggs must be balanced with the need for adequate storage equipment, considering that the import of eggs is carried out in large quantities and is not directly absorbed by the market. Poor egg storage results in many eggs dying, hatching not being uniform, and the larvae of the following generation being weak [77,97].

Another problem with silkworm eggs is related to the limitations of the center and breeders. Up until now, there has been only one breeder operating (KSPA Soppeng) with a limited capacity. The limited number of breeders impacts the use of silkworm eggs in rearing locations that are far from the nursery location. With the lack of activity of the Candiroto Silkworm Breeding Center (PPUS), silkworm farmers in West Java and surrounding areas obtain silkworm eggs from Soppeng's KPSA in South Sulawesi, thus requiring a lengthy shipping process. The time the eggs are in transit increases the risk of egg damage.

The PS 01 hybrid production is a solution for the existence of superior silkworm eggs. Unfortunately, PS 01 silkworm eggs are not yet on the market [98]. The Directorate of Social Forestry and Customary Forest Business Development (Bina Usaha Perhutanan Sosial & Hutan Adat/BUPSHA), part of the General Directorate of Social Forestry and Environmental Partnership (PSKL) in collaboration with the Center for Forest Research and Development of Forest, has created Bina Mandiri Farmer Group (BMFG) to become the PS 01 breeder, functioning as a supplier of silkworm eggs for the Sukabumi Regency and surrounding areas [98]. However, until now, it has not been produced optimally.

2.2.3 Silkworm Rearing

Silkworm rearing is a series of activities to produce cocoons using silkworm eggs with certain cultivation standards. The success of cocoon production is highly dependent on compliance with its SOPs, temperature, and humidity, and the condition of the admin- istered leaves [99]. Sericulture farmers in Sukabumi have good knowledge of silkworm rearing standards [100], but many do not apply them, resulting in low or even negative income [79]. Support from Sericulture Laboratory technicians of the Forest Research and Development Center for applying good silk-rearing standards in Forest Management Unit/FMU (Kesatuan Pemangkuan Hutan/KPH) Boalemo has resulted in higher cocoon productivity than that of FMU Yogyakarta without assistance [101]. Therefore, efforts are needed to increase farmers' adoption of good rearing standards by providing consistent, accessible, relevant, and practical information with more personal means of communica- tion and stakeholder collaboration, using an interdisciplinary approach including social scientists, communication specialists, and marketing experts [102].

People who still cultivate silkworms at home can potentially interfere with silkworm cultivation activities [41]. *B. mori* L. is very delicate, highly sensitive to environmental fluctuations, and unable to survive extreme conditions [103]. Thus, traditional rearing activities carried out at home will result in low cocoon productivity.

Complete and appropriate infrastructure facilities (separate rearing houses for first and last instar silkworms, mulberry fields) are important for silkworm rearing. They will affect the number of rearing cycles in a year. Silk cultivation in FMU Boalemo can be done 12 times per year with appropriate facilities and infrastructures in one location. This differs from the BMFG in Sukabumi, which can only cultivate eight cycles in one year and keeps young silkworms and grown silkworms separate. More rearing cycles will increase yield productivity. Silkworm rearing in Indonesia can be seen in Figure 4.

(a) (b)

Figure 4. Silkworm rearing in Indonesia. (a) South Sulawesi Province; (b) West Java Province.

2.2.4 Pests and Diseases

Sericulture in Indonesia has experienced two waves of epidemics of pebrine disease and Nuclear Polyhedrosis Virus (NPV) infection, causing a drastic decline in sericulture pro- duction [21]. Both mulberries and silkworms are infested by a number of pests, which affect the cocoon quality and productivity, resulting in economic losses for farmers [104]. Among the silkworm diseases that cause economic damage, the highest crop loss is attributed to viral diseases, accounting for 70% of the total loss of the crop every year [105].

Diseases that often attack silkworms in Indonesia are NPV, Infectious Flacherie (FV), Aspergillus, Muscardine, pebrine, bacteria, and poisoning from agricultural chemicals [21,106,107]. The causes of these diseases include bacteria, fungi, viruses, and physical and chemical factors as well as environmental factors that result in abnormal conditions and death [108].

2.3 Silkworm Products

2.3.1 Main Products

The main products of sericulture are cocoons, yarns, and material fabrics (Figure 5). The trend of cocoon production [72] and yarn production [109] continues to decline. In the 1990s, Indonesia's silk yarn production reached 120–140 tons/year. Nowadays, it is only around 2.5 tons/year [20]. The current national demand for silk yarn is 900 tons per year [110], and it is predicted to increase by 12.2% annually [72]. The solution to the scarcity of silk yarn supply is generally importing. As stated by Antaranews [110], 95% of Indonesia's silk yarn is still imported from China, India, and Japan. Therefore, Indonesia has changed its sericulture strategy for the export market to meet domestic needs [111].

(a) (b) (c)

Figure 5. The main products of sericulture. (a) cocoons; (b) yarns; (c) material fabrics.

◉ Cocoons

Cocoons are the final product of silkworm rearing. The quality of the cocoons is determined by the hereditary nature of the silkworm strains and environmental conditions during rearing, spinning stage, etc. [112]. The requirements of a perfect cocoon are healthy (no defects), clean (clean white), the inside (pupa) not damaged or crushed, and a hard silk shell (layer of silk fibers) resistant to finger pressure. Several parameters of cocoon quality affect the quality of the raw materials for silk fiber and determine the quality, quantity, and efficiency of the process [113]. Cocoon characteristics, such as cocoon weight, cocoon shell weight, and cocoon shell ratio, are the most important economic determinants of silkworm rearing [114]. Different shapes and sizes of cocoons will produce various sizes of silk fiber and yarn quality.

The quality of the C301 hybrid showed higher yields compared with hybrids from China with respect to these parameters: percentage of normal cocoons, the weight of cocoons, weight of cocoon shells, and proportion of cocoon shells [77]. China's normal hybrid cocoons are 80.08–90.08%, while the C301 hybrid is 86.75–93.25%. The weight of the Chinese hybrid cocoon is 1.44–1.57 g, while that of the C301 hybrid is 1.7–2.04 g. The weight of the Chinese hybrid cocoon shell is 0.3–0.32 g, while the C301 hybrid is 0.41–0.46 g. The percentage of Chinese hybrid cocoon skin is 20.58–20.96%, while that of the C301 hybrid is 22.94–23.97% [77].

Hybrid PS 01 (Chinese race male and Japanese race female—(804 × 927)) has an egg hatching percentage of 97.00% compared with hybrid C301, which is only 96.01%. Hybrid PS 01 produces a cocoon weight of 1.6 g, cocoon shell weight of 0.29 g, and a cocoon shell ratio of 20.35–21.36%, the hybrid C301 cocoon weighs 1.27 g, the cocoons shell weighs 0.2 g, and the cocoon shell ratio is 17.9–19.54% [39].

The cocoon shell percentage is one of the benchmark measures for determining the selling price of cocoons. The highest level (grade of cocoon quality) is a percentage of cocoon shells of 22–25% [115].

High competition with producers in other countries, low government support for this industry, lagging technology, and poor handling of pests and diseases are thought to hinder the development of silkworm cultivation in Indonesia. In 2020, globally, Indonesia's silk production was only twelfth behind China, India, Brazil, Thailand, and Japan [20]. From year to year, Indonesia has experienced a decline in cocoon production.

- Silk Yarn

The quality of silk fiber is very important in the reeling process because it will affect the silk yarn obtained. These qualities include the dry cocoon weight, number of fiber breaks, remaining cocoon, percentage of fiber weight, and reelability (fiber unwinding). Reelability is the ability of silk fibers to unravel and coil when the cocoon is spun. Cocoon is highly dependent on the silkworm variety, temperature, and humidity during spinning. The cocoon mountages also affect the cocoon unraveling capacity [60].

Silkworm eggs from KSPA Soppeng and PPUS Candiroto, when maintained in conditions of high temperature and low humidity, did not show any difference in a produced cocoon. Still, there was a difference in the filament yield. Longer filaments were produced by silkworm eggs from KSPA Soppeng compared with the eggs from PPUS Candiroto [35]. The quality of the C301 hybrid filament is better than that of the Chinese hybrid for the parameters of filament length, percentage of filament, denier thickness, and reelability. The China hybrid filament length is 889.08–994.84 m; the filament percentage is 17.91–19.14%, the denier thickness is 2.37–2.97, the reelability is 85.57 –95.56, and the yarn weight per 50 cocoons is 8.87–10.7. The C301 hybrid produces a filament length of 1026.05–1127.83 m, filament percentage of 21.24–21.49%, denier thickness of 2.71–3.63, reelability of 76.11–89.58, and yarn weight per 50 cocoons of 9.51–13.66 [77]. Hybrid PS 01 (Chinese race male and Japanese race female—(804 × 927)) produced a filament length of 1003.67–1160 m, 32% longer than the C301 hybrid, which only had a filament of 755.67–883.33 m [39].

To encourage the development of the national silkworm industry, the Government of Indonesia (GOI), through the Ministry of Forestry, Ministry of Industry, and Ministry of Co-operatives and Small and Medium Enterprises—issued Joint Regulation No. P47/Menhut-II/2006, No. 29/M-IND/PER/6/2006 and No. 07/PER/M. KUKM/VI/2006 concerning the Guidance and Development of National Natural Silk with a Cluster Approach [27]. Through this joint regulation, the government plans to build an integrated silk industry cluster from upstream to downstream, including cultivating mulberries, silkworms rearing, and cocoons for the yarn and silk reeling industry. Furthermore, to encourage the above efforts, the Ministry of Forestry issued Ministerial Regulation no. P.56/Menhut-II/2007 concerning the

Procurement and Circulation of Silkworm Eggs.

Ironically, when the cluster effort was launched (2006), the annual national silk yarn production fell to less than 32 tons and the yearly demand for domestic industry to around 700 tons [116]. On the other hand, at the global level, the market for silk yarn in 2005 continued to increase, reaching the demand of 100,000 tons. Meanwhile, to fill the shortage of domestic silk yarn, the government allowed imports of silk yarn of more than 600 tons from China, India, Thailand, and other ASEAN countries [116,117]. The imports from silk yarn-producing countries turned out to harm the national silk yarn industry. Domestic artisans prefer imported yarn, especially from China, because the price is lower. For silk entrepreneurs, the presence of spikes yarn or synthetic yarn at a low price affects the selling power of natural silk fabrics. Thus, lower-class people especially prefer it because the price is low [118].

- Material Fabrics

(a) (b) (c) (d)

Figure 6. Process and material fabrics. (a) spinning; (b) weaving; (c) Sulawesi woven fabrics; (d) West Java woven fabric.

Spinning and weaving will produce material fabric. Spinning is the process of extruding cocoon fibers into spun strands. There are two categories of spinning, modern spinning and people spinning. Modern spinning is spinning that uses machines, both fully automatic and semi-automatic. People's spinning machines in South Sulawesi have three stages in the spinning process, and all stages affect the quality of the yarn. The three stages are the process of cocoon handling, spinning, and yarn handling spinning result [40]. The patterns of woven fabrics are produced based on regional characters. The results of woven fabrics from two areas of silk development in Indonesia can be seen in Figure 6.

2.3.2 Diversification Product

Despite the aforementioned challenges, the untapped potential of silk-based products makes us optimistic. Beyond conventional textiles, there are many opportunities for product diversification. This is due to the fact that silkworm cocoons and pupae can be exploited as sericin [119], regenerated silk fibroin [120], and chitin/chitosan [121]. These are high-value biomaterials with potential applications in many fields. In the last couple of decades, progress around the science of silk has enabled the utilization of silk- based materials in exciting areas such as the pharmaceutical/medical field [121–124], cosmetics [125,126], electronics [127–129], the energy sector [130,131]. Although these are niche markets, product diversification efforts may lead to renewed interest from investors and developers in the sericulture industry. To advance in this direction, it is important for the scientific community to collaborate with farmers in the hope that high-value niche products may give some additional competitive advantage against other commodities at both national and global scales.

2.4 Social, Economic, Cultural, and Institutional Factors

2.4.1 Social, Economic, and Cultural Factors

Sericulture plays an important role, especially in rural communities [109,132]. The sericulture industry involves minimal investments, a short cultivation period, high em- ployment potential, and very profitable returns that are suitable for the economy of a rural agrarian community [133,134].

Natural silk is a labor-intensive industry and provides jobs for many people [135]. The proportion of labor costs for silk cultivation reaches 73.14% [85]. Sericulture business actors, especially the upstream sector in Indonesia, are still dominated by people with low levels of education, mostly at an elementary school level [17,44,79,135]. However, some of them have a high school or undergraduate education [41,136], and the majority are aged between 15 and 64 years [137]. Even though education is not the main requirement [44], the sustainability of silk needs to be supported by an adequate level of education to update technical knowledge, management, and mindset [138] in the adoption of technology for a commercial business. In its implementation, the upstream sector involves a lot of family labor and does not use wage labor, so it is more of commercial household business.

Of 9223 Business Farmer Groups in the Social Forestry unit, only 27 focus on sericulture development (West Java, North of Kalimantan, and South Sulawesi); 23 focus on mulberry and cocoon production, and 4 on silk processing [139]. Farmers carry out sericulture cultivation as a primary or secondary job [44]. The average income of cocoon farmers is Rp. 2,383,969 (USD 156.56) per head of family/year, while the average income of yarn farmers is Rp. 8,004,500 (USD 525.68) per family head per year [140]. The income showed the shrink of the business and led these activities to be part of side occupations which contribute to the additional income for the family.

There is a decreasing number of silk farmers as they switch to other commodities [40]. An insufficient number of eggs was one of the main problems. On the other hand, if enough eggs are produced, the quality is sometimes not so good. Hence, the production will fail and cause uncertainty in the cocoon yields. The minimal number of cocoons produced affected the availability of the raw materials for weavers [141].

Women play an important role in natural silk cultivation [142,143]. According to [17,143,144]. silk activities involve distributed gender roles through internal agree- ments [145,146], and this condition is different in each region [40]. Men are more focused on mulberry cultivation and leaf harvesting, which require physical energy, while women are dominant in rearing, feeding, harvesting cocoons, weaving, and reeling, which require patience [142,147]. Some rural women consider silk activities as their primary or secondary livelihood. The distribution of internally agreed roles involves substantial social capital in supporting the future development of labor-intensive sericulture [147].

Processing cocoons into yarn and silk is characterized as a traditional household business [148]. The craft of silk weaving was initially to meet the family's needs and turned into a commercial enterprise with a cluster of handicraft industries. However, traditional characteristics remain attached to simple reeling (non-machine looms). In South Sulawesi, there are around 4900 silk woven fabric artisans that produce 99,000 sarongs per year; 227 of them are non-machine loom craftsmen with a fabric production of 1,580,000 m per year [22]. Thus, the silk business is a cultural asset of the Indonesian nation that must be maintained and preserved.

The foremost opportunity for the development of Indonesian silk is the culture of the people in the use and management of silk production [136]. Silk cloth became a traditional costume and a symbol of social status. Silk is used at traditional, religious, and official national events [40,44] and is a mainstay commodity and souvenir typical of South Sulawesi [41]. Silk motifs and creations also represent the wearer's position, happiness, and safety [149].

2.4.2 Sericulture Development Policy

The policy of developing silk in Indonesia aims to encourage agribusiness to produce silk products of good quality in large quantities to meet the demands and standards of the domestic and export markets [24,25]. Sericulture was first developed in South Sulawesi Province through the cooperation of the Directorate General of Forestry, Ministry of Agriculture, with the Japan International Cooperation Agency (JICA).

In 2002 through Forestry Minister Decree No. 664/Kpts-II/2002, a Natural Silk Center was formed to support the development of silk in Indonesia. Balai Persuteraan Alam (BPA) has a vast working area, covering almost the entire territory of Indonesia with natural silk development activities, starting from institutions arrangement and skills development through training to data collection from upstream to downstream. Silk development refers to the Joint Regulation of the Minister of Forestry Number P.47/MENHUT-II/2006; the Ministry of Industry Number 29/M-IND/PER/6/2006, and the State Ministry of Coopera- tives and Small-Medium Enterprises Number 07/PER/M.UMKM/VI/2006 concerning the Guidance and Development of National Natural Silk through a cluster approach. Business development with a cluster system (upstream-downstream) was developed through a partnership scheme. In this scheme, business groups engaged in the sericulture business (mulberry farmers, silkworm cultivators, and silk yarn industries) and related institutions are geographically close together.

At the provincial level, the government of South Sulawesi launched the "Restore the Glory of Silk" program to recreate the level of Indonesian silk production achieved in the 1970s. In order to support this agenda, the South Sulawesi Regional Government has developed various programs to strengthen the upstream and downstream sectors, including technical assistance, equipment, marketing, farmer empowerment, and capi- tal [41]. To improve the quality and quantity of silkworm eggs and ensure the quality and availability of silk cocoons, in 2017, the Ministry of Forestry launched MoF number P.37/MenLHK/Setjen/Kum.1/6/2017, which regulates the procurement and circulation of silkworm eggs.

In 2016, there was an institutional change at BPA to become the Center for Social Forestry and Environmental Partnerships (Balai Perhutanan Sosial dan Kemitraan Lingkun-gan/BPSKL) through the Regulation of the Minister of Environment and Forestry of the Republic of Indonesia Number P. 14 /Menlhk/Setjen/OTL.O/I/2016.

The development of the silk agro-industry in Indonesia is influenced by changes in policies related to institutions. Starting in 2016, the silk agro-industry development activities became part of the social forestry (PS) activities supported by the Decree of the Minister of Environment and Forestry No. 83/2016 concerning Social Forestry, which provides access rights for communities in forest management areas. However, silk activities have become limited because they are only a small part of social forestry. They are, therefore, only supported by 1.2–1.5% of the total budget of BPSKL.

Fortunately, in 2020 the GOI stipulated Law Number 11 of 2020 concerning Job Creation and Government Regulation (PP) and its derivatives Number 23 of 2021 concerning forestry administration which opened up opportunities for forest areas to permit holders to develop multiple forestry businesses, including the development of silk cultivation through Forest Utilization Business Permits (PBPH). The dynamics of government policies that affect the development of the silk agro-industry in Indonesia can be seen in Supplementary Table S2.

2.4.3 Institutional Factors

Sericulture agribusiness is carried out by three leading players, namely the public sector (government institutions, both central and regional), the private sector (legal entity organizations), and farmer groups [150]. The public sector has a role in setting regula- tions [150] to ensure mulberry availability and quality, the provision of silkworm technical guidance, evaluating facilities and infrastructure, increasing human resources, and updating databases [40]. Private sector groups play a role in terms of providing capital, distributing various inputs needed in mulberry and silkworm cultivation (fertilizer, mulberry, and silkworm planting material), the yarn spinning industry, and marketing the products. The farmer group has a role as a mulberry and silkworm cultivation actor in producing cocoons.

These three stakeholders are involved in the upstream and downstream activities of sericulture. The parties involved are as follows: (a) Upstream Sector: public groups (BPA/BPSKL, Provincial Forestry Service, District Forestry Service), private groups (Perum Perhutani and BMFG, which produces silkworm), mulberry farmer groups and silkworm farmers, (b) Downstream Sector: public groups (Provincial Industry and Trade Office, Regency Industry and Trade Office), Business Groups (PT. Kokon Sutera Sulawesi*)*; and (c) Supporters (financial institutions, Forest Research and Development Agency, Industry & Trade Research Institute, colleges).

The parties involved in silk activities have different organizational structures and rules [151]. From 2006 until 2012, natural silk still relied on cluster policies [27]. During this period, institutional development was carried out by the Ministry of Environment and Forestry, the Ministry of Industry, and the Ministry of Cooperatives and Small Enterprises. However, in reality, these three institutions are still running independently [27]. This shows that there is a fragmentation between upstream and downstream in the management of Indonesia's silk, which has the potential to weaken efforts toward improvement.

The motivation for establishing silk institutions at the upstream level was different in the silk centers of Indonesia. The establishment of silk institutions in West Java and Central Java was generally initiated by the national reforestation program. On the other hand, the silk farmers 'institutions in South Sulawesi were formed on the motivation of the socio-cultural attachment of the community to the use of silk fabrics. This has induced Wajo farmers to be more independent. This condition is different from silk farmers in West Java, who still depend heavily on assistance from other parties [100].

The institutional capacity of the sericulture business in Indonesia is generally at a small scale, even at the household level, and only reaches the silkworm cocoon stage. Small-scale development, only up to the cocoon production stage, causes the bargaining position of farmers to be relatively weak, with prices determined by buyers [85]. Widiarti et al., [152] found that the basic problems of small businesses such as sericulture are a lack of management skills and professionalism, as well as limited access to capital, technological innovation, and marketing networks. These challenges have hampered institutional performance, especially at the farmer level, and often led to the reluctance of large companies to cooperate.

The change in the centralization of the forestry sector from regencies to provinces has also disrupted the partners of natural silk farmer groups in the field [40]. The ongoing changes in government institutions hamper the sustainability of the silk program, such as the lack of control in the development of techniques, infrastructure, human resource development, and a database that supports decision-making. In addition, public sector institutions that handle silk often change their functions. The Natural Silk Center, as one of the technical implementing units of the Ministry of Environment and Forestry that is specialized in handling natural silk, has turned into a Social Forestry and Environmental Partnership Center (BPSKL), whose main task is not only dealing with silk. This change aims to simplify the structure but is affected functionality. However, the problem of silk is very complex (upstream and downstream, from mulberry to yarn), so organizational changes are not sufficient. One of the consequences of these institutional changes is the loss of information (in the term of database) on silk production, such as production potential, demand, the number of parties involved, the area used for mulberry cultivation, etc.

3. THE WAY FORWARD

3.1 Moriculture

Based on the existing conditions of natural silk related to the fulfillment of silkworm (*B.mori* L.) feed sources, the main problem was the low and discontinuous productivity of mulberries. The food availability factor is important since silkworms are monophagous; hence their entire life cycle depends on mulberry leaves [153].

However, the tropical climate in Indonesia causes the productivity and supply of mulberry leaves to be discontinuous [28,66].

Efforts that can be made to overcome the problems with mulberry productivity as a source of silkworm feed include:

1. Enhance the commitment and compliance of all related stakeholders to apply standard operating procedures in cultivating mulberries (selection of suitable locations for mulberry cultivation, cultivation techniques, intensive rearing, appropriate fertilization, and preventive measures to anticipate plant pests and diseases).
2. Superior mulberry seeds with high productivity that can be planted under various growing conditions. Indonesia, through the Ministry of Environment and Forestry, has initiated and produced a mulberry hybrid with productivity of 52.35 tons/ha/year [39].
3. Integrating mulberry agroforestry with other plants to increase land productivity and mulberry production in Indonesia. Agroforestry that integrates mulberry plantations with agricultural crops provides multidimensional benefits, including the efficient use of land area, increased production and total leaf yield, improved nutrition, efficient use of light and water from the mulberry plant, and its inherent ability to combat pests and diseases [46]. Expanding the mulberry plantation to the potential of the Social Forestry areas of 5.03 million Ha [139].
4. Managing mulberry blocks to ensure the continuity of mulberry production. A good field division will optimize feed production so that, in addition to optimal leaf quality and productivity, it can also increase the intensity of silkworm rearing per year. The separation of young and grown silkworms into different facilities with the different dedicated mulberry fields has the advantages of intensive maintenance, pest control, and feed productivity prediction, considering that young silkworms need one-month-old mulberry feed. In contrast, grown silkworms require two months old mulberry feed. Otherwise, it will cause losses due to separation, which will result in high maintenance operational costs and will require a wider allocation of the mulberry plantation area.

3.2 Silkworm Cultivation (Sericulture)

One of the main obstacles to sericulture development in Indonesia is the absence of quality silkworm eggs and the application of silkworm rearing that does not follow the appropriate procedures. Silkworm egg producers are still limited with low production. Some of the efforts that can be made include:

1. Creation of silkworm breeding centers as silkworm egg producers with superior strains of silkworm eggs. The scheme for establishing a silkworm nursery center in collaboration with related stakeholders in Sukabumi can

be an example to be implemented in other areas, especially in South Sulawesi.

2. Develop superior silkworm hybrids. The Ministry of Environment and Forestry has initiated the distribution of the superior silkworm hybrid strains PS 01, which has 35–40 kg/box cocoon production. This strain can be developed more widely to increase the productivity of Indonesian sericulture.
3. Enforcement in implementing SOP for silkworm rearing. The SOP for maintaining silkworms that are not carried out properly has an impact on the productivity of co- coons and yarns. Adherence to selecting a good and suitable location (effectiveness of feed sources and minimal air pollution), carrying out room and silkworm disinfection procedures, and paying attention to the timing and rearing procedures correctly will be steps to ensure optimal cocoons and yarns.
4. The separation of young and grown silkworm farmers can increase productivity. Young silkworms require intensive care and are susceptible to disease. Productivity will increase if skilled farmers carry out the maintenance of young silkworms. Only then can the maintenance of grown silkworms be carried out by novice farmers. The advantages of this separation are the effective use of the rearing room, and the productivity is maintained because young silkworms are susceptible to disease, so they will only be kept by skilled farmers. Good cooperative arrangements can increase the intensity of maintenance. Disadvantages will arise if communication is established between young and grown silkworm farmers; high mobility is also required because of the distance between the two.

3.3 Products

The main products of sericulture are cocoons and yarns. So far, the quality of the cocoons produced does not match the market or consumer standard, which has an impact on yarn imports and causes competition with synthetic yarns such as rayon and cupro. A strategic step to overcome this problem is the selection of cocoons at the time of silkworm rearing so that the quality is optimal and the price can be higher. Supposing the requirement for quality silk yarn can be appropriately fulfilled. The problem of competition with synthetic yarns can be avoided, considering that the existence of synthetic yarns is due to the unavailability of raw materials for silk yarns. Silk's loyal consumers have a different market share; hence, they will not be affected by synthetic yarns. Moreover, the cultural factor in South Sulawesi, which requires the use of silk, means the silk market can be maintained.

3.4 Social, Economic, and Cultural Factors

Sericulture can increase people's income, absorb labor, and be applied in rural areas. Livelihoods in the sericulture sector are discontinuous due to the absence of eggs causing many people to turn to other commodities. Some efforts can be made to tackle the situation:

1. Build production networks and cooperate with stakeholders, facilitated by the government to ensure the sustainability of the production of superior eggs and the propagation of silkworm egg producers.
2. Internalization of silkworm rearing (sericulture) through both formal and informal education involving various stakeholders, including business stakeholders, to maintain socio-cultural values and develop the value of silk entrepreneurship as part of the nation's identity.
3. Increasing the community's capacity for silkworm rearing in terms of technical aspects, farm management, farmer group organization management, and improved access to markets. The yield of silk products is low due to a lack of assistance regarding proper rearing procedures. Efforts are needed to increase farmers' adoption of good rearing standards by providing consistent, accessible, relevant, and practical information via more personal means of communication and coupling with stakeholders by a transdisciplinary approach, including social scientists, communication specialists, entrepreneurs, and marketing experts. The role of the Forest Management Unit/FMU, which is the implementation of management at the site level, needs to be more intensive for facilitating sericulture development.

3.5 Institutional Factors

One important factor that needs to be considered in the institutional system is the arrangement between rights and obligations contained in a mutually binding agreement. One example is the collaboration between silk farmer groups in Sukabumi with PT. Be-gawan Sutera Nusantara (PT. BSN) in terms of joint utilization of production activities that can increase competitive advantage through capacity utilization and product diversification [85,152]. Shared use arises from the opportunity to share activities (PT BSN, which carries out the egg production process and raises young silkworms, is a producer of eggs for young silkworms, while farmers are rearing grown silkworms and harvesting cocoons that are taken and paid for by PT BSN), similarities in technology (assistance is provided by the company), distribution or other factors. The joint use of production activities can increase utilization capacity and product diversification and, in turn, achieve economies of scale. This will increase the productivity and competitiveness of sericulture. Improving the ability to build networks, increasing social capital within the group, and improving the organizational management and negotiation skills of farmers are important factors in supporting silk farming institutions. In addition, local governments can provide formal support through local regulations, policies, and silk development programs at the local level.

4. CONCISE SUMMARY

Sericulture in Indonesia include the cultivation of non-timber forest products and has been a longstanding practice among indigenous communities for their sustenance. Nonetheless, there has been a decline in this activity over the past 5 to 10 years. Numerous intricate issues from upstream to downstream have contributed to this terrible development. Moriculture and sericulture techniques, product quality and quantity, socio-economic factors, institutional frameworks, and community engagement are interrelated. They have created a difficult environment for the advancement of sericulture. Conversely, there exist potential resources, like the investigation of proficient mulberry cultivation and high-quality silkworm production via research and development, significant cultural elements, and prospective stakeholders for fostering network participation. The engagement, collaboration, and initiative of all stakeholders are essential for advancing sericulture development in Indonesia. The central government may significantly facilitate multi-stakeholder collaborations in the integrated development of sericulture in Indonesia.

REFERENCES

1. Sparrow, R.; Dartanto, T.; Hartwig, R. Indonesia Under the New Normal: Challenges and the Way Ahead. *Bull. Indones. Econ. Stud.* **2020**, *56*, 269–299. [CrossRef]
2. Halimatussadiah, A.; Cesarina, A.; Siregar, A.A.; Hanum, C.; Wisana, D.; Rahardi, F.; Bintara, H.; Rezki, J.F.; Husna, M.; Azar, M.S.; et al. *Thinking Ahead: Indonesia's Agenda on Sustainable Recovery from COVID-19 Pandemic*; LPEM FEB UI dan BAPPENAS: Jakarta, Indonesia, 2020.
3. United Nations. Ending Poverty. Available online: https://www.un.org/en/global-issues/ending-poverty (accessed on 15 November 2020).
4. Golar, G.; Malik, A.; Muis, H.; Herman, A.; Nurudin, N.; Lukman, L. The social-economic impact of COVID-19 pandemic: Implications for potential forest degradation. *Heliyon* **2020**, *6*, e05354. [CrossRef] [PubMed]
5. Djalante, R.; Lassa, J.; Setiamarga, D.; Sudjatma, A.; Indrawan, M.; Haryanto, B.; Mahfud, C.; Sinapoy, M.S.; Djalante, S.; Rafliana, I.; et al. Review and analysis of current responses to COVID-19 in Indonesia: Period of January to March 2020. *Progress in Disaster Science* **2020**, *6*, 100091. [CrossRef]
6. Suryahadi, A.; Al Izzati, R.; Suryadarma, D. *The Impact of COVID-19 Outbreak on Poverty: An Estimation for Indonesia (Draft)*; SMERU Working Paper; The SMERU Research Institute: Jakarta, Indonesia, 2020; pp. 1–20.
7. FAO. *The impact of disasters and crises on agriculture and food security: 2021*; Food and Agriculture Organization of The United Nations: Rome, Italy, 2021.
8. Cahyadi, H.S.; Newsome, D. The post COVID-19 tourism dilemma for geoparks in Indonesia. *Int. J. Geoheritage Park.* **2021**, *9*, 199–211. [CrossRef]

9. Supriyanto, B. *Cara Belajar Baru Petani Dalam Buku Salam Lima Jari*; Direktorat Penyiapan Kawasan Perhutanan Sosial: Jakarta, Indonesia, 2020.

10. Shackleton, C.; Delang, C.O.; Shackleton, S.; Shanley, P. Non-timber Forest Products: Concept and Definitions. In *Non-Timber Forest Products in the Global Context*; Shackleton, S., Shackleton, C., Shanley, P., Eds.; Springer: Berlin/Heidelberg, Germany, 2011; pp. 3–21.

11. Sardeshpande, M.; Shackleton, C. Wild Edible Fruits: A Systematic Review of an Under-Researched Multifunctional NTFP (Non-Timber Forest Product). *Forests* **2019**, *10*, 467. [CrossRef]

12. Simangunsong, B.C.H.; Manurung, E.G.T.; Elias, E.; Hutagaol, M.P.; Tarigan, J.; Prabawa, S.B. Tangible economic value of non-timber forest products from peat swamp forest in Kampar, Indonesia. *Biodiversitas* **2020**, *21*, 5954–5960. [CrossRef]

13. Wahyudi, W. Non-Timber Forest Product (Ntfp) Commodities Harvested and Marketed By Local People At the Local Markets in Manokwari—West Papua. *Indones. J. For. Res.* **2017**, *4*, 27–35. [CrossRef]

14. Kartodihardjo, H.; Suwito. *Survei dan Indeks Perhutanan Sosial: Jalan Menuju Kesejahteraan Rakyat dan Kelestarian Hutan*; Katadata Insight Center: Jakarta, Indonesia, 2020.

15. Barcelos, S.M.B.D.; Salvador, R.; Guedes, M.D.G.; de Francisco, A.C. Opportunities for improving the environmental profile of silk cocoon production under Brazilian conditions. *Sustainability* **2020**, *12*, 3214. [CrossRef]

16. Agustarini, R.; Andadari, L.; Minarningsih; Dewi, R. Conservation and breeding of natural silkworm (*Bombyx mori* L.) in Indonesia. *IOP Conf. Ser. Earth Environ. Sci.* **2020**, *533*, 012004. [CrossRef]

17. Kasi, E. Role of Women in Sericulture and Community Development. *SAGE Open* **2013**, *3*(3), 2158244013502984. [CrossRef]

18. Hogeboom, R.J.; Hoekstra, A.Y. Water and land footprints and economic productivity as factors in local crop choice: The case of silk in Malawi. *Water (Switzerland)* **2017**, *9*, 802. [CrossRef]

19. Harbi, J.; Nurrochmat, D.R.; Kusharto, C.M. Pengembangan Usaha Persuteraan Alam Kabupaten Wajo, Sulawesi Selatan. *Risal. Kebijak. Pertan. Dan Lingkung. Rumusan Kaji. Strateg. Bid. Pertan. Dan Lingkung.* **2015**, *2*, 128. [CrossRef]

20. ISC. Global Silk Industry. Available online: https://inserco.org/en/statistics (accessed on 21 June 2021).

21. Nuraeni, S. Gaps in the thread: Disease, production, and opportunity in the failing silk industry of south sulawesi. *For. Soc.* **2017**, *1*, 110–120. [CrossRef]

22. Rusniati. *Analisis Pengendalian Proses Produksi LIPA SABBE' (Sarung Sutera) Sengkang di Kecamatan Tanasitolo Kabupaten Wajo*; Universitas Muhammadiyah Makassar: Makassar, Indonesia, 2018.

23. Minister of Forestry. *Peraturan Menteri Kehutanan Republik Indonesia No.P.21/Menhut-II/2009 tentang Kriteria dan Indikator Penetapan Jenis Hasil Hutan Bukan Kayu Unggulan*; Ministry of Environment and Forestry of The Republic of Indonesia: Jakarta, Indonesia, 2009.

24. Mutiara, F.; Asnani, D. Strategi pengembangan agribisnis Ulat Sutera Pemakan Daun Singkong di Kabupaten Malang. *J. Ilmu-Ilmu Peternak.* **2017**, *27*, 24–38. [CrossRef]

25. Tenriawaru, A.N.; Fudjaja, L.; Jamil, M.H.; Rukka, R.M.; Anisa, A.; Halil. Natural silk agroindustry in Wajo Regency. *IOP Conf. Ser. Earth Environ. Sci.* **2021**, *807*, 012037. [CrossRef]

26. Hartati. *Analisis Fenotip Ulat Sutera (bombyx mori L.) Hasil Persilangan Ras Jepang, Chna, dan Rumania*; Global Research and Consulting Institute (Global-RCI): Makassar, Indonesia, 2015; pp. 21–23.

27. Sadapotto, A. Proses Kebijakan Persuteraan Alam di Sulawesi Selatan. *J. Parennial* **2012**, *8*, 1–5.

28. Nurjayanti, E.D. Budidaya Ulat Sutera Dan Produksi Benang Sutera Melalui Sistem Kemitraan Pada Pengusaha Sutera Alam (PSA) Regaloh Kabupaten Pati. *Mediagro* **2011**, *7*, 1–10.

29. Zambrano-Gonzalez, G.; Ramirez-Gonzalez, G.; Almanza-P, M.I. The evolution of knowledge in sericultural research as observed through a science mapping approach. *F1000Research* **2017**, *6*, 2075. [CrossRef]

30. Sathyanarayana, K.; Saratchandra, B. *Carbon trading through afforestation programmes with host plants of tasar silkworm (Antheraea mylitta Drury) in private waste lands: A Value Add for Tribals*; Council of Research and Experiments in Agriculture, Apiculture and Sericulture Unit of Bologna: Padua, Italy, 2013.

31. Rohela, G.K.; Shukla, P.; Muttanna; Kumar, R.; Chowdhury, S.R. Mulberry (*Morus* spp.): An ideal plant for sustainable development. *Trees For. People* **2020**, *2*, 100011. [CrossRef]

32. Deepa, K.B.; Vishaka, G.V.; Kumar, N. Mulberry as a avenue plant. *J. Pharmacogn. Phytochem.* **2020**, *9(4S)*, 135–137.

33. McNair, P.D. Reviving the Sericulture Industry. In Proceedings of the 5th BACSA International Conference "Sericulture for Multi Products—New Prospects for Development" SERIPRODEV, Bucharest, Romania, 11–15 April 2011.

34. Tarigan, D.; Fauzi, A.M.; Suryani, A.; Kaomini, M.; Tarigan, D.; Fauzi, A.M.; Suryani, A.; Kaomini, M. Strategi pengembangan agroindustri sutera alam melalui pendekatan klaster a strategy for the development of silk agroindustry using cluster approach. *J. Teknol. Ind. Pertan.* **2019**, *20*, 39–47.

35. Nursita, I.W. Perbandingan produktifitas ulat Sutra dari dua tempat pembibitan yang berbeda pada kondisi lingkungan pemeliharaan panas. *J. Ilmu Peternak.* **2011**, 3, 10–17.

36. Muin, N.; Suryanto, H.; Minarningsih. Ujicoba Hibrid Morus Khunpai dan Morus Indica Sebagai Pakan Ulat Sutera (*Bombyx Mori.* Linn). *J. Penelit. Kehutan. Wallaceae* **2015**, *4*, 137–145. [CrossRef]

37. Pudjiono, S. *Ulat Sutra*; Badan Litbang dan Inovasi: Jakarta, Indonesia, 2015; pp. 42–59.

38. Shantibala, T.; Lokeshwari, R.K.; Gusheinzed, W.; Agarwala, B.K. *Documentation of Ethno-Entomophagy Practices in Ethnic Communities of Manipur, North-East India. Ancestral Knowledge in Agri-Allied Science of India*; India Publishing Agency: New Delhi, India, 2013.

39. Andadari, L. Pemilihan jenis hibrid ulat sutera yang optimal untuk dikembangkan di dataran tinggi dan/atau dataran rendah. *J. Penelit. Hutan Tanam.* **2016**, *13*, 13–21. [CrossRef]

40. Sadapotto, A.; Palulungan, L.; Riwu, M.; Sahide, M.A.K.; Sirimorok, N.; Parenreng, S.M.; Salindeho, Y.M.; Yahya, A.F.; Mulyani, A.; Said, R.D.; et al. *Laporan Kajian Final Rantai Nilai Komoditas Sutra Sulawesi Selatan*; Bappelitbangda Provinsi Sumatera Selatan: Makassar, Indonesia, 2021.

41. Muin, N.; Bisjoe, A.R.H. Budidaya Ulat Sutera Di Desa Sudu, Kecamatan Alla, Kabupaten Enrekang, Sulawesi Selatan. *J. Penelit. Hutan Tanam.* **2013**, *10*, 229–239. [CrossRef]

42. Hadi, P.; Rustiono, D. Silkworm Agribusiness in Bejen Village Temanggung. *Int. J. Agric. Innov. Res.* **2015**, *3*, 1592–1594.

43. Maturidy, A.F. Preferensi pelaku usaha dalam pengembangan persuteraan alam di Provinsi Sulawesi Selatan. *J. Pasca Sarj. Univ. Hasanuddin* **2011**, 1–11.

44. Isnan, W. Perception and Motivation of Farmers in the Development of Natural Silk Business in Soppeng Regency Sulawesi Selatan. *J. Wasian* **2019**, *6*, 1–10. [CrossRef]

45. Sharma, A.; Krishna, V.; Kaur, P.; Rayal, R. Characterization and Screening of Various Mulberry Varieties Throgh Morpho- Biochemical Characteristics. *J. Glob. Biosci.* **2015**, *4*, 1186–1192.

46. Mir, M.R.; Khan, I.L.; Baqual, M.F.; Banday, M.; Raja, R.; Khursheed, S. *Mulberry-Based Agroforestry System: An Effective Way of Maintaining Livelihood Security and Climate Change Mitigation*; Apple Academic Press: Palm Bay, FL, USA, 2019.

47. Andadari, L.; Minarningsih; Dewi, R. The Effect of Mulberry Types on the Productivity of Cocoon of Two *Bombyx mori* L. Silkworm Hybrids. *Widyariset* **2017**, *3*, 119. [CrossRef]

48. Sasmita, N.; Komara, L.L. Species Test of Morus alba and Morus cathayana in Indonesia. *Adv. Biol. Sci. Res.* **2021**, *11*, 246–252. [CrossRef]

49. Daulay, N.S. *Pengaruh Kualitas Daun Murbei Moruscathayana Terhadap Indeks Nutrisi Ulat Sutera Bombyx mori L. (Lepidoptera: Bombicidae)*; Repositori Institusi Universitas Sumatera Utara: Sumatera Utara, Indonesia, 2013.

50. Triandi, A.; Mubin, N.; Nurulalia, L. Identification of Mulberry Pests and Its Natural Enemies at "Rumah Sutera Alam" Ciapus, West Java. *Akta Agrosia* **2021**, *24*, 9–18. [CrossRef]

51. Sulthoni, A. *Peranan Entomologi Hutan dan Tantangannya Menghadapi Pembangunan Hutan Tanaman Industri. Pidato Pengukuhan Guru Besar dalam Ilmu Hama Tanaman*; Universitas Gadjah Mada: Yogyakarta, Indonesia, 1991.

52. Pudjiono, S.; Na'iem, M. Pengaruh Pemberian Pakan Murbei Hibrid Terhadap Produktivitas dan Kualitas Kokon. *J. Pemuliaan Tanam. Hutan* **2007**, *1*, 81–87. [CrossRef]

53. Singh, B.; Makkar, H.P.S. The Potential of Mulberry Foliage as Feed Supplement in India. In Proceedings of the FAO Electronic Conference on Mulberry for Animal Production (Morus I-L), Rome, Italy, August 2020; pp. 1–20.

54. Yongkang, H. Mulberry cultivation and utilization in China. In Proceedings of the Electronic Conference, 2000.

55. Doss, S.G.; Chakraborti, S.P.; Roychowdhuri, S.; Das, N.K.; Vijayan, K.; Ghosh, P.D. Development of mulberry varieties for sustainable growth and leaf yield in temperate and subtropical regions of India. *Euphytica* **2012**, *185*, 215–225. [CrossRef]

56. Bhatia Narendra, K.; Yousuf, M.; Nautiyal, R. Yield Gap Analysis of Mulberry Sericulture in Northwest India. *Int. J. Ind. Entomol.* **2013**, *27*, 131–141. [CrossRef]

57. Sori, W.; Gebreselassie, W. Evaluation of mulberry (*Morus* spp.) genotypes for growth, leaf yield and quality traits under Southwest Ethiopian condition. *J. Agron.* **2016**, *15*, 173–178. [CrossRef]

58. Nursyamsi. Propagasi Tiga Varietas Murbei Melalui Teknik Kultur Jaringan. *J. Penelit. Hutan Tanam.* **2012**, *9*, 75–82. [CrossRef]

59. Pudjiono, S.; Septina, S. Morfologi Tanaman Hibrid Murbei di Purwobinangun Yogyakarta. *J. Pemuliaan Tanam. Hutan* **2008**, *2*(1), 163–171. [CrossRef]

60. Atmosoedarjo, S.; Kartasubrata, J.; Kaomini, M.; Saleh, W.; Murdoko, W. *Sutera Alam Indonesia*; Yayasan Sarana Wana Jaya: Jakarta, Indonesia, 2000.

61. Andadari, L.; Pudjiono, S.; Suwandi; Rahmawati, T. *Budidaya Murbei dan Ulat Sutra*; Forda Press: Bogor, Indonesia, 2013.

62. Santoso, B.; Budisantoso, H. Adaptasi Varietas Murbei Hasil Silangan. In Proceedings of the Prosiding Ekspose Hasil Penelitian dan Pengembangan Kehutanan Ujung Pandang, Makassar, Indonesia, 1999.

63. Andadari, L.; Dewi, R.; Pudjiono, S. Adaptation Test of Five New Hybrid Mulberry to Improve Natural Silk Productivity. *Widyariset* **2016**, *2*(2), 96–105. [CrossRef]

64. Pudjiono, S.; Andadari, L.; Darwo. Pemilihan Jenis Hibrid Murbei Untuk Dikembangkan Di Dataran Tinggi. *J. Penelit. Hutan Tanam.* **2016**, *13*, 133–138.

65. Gupta, S.K.; Dubey, R.K. Impact of heavily polluted air and acid rain on mulberry silkworm and cocoon production of Bombyx mori Linn. *Int. J. Fauna Biol. Stud.* **2021**, *8*, 26–30. [CrossRef]

66. Yulistiani, D. Tanaman murbei sebagai sumber protein hijauan pakan domba dan kambing. *Wartazoa* **2012**, *22*, 46–52.

67. Mahadeva, A. Insect Pest Infestation, an Obstacle in Quality Mulberry Leaves Production. *Asian J. Biol. Sci.* **2018**, *11*, 41–52. [CrossRef]

68. Prayudyaningsih, R.; Tikupadang, H.; Santoso, B. Hama Dan Penyakit Jenis Murbei Eksot Dan Tingkat Kehilangan Daunnya Pada Akhir Musim Kemarau. *J. Penelit. Hutan dan Konserv. Alam* **2006**, *3*, 429–435. [CrossRef]

69. Dewi, R.; Andadari, L.; Maharani, K.E. Tinjauan Bioekologi dan Pengendalian Hama Kutu Kebul (*Bemisia tabaci* Genn.). In Proceedings of the Prodising Seminar Nasional PEI, Jatinangor, Bandng, 25–26 October 2017; pp. 40–45.

70. Mukhopadhyay, S.K.; Trivedy, K. Intregated management and forecasting of insect pests of mulberry (*Morus Alba* L.) for Eastern and Northeastern regions of India. *J. Environ. Sociobiol.* **2016**, *13*, 137–145.

71. Soliman, N.H.; Sherif, D.F.E. rovided for non-commercial research and education use. Not for reproduction, distribution or commercial use. Hepatoprotective Agents in Rats. *Egypt. Acad. J. Biol. Sci.* **2019**, *11*, 91–95.

72. Maru, R.; Badwi, N.; Abbas, I.; Nurfadillah; Nur, M.M.; Basram, N.F. Opportunities and Challenges of SilkWorm Cultivation Development in Geography Perspectives. *La Geogr.* **2021**, *19*, 201–210.

73. Shi, M.; Li, G.; Zheng, X.; Liu, F.; Zhao, S.; Cai, M.; Zhu, Y. Effects of dimethoate on the gonads and oxidative stress response in the silkworm, Bombyx mori. *Acta Entomol. Sin.* **2019**, *62*, 1186–1196.

74. BPA. *Laporan Bulanan. Pelaksana Harian Balai Persuteraan Alam (BPA), Kabupaten Soppeng*; Balai Persuteraan Alam (BPA): Makassar, Indonesia, 2014.

75. Fambayun, R.A.; Agustarini, R.; Andadari, L. Cultivation and Breeding Techniques for Increase Silk Productivity in Indonesia. *IOP Conf. Ser. Earth Environ. Sci.* **2022**, *995*, 012055. [CrossRef]

76. Deni; Diba, F.; Tavita, G.E. Deni, Farah Diba, Gusti Eva Tavita. *J. Hutan Lestari* **2019**, *7*, 874–883.

77. Andadari, L.; Kuntadi. Comparison of Chinese-Origin and Local Silkworm (L.) Hybrids in South Sulawesi. *J. Penelit. Hutan Tanam.* **2014**, *11*, 173–183. [CrossRef]

78. Djabar, M.; Utiarahman, N. Kelayakan Usaha Budidaya Ulat Sutera (*Bombix Mori* L.) Berdasarkan Aspek Non Finanasial Kabupaten Boalemo. *J. Penelit. Kehutan. Bonita* **2019**, *1*, 15–22. [CrossRef]

79. Syam, D. *Analisis Pendapatan Pemelihara Ulat Sutera Pada Pemeliharaan Konvensional di Desa Sering, Kecamatan Donri-Donri, Kabupaten Sorong (Studi Kasus Kelompok Tani Batu Tungke'E)*; Universitas Hasanuddin: Makassar, Indonesia, 2017.

80. Watanabe, H. Genetic resistance of the silkworm, Bombyx mori to viral diseases. *Curr. Sci.* **2002**, *83*, 439–446.

81. Potrich, M.; Alves, L.F.A.; Brancalhão, R.C.; Dalcin, G. Entomopatógenos Associados a Lagartas De *Bombyx Mori* L. (Lepidoptera: Bombycidae) No Estado Do Paraná. *Arq. Do Inst. Biológico* **2007**, *74*, 363–367. [CrossRef]

82. Nuraeni, S. *Kuantitas dan kualitas produksi F1 ulat sutera (Bombyx mori L.) hasil persilangan antara Ras Rusia dengan Ras China*; Fakultas Pertanian universitas Hasanuddin: Makassar, Indonesia, 1993.

83. Budisantoso. Pengeringan Kokon dengan Alat Pengering Tenaga Matahari dan pengaruhnya Terhadap Kualitas Serat Sutra. *Bul. Penelit. Kehutan.* **2001**, *7*(1), 24–36.

84. Kasip, M. *Pembentukan Galur Baru Ulat Sutera (Bombyx mori L) Melalui Persilangan Ulat Sutera Bivoltin dan Polivoltin*; IPB University: Bogor, Indonesia, 2001.

85. Yuniati, D.; Suharti, S.; Widiarti, A.; Andadari, L.; Heryati, Y.; Agustarini, R. Business feasibility of several PS-01 hybrid silkworms (*Bombix mori* L.) cultivation scheme. *IOP Conf. Ser. Earth Environ. Sci.* **2021**, *917*, 012031. [CrossRef]

86. Estetika, Y.; Endrawati, Y.C. Produktivitas Ulat Sutera (*Bombyx mori* L.) Ras BS-09 di Daerah Tropis. *J. Ilmu Prod. Dan Teknol. Has. Peternak.* **2018**, *6*, 104–112. [CrossRef]

87. Minister of Forestry. *Keputusan Menteri Kehutanan Republik Indonesia No. SK. 794/Menhut-II/2013 Tentang Pelepasan Bibit Ulat Sutera (Bombyx mori L.) Hibrid PS-01*; Ministry of Environment and Forestry of The Republic of Indonesia: Jakarta, Indonesia, 2013.

88. Ministry of Environment and Forestry. *Decree of the Minister of Environment and Forestry of the Republic of Indonesia No: SK.300/Menlhk/Setjen/KUM.1/4/2019 Dated 24 April 2019*; Ministry of Environment and Forestry of The Republic of Indonesia: Jakarta, Indonesia, 2019.

89. Cholis, N. Studi Penampilan Produksi Ulat Sutera F1 Hibrid Hasil Persilangan RAS Jepang dan RAS Cina yang Berasal dari Pusat Pembibitan Soppeng dan Temanggung. *J. Ternak Trop.* **2014**, *15*, 72–76.

90. Cholis, N. Studi tentang polimorfisme ulat sutera F1 hibrid hasil persilangan ras Jepang dan ras Cina yang berasal dari pusat pembibitan Soppeng dan Temanggung dengan menggunakan enzim restriksi Pst1 dan EcoR1. *J. Ilmu-Ilmu Peternak.* **2015**, *25*, 61–65. [CrossRef]

91. Minister of Forestry. *Keputusan Menteri Kehutanan Indonesia No. SK. 369/Menhut-II/2004 Tentang Pelepasan Bibit Ulat Sutera (Bombyx mori L.) Hibrid BS-08 dan BS-09*; Ministry of Environment and Forestry of The Republic of Indonesia: Jakarta, Indonesia, 2004.

92. Rahmathulla, V.K. Management of Climatic Factors for Successful Silkworm (*Bombyx mori* L.) Crop and Higher Silk Production: A Review. *Psyche* **2012**, *11*, 1–12. [CrossRef]

93. Mubarak, H. *Produktivitas Ulat Sutera (Bombyx Mori L.) Candirotobs- 09dan Hibrid PS.01 Yang Diberi Pakandaun Murbei (Morus Multicaulis)*; Institut Pertanian Bogor: Bogor, Indonesia, 2017.

94. Nuraeni, S.; Putranto, B. Aspek Biologis Ulat Sutera (*Bombyx mori* L.) dari Dua Sumber Bibit di Sulawesi Selatan. *J. Perenn.* **2007**, *4*, 10–17. [CrossRef]

95. Sadapotto, A.; Kartodihardjo, H.; Triwidodo, H.; Darusman, D.; Sila, M. Penataan institusi untuk peningkatan kinerja persuteraan alam di Sulawesi Selatan. *J. Forum Pascasarj.* **2010**, *33*, 133–140.

96. Andadari, L.; Minarningsih, S. The effect of feeding various species of mulberry (*Morus* spp.) on the growth of silkworm and quality of cocoon hybrid BS 09. *IOP Conf. Ser. Earth Environ. Sci.* **2021**, *914*, 012017. [CrossRef]

97. San-Ming, W. *Silkworm Egg Production*; FAO: Rome, Italy, 1989; p. 64.

98. Yuniati, D.; Widiarti, A.; Suharti, S. Pengembangan Persutraan Alam di Kabupaten Sukabumi. In *Pembelajaran Dunia Riset untuk Pengembangan HHBK*; Widiarti, A., Irma, Y., Minarningsih, A.R., Eds.; IPB Press: Bogor, Indonesia, 2020.

99. Prihatin, J. Pemeliharaan dan Pengokonan Ulat Sutera Bombyx mori L. Menggunakan Seriframe Standar dan Seriframe Modifikasi. *Bioedukasi* **2009**, *7*, 31–38.

100. Fauziyah, E.; Wijayanto, N. Strategi pengembangan usaha persuteraan alam di kabupaten sukabumi, propinsi jawa barat (Development Strategy of Striving Nature Silk in Kabupaten Sukabumi, Propinsi Jawa Barat). *J. Manaj. Hutan Trop.* **2007**, *13*, 59–77.

101. Agustarini, R.; Andadari, L. Implementasi Persuteraan Alam dalam Mendukung Perhutanan Sosial. In Proceedings of the Prosiding Seminar Nasional Ekosistem Unggul Membangun Hutan Sebagai Ekosistem Unggul Berbasis Das: Jaminan Produksi, Pelestarian, dan Kesejahteraan, Yogyakarta, Indonesia, 23 August 2018; pp. 151–162.

102. Ritter, C.; Jansen, J.; Roche, S.; Kelton, D.F.; Adams, C.L.; Orsel, K.; Erskine, R.J.; Benedictus, G.; Lam, T.; Barkema, H.W. Invited review: Determinants of farmers' adoption of management-based strategies for infectious disease prevention and control. *J. Dairy Sci.* **2017**, *100*, 3329–3347. [CrossRef] [PubMed]

103. Sarkhel, S.; Shrivastava, S.; Pouranik, M. The Effective Influence of Temperature on the Varied Characteristic of Silkworm: A Review. *Asian J. Exp. Sci.* **2017**, *31*, 31–37.

104. Sakthivel, N.; Kumaresan, P.; Qadri, S.M.H.; Ravikumar, J.; Balakrishna, R. Adoption of integrated pest management practices in sericulture—A case study in Tamil Nadu. *J. Biopestic.* **2012**, *5*, 212–215.

105. Sharma, A.; Sharma, P.; Thakur, J.; Murali, S.; Bali, K. Viral diseases of Mulberry Silkworm, *Bombyx mori* L. —*A Review.* **2020**, *9*, 415–423.

106. Nuraeni, S.; Muin, N.; Sanusi, D. Respon Empat Ras Ulat Sutera Terhadap Bombyx mori Nuclear Polyhedrosis Virus (BmNPV). In Proceedings of the Prosiding Seminar Nasional Mikrobiologi Kesehatan dan Lingkungan, Makassar, Indonesia, 29 January 2015.

107. Muin, N. *Pengaruh Pakan Pada Resistensi Ulat Sutra (Bombyx mori L.) Terhadap Penyakit Grasserie*; IPB University: Bogor, Indonesia, 2009.

108. Guo-Ping, K.; Xi-Jie, G. Overview of silkworm pathology in China. *Afr. J. Biotechnol.* **2011**, *10*, 18046–18056. [CrossRef]

109. Nuraeni, S.; Baharuddin. The adaptability of the commercial seeds of silkworm (Bombyx Mori L) on different altitude preservation. *IOP Conf. Ser. Earth Environ. Sci.* **2019**, *270*, 012034. [CrossRef]

110. Antaranews. *Ulat "Sinar" Diharapkan Penuhi Kebutuhan Benang Sutera Lokal*; Antaranews: Jakarta, Indonesia, 2019.

111. ASSIA. World Sericulture in Industry 4.0. In Proceedings of the ASEAN Country Member National Brief on Silk and Sericulture: Indonesia, Bangkok, Thailand, 28–29 July 2019.

112. Hussain, M.; Khan, S.A.; Naeem, M.; Aqil, T.; Khursheed, R.; ul Mohsin, A. Evaluations of silkworm lines against variations in temperature and RH for various parameters of commercial cocoon production. *Psyche* **2011**, *2011*, 1–11. [CrossRef]

113. Gowda, B.N.; Reddy, N.M. Influence of different environmental conditions on cocoon parameters and their effects for reeling performance of bivoltine hybrids of silkworm, Bombyx mori L. *Int. J. Indust. Entomol* **2007**, *14*, 15–21.

114. Mirhosseini, S.Z.; Nematollahian, S.; Ghanipoor, M.; Seidav, A. Comparison of Phenotypic and Genetic Performance of Local Silkworm Groups and Two Commercial Lines. *Biol. Res.* **2010**, *43*, 411–416. [CrossRef]

115. Kim, S.E. Silkworm Breeding. In *Principle and Practices in Sericulture*; NSERI: City, Korea, 1998.

116. BPS. *Data dan Informasi Kemiskinan 2005 2006, Buku 1: Provinsi [Poverty Data and Information 2005 2006, Book 1: Province]; BPS:* Jakarta, Indonesia, 2007.

117. Rochmawati, R. *Kualitas Kokon Hasil Silangan Ulat Sutera (Bombyx mori L.) Ras Cina dengan Ras Jepang Secara Resiprocal*; IPB University: Bogor, Indonesia, 2011.

118. Sadapotto, A. Strategi Pengembangan Sutra Alam di Kabupaten Poliwali Mandar, Sulawesi Barat. In Proceedings of the Seminar Nasional MAPEKI XVII, Medan, Indonesia; 2014.

119. Martinez-Mora, C.; Mrowiec, A.; Garcia-Vizcaino, E.M.; Alcaraz, A.; Cenis, J.L.; Nicolas, F.J. Fibroin and sericin from Bombyx mori silk stimulate cell migration through upregulation and phosphorylation of c-Jun. *PLoS ONE* **2012**, *7*, e42271. [CrossRef]

120. Rockwood, D.N.; Preda, R.C.; Yucel, T.; Wang, X.; Lovett, M.L.; Kaplan, D.L. Materials fabrication from Bombyx mori silk fibroin. *Nat. Protoc.* **2011**, *6*, 1612–1631. [CrossRef]

121. Battampara, P.; Nimisha Sathish, T.; Reddy, R.; Guna, V.; Nagananda, G.S.; Reddy, N.; Ramesha, B.S.; Maharaddi, V.H.; Rao, A.P.; Ravikumar, H.N.; et al. Properties of chitin and chitosan extracted from silkworm pupae and egg shells. *Int. J. Biol. Macromol.* **2020**, *161*, 1296–1304. [CrossRef] [PubMed]

122. Lin, M.J.; Lu, M.C.; Chan, Y.C.; Huang, Y.F.; Chang, H.Y. An Insulin-like Growth Factor-1 Conjugated Bombyx mori Silk Fibroin Film for Diabetic Wound Healing: Fabrication, Physicochemical Property Characterization, and Dosage Optimization In Vitro and In Vivo. *Pharmaceutics* **2021**, *13*(9), 1459. [CrossRef]

123. Zhang, Y.; Liu, Y.; Jiang, Z.; Wang, J.; Xu, Z.; Meng, K.; Zhao, H. Poly(glyceryl sebacate)/silk fibroin small-diameter artificial blood vessels with good elasticity and compliance. *Smart Mater. Med.* **2021**, *2*, 74–86. [CrossRef]

124. Veiga, A.; Castro, F.; Rocha, F.; Oliveira, A.L. Protein-Based Hydroxyapatite Materials: Tuning Composition toward Biomedical Applications. *ACS Appl. Bio Mater.* **2020**, *3*, 3441–3455. [CrossRef] [PubMed]

125. Su, D.; Ding, S.; Shi, W.; Huang, X.; Jiang, L. *Bombyx mori* silk-based materials with implication in skin repair: Sericin versus regenerated silk fibroin. *J. Biomater. Appl.* **2019**, *34*, 36–46. [CrossRef]

126. Ulfa, M.; Yahya, P.T.; Angriani, P.D.; Nur, A.A. Formulasi dan Uji Efektivitas Krim Limbah Air Kokon Ulat Sutera (*Bombyx mori*) Asal Kabupaten Soppeng sebagai Pelembab Kulit. *J. Farm. UIN Alauddin Makassar* **2020**, *8*, pagination. [CrossRef]

127. Hwang, S.W.; Tao, H.; Kim, D.H.; Cheng, H.; Song, J.K.; Rill, E.; Brenckle, M.A.; Panilaitis, B.; Won, S.M.; Kim, Y.S.; et al. A physically transient form of silicon electronics. *Science* **2012**, *337*, 1640–1644. [CrossRef]

128. Kim, D.-H.; Viventi, J.; Amsden, J.J.; Xiao, J.; Vigeland, L.; Kim, Y.-S.; Blanco, J.A.; Panilaitis, B.; Frechette, E.S.; Contreras, D.; et al. Dissolvable films of silk fibroin for ultrathin conformal bio-integrated electronics. *Nat. Mater.* **2010**, *9*, 511–517. [CrossRef] [PubMed]

129. Aramwit, P.; Kanokpanont, S.; Nakpheng, T.; Srichana, T. The Effect of Sericin from Various Extraction Methods on Cell Viability and Collagen Production. *Int. J. Mol. Sci.* **2010**, *11*, 2200–2211. [CrossRef]

130. Xiang, M.; Wang, Y.; Wu, J.; Guo, Y.; Wu, H.; Zhang, Y.; Liu, H. Natural Silk Cocoon Derived Nitrogen-doped Porous Carbon Nanosheets for High Performance Lithium-Sulfur Batteries. *Electrochim. Acta* **2017**, *227*, 7–16. [CrossRef]

131. Ho, J.; Cao, C.; Idrees, F.; Ma, X. Hierarchical Porous Nitrogen-Doped Carbon Nanosheets Derived from Silk for Ultrahigh- Capacity Battery Anodes and Supercapacitors. *ACS Nano Publ.* **2015**, *9*, 2556–2564. [CrossRef]

132. Iwang, B.; Sudirman. Peranan Pemerintah dalam Memajukan Perusahaan Sutera di Sulawesi Selatan, Indonesia. *Southeast Asian Soc. Sci. Rev.* **2020**, *5*, 103–132.

133. Chanotra, S.; Bali, K.; Bali, R. Sericulture: An opportunity for the upliftment of rural livelihood. *J. Entomol. Zool. Stud.* **2019**, *7*, 1100–1103.

134. Pratama, A.G.; Supratman, S.; Makkarennu, M. Examining forest economies: A case study of silk value chain analysis in Wajo District. *For. Soc.* **2019**, *3*, 22–33. [CrossRef]

135. Dewangan, S.K.; Sahu, K.R.; Soni, S.K. Breaking of poverty through sericulture among the tribe- A Socio-Economic study of Dharamjaigarh block. *Res. J. Recent Sci.* **2012**, *1*, 371–374.

136. Ibrahim, H. *Pemberdayaan Pengrajin Ekonomi Kreatif Kerajinan Sutera Di Perdesaan Provinsi Sulawesi Selatan*; IPB Press: Bogor, Indonesia, 2014.

137. BPS. *Sulawesi Selatan Dalam Angka 2020*; Badan Pusat Statistik Provinsi Sulawesi Selatan: Makassar, Indonesia, 2020.

138. Sari, R.P.; Holilulloh; Yanzi, A. Faktor-faktor Yang Mempengaruhi Pola Pikir Masyarakat Terhadap Pentingnya Pendidikan di Desa Cugung. *J. Kult. Demokr.* **2015**, *3*, pagination.

139. goKUPS. Perhutanan Sosial dan Kemitraan Lingkungan. Available online: http://sinav.usahahutan.id/index.php/frontend/ analisa/kups (accessed on 20 August 2022).

140. Haruna, A.; Ridwan Sadapotto, A. Nilai Tambah Produk Persutraan Alam Dari Produk Intermediate Kokon ke Produk Akhir Benang di Kelurahan Walannae Kecamatan Sabbangparu Kabupaten Wajo. *J. Eboni* **2021**, *3*, 24–38.

141. Muin, N.; Hayati, N. *Penerapan Iptek Persuteraan Alam di Sulawesi Selatan [Laporan Hasil Penelitian]*; Balai Penerapan Standarisasi Instrumen Kementrian Lingkungan Hidup dan Kehutanan: Makassar, Indonesia, 2017.

142. Alam, A.M.K. *Peran Perempuan pada Usaha Persuteraan Alam di Desa Pising Kecamatan Donri-Donri Kabupaten Soppeng*; Universitas Hasanuddin: Makassar, Indonesia, 2016.

143. Bukhari, R.; Kour, H.; Aziz, A. Women and the Indian Sericulture Industry. *Int. J. Curr. Microbiol. Appl. Sci.* **2019**, *8*, 857–871. [CrossRef]

144. Geetha, G.S.; Indira, R. Silkworm Rearing by Rural Women in Karnataka: A Path to Empowerment. *Indian J. Gend. Stud.* **2011**, *18*, 89–102. [CrossRef]

145. Hanum, F. *Kajian dan Dinamika Gender*; Instrans Publishing: Malang, Indonesia, 2018.

146. Hubeis, A.V.S. *Pemberdayaan Perempuan dari Masa Ke Masa*; IPB Press: Bogor, Indonesia, 2010.

147. Muin, N.; Bisjoe, A.R.H. Strengthening Gender Role in Managing Private Forests in South Konawe, South East Sulawesi Province. *J. Penelit. Sos. Dan Ekon. Kehutan.* **2019**, *16*, 127–135. [CrossRef]

148. Harbi, J. Kajian teknis usaha persuteraan alam kelurahan walennae, kec. sabbangparu, kab. wajo, sulawesi selatan. *J. Penelit. Ilmu-Ilmu Kehutan.* **2016**, *5*, 45–53.

149. Muhammad, S.; Dharmawan, A.; Unito, S.; Damanhuri, D. Kearifan Lokal dalam Sistem Sosial Ekonomi Masyarakat Penenun Bugis-Wajo. *Mudra J. Seni Budaya* **2013**, *28*(2), 129–142.

150. Kumar, S.N. *Strategies for Adoption of Appropriate Innovations and Technologies for the Development of Sericulture in SAARC Countries*; SAARC Agriculture Center: Dhaka, Bangladesh, 2015.

151. Anantanyu, S. Kelembagaan Petani: Peran Dan Strategi Pengembangan Kapasitasnya. *SEPA* **2011**, *7*, 102–109.

152. Widiarti, A.; Andadari, L.; Suharti, S.; Heryati, Y.; Yuniati, D.; Agustarini, R. Partnership model for sericulture development to improve farmer's welfare (a case study at bina mandiri farmer group at Sukabumi Regency). *IOP Conf. Ser. Earth Environ. Sci.* **2021**, *917*, 012009. [CrossRef]

153. Kanafi, R.R.; Ebadi, R.; Mirhosseini, S.Z.; Seidavi, A.R.; Zolfaghari, M.; Etebari, K. A review on nutritive effect of mulberry leaves enrichment with vitamins on economic traits and biological parameters of silkworm Bombyx mori L. *Invertebr. Surviv. J.* **2007**, *4*, 86–91.

3
Chapter

BIBLIOMETRIC EXAMINATION OF TRENDS IN MULBERRY AND SILKWORM RESEARCH PERTAINING TO SILK PRODUCTION AND ITS BY-PRODUCTS.

1. INTRODUCTION: AN OVERVIEW

Sericulture is the agricultural practice that involves the cultivation of mulberry trees (Morus sp.pl.) to produce leaves utilized for feeding silkworms (Bombyx mori L.), which are raised for silk production. In a historical epoch, silk trade constituted a significant endeavor that facilitated an early kind of globalization, linking Europe to Asia via the "Silk Road" approximately 2000 years ago. Currently, silk production accounts for approximately 0.2% of the total global textile output value and is distributed across over 60 nations. The principal nations for silk production are China, India, Uzbekistan, Vietnam, Thailand, and Brazil. [5].

The term "silk" refers to a class of natural structural proteins produced and spun as long fibrous filament by different Arthropoda, for example Lepidoptera and spiders [6–9]. The most famous and studied type of silk is produced by the domestic silkworm *B. mori* L. [6].

From the beginning of silk manufacturing, silk has been used essentially for textile and medical applications, since efforts in medical research demonstrated how silk could be successfully applied to surgery due to its biocompatibility [10–12]. Over the last few years, there has been a growing interest in the engineering of biomaterials capable of mimicking the structure and characteristics of tissues and several studies have focused on the use of natural silk fibroin [10,13–15], due to its high biocompatibility when separated from sericin [13]. In recent years even sericin acquired a high commercial value [16] thanks to a series of properties that make it suitable for application in the pharmaceutical, cosmetic and food sector [17]. In fact, sericin has been shown to have several biological activities, such as antioxidant and anti-tyrosinase properties [17–20] and it is used for controlled drug-releasing biomaterials to promote stability and prolonged administration of drugs, enzymes and insulin [20–22].

To produce commercial silkworm eggs male and female larvae of pure strains are reared in a dedicated environment to create the parental lines that are then crossed among them [23]; the most common parental lines belong to Chinese and Japanese strains [24,25]. Hybrids and polyhybrids are in general more productive than pure strains and more resistant to biotic and abiotic stress; Nevertheless the management of environmental parameters, as air temperature and relative humidity, is fundamental for the optimal development of reared larvae since these factors heavily affects larval metabolism [26].

Traditionally, *B. mori* larvae feed on fresh leaves of the white mulberry (*Morus alba* L.) [27,28], although even leaves of other species belonging to *Morus* genus can be used [29]. Leaves of *M. alba* contain the optimal amount of nutrients and compounds that are fun- damental for the growth and development of *B. mori* larvae and for their silk cocoon spinning. Besides being the only feeding source for silkworms, several studies have also been focused on alternative uses of by-products from *Morus*: leaves, roots, stems, and fruits of different species belonging to the *Morus* genus (in particular, *M. alba* and *nigra* L.), contain relatively high amounts of phenols [30–32], in particular flavonoids [32–34] with predominant quercetin and kaempferol glycosides as well as anthocyanins [35]. Different experiments showed how mulberry cultivation can represent an interesting and sustainable source of these compounds for nutraceuticals, which are usually contained in the mulberry leaf powder and mulberry fruit juice [36,37].

The previous paragraphs offer a brief overview about the whole compart of the silk production and show how the theme is wide and linked to several other topics, as engineering, medicine, and nutrition. In this complex framework, a text-mining analysis as proposed by Cogato et al. [38] and Ferrari et al. [39] seems to be appropriate to appreciate the interconnection among the hotspots of the research over the past two decades, from 2000 to 2020. This kind of analysis, in fact, allows to objectively identify and weigh the most important topics in a specific research field and to study how they interact.

The objective of this analysis is to deliver a thorough and comprehensive overview of the current literature on sericulture within the domains of agriculture and biological sciences. The specific objectives of this study were to: (i) delineate the temporal trend of publications over the past two decades; (ii) identify the predominant topics that have directed research in the field of sericulture; and (iii) analyze the significant interconnections among these topics, which is crucial as sericulture relies on the interactions of various production chains and processes. To reach these results, the authors opted for a quantitative analysis, the most effective and objective methodology to fulfill the specified objectives.

2. MATERIALS AND METHODS

As first step for the bibliometric analysis, the authors downloaded a collection of documents indexed by the SCOPUS database, taking advantage of the *advanced search* to precisely define the topics. The analysis of scientific documents contained in the SCOPUS database allowed the authors to understand how the interests of researchers developed and changed in the past years.

The authors performed a *text mining* process of the selected documental collection, in order to derive significant numeric indices and information, by analyzing unstructured (textual) information. The statistical analysis of these indices provides a key to the text interpretation, obtaining high-quality information useful for content interpretation. The words appearing in the title, keywords, and abstract of all articles were first pre-processed in the Microsoft® pre-installed Notepad environment and used for the creation of basic graphs in Microsoft® Excel®; the more elaborated graphical representations were obtained using Gephi (Gephi® Consortium, Compiegne, France), an open-source software developed to perform network analysis.

2.1 Article Selection

The analysis was based on the words "*silkworm*" and "*mulberry*" both as parts of the process that leads to silk production. Moreover, to include articles containing the derived forms of those two nouns, the scripts "silkworm * " and "mulberry*" were used for the initial research in SCOPUS, since the asterisk "*" indicate any word-related declination: e.g., "mulberry", "mulberries", "mulberroside" and so on (Table 1). With the initial examination, the program selected the articles that contain the string "silkworm" and "mulberr" and their derived terms in the title, keywords, or abstract. The authors applied some filters (in Table 1, Filter application n.1) for a more pertinent selection of the articles: the field was limited to "Agricultural and Biological Sciences", the language was limited to "English" and the type of document was limited to "Scientific Article"; the time span considered was 2000–2020. According to Author's expectations, this analysis resulted in finding many articles (more than five thousand) even with the application of the described filters.

Table 1. Scripts for the extraction of research papers in the SCOPUS database.

Step	Script	Number of Papers[1]
Initial	TITLE-ABS-KEY (mulberry *) OR TITLE-ABS-KEY (silkworm *)	23,580
Filter application n.1	TITLE-ABS-KEY (mulberry *) OR TITLE-ABS-KEY (silkworm *) AND (LIMIT-TO (DOCTYPE, "ar")) AND (LIMIT-TO (SUBJAREA, "AGRI")) AND (LIMIT-TO (LANGUAGE, "English"))	5184

Filter application n.2	TITLE-ABS-KEY(mulberry*)OR TITLE-ABS-KEY (silkworm *) AND NOT TITLE-ABS-KEY (Antheraea)) AND (LIMIT-TO (DOCTYPE, "ar")) AND(LIMIT-TO(SUBJAREA, "AGRI")) AND(LIMIT-TO (LANGUAGE,"English"))	4598

[1] Referred to years from 2000 to 2020.

The authors then applied a second series of filters to reduce misleading results. In particular, the word "*Antheraea*" was excluded from the research since it is linked to the topic of silk production but referred to a wild Lepidopteran rather than to *B. mori*. Articles collected by popular and dissemination journals were also excluded by the research. The new script is shown in Table 1 at Filter application n.2 row. With the applications of those new filters, the number of articles dropped to about 4600.

The download procedure from the SCOPUS database was repeated for each year between 2000 and 2020; the .csv format was chosen for the downloads.

2.2 Article Elaboration

The results of the research were converted and saved as .txt files to have all the data stored in a file format that could be easily processed. The first step of the data pre-processing was the so-called *tokenization*, a procedure through which the sentences are divided into their essential components (i.e., single words). All the other parts of the text as punctuation marks, hyphens, brackets, and others special characters were removed. The authors run a further elaboration to convert all capital letters into lowercase and to identify and convert singular/plural nouns (e.g., mulberry/mulberries or leaves/leaf) and synonyms (e.g., white mulberry and *M. alba*).

All .txt files obtained at the end of the previously described process were imported into Microsoft® Excel® (Microsoft 365 MSO Version 2111 Build 16.0.14701.20254 64-bit software, Microsoft Corporation, Redmond, WA, USA) to order all the single selected words and to count how many times each one appeared. This elaboration was necessary for the identification of the more frequent words in each considered year of the past two decades. Using Excel, the 100 most relevant words were identified and used for the subsequent analysis. Another interesting function provided by the Excel® software is the so-called *sparkline graphs* that creates a small graph into a single cell; the authors used this function to create time series graphs for all the analyzed nouns, thus recognizing in a fast way which words increased in importance during the analyzed time. As final step, the results were imported as .csv dedicated files in the Gephi software [40], which is a free open-source software that allows for the creation of complex graphical representation of word associations. To create a connection graph in Gephi, the user has to import into the software a .csv file containing information about the *nodes* (here, the most frequent nouns derived from the previous elaboration) and another .csv file containing information about the *edges*, the connection thanks to which each node is related to the others. Gephi considers the edges as vectors, directed or undirected, each one with its specific *weight*, which is generated as described in Section 2.2.1.

2.2.1 Combination Matrix

A word–word connection matrix was built as detailed by Ferrari et al. [39], based on the list of the 100 most recurrent words. The matrix had 100^2 cells and, that was the result of a total of 10,000 couples. Starting from this matrix, the authors built a dedicated connection matrix in which, for each couple of words (e.g., *mulberry-silkworm*) the number of articles that contained both the nouns was indicated; the connections of words are not directional, namely the order of the words forming the couple was not considered (e.g., *mulberry-silkworm* is equivalent to *silkworm-mulberry*) in order to avoid duplications. As a result of the matrix, 4950 couples of nouns were obtained.

2.2.2 Cluster Definition

Besides the analysis for the combination matrix and therefore for studying how the nouns are interconnected, the authors performed a so-called *cluster analysis*. The aim of the cluster analysis was to define a rule (or a feature, or a set of features) that allows to create groups containing a set of objects (the words in this case) and sharing one or more previously defined features: in this precise kind of cluster analysis the aim is grouping words according to the thematic area they belong to. When the main topic can be subdivided in several interrelated parts, as the one here-considered, cluster analysis allows to study and characterize the relationship among the different sections of the main topic: for example, how the cultivation of mulberry trees connects to silkworm rearing and to silkworm applications. As this analysis is only qualitative, by further differentiating results for the years or for the geographical area, it is possible to describe in a precise way the evolution of the relationship among clusters. It is worth noting that there are no fixed rules to define clusters, but the experience of the authors and the definition of precise rules, useful to minimize equivocations, allow the authors to discriminate the clusters. In the present work, the authors propose the different activities of the sericultural chain as criteria to define the clusters and, in this way, five clusters were identified: Cultivation, Silkworm, Rearing, Process and Product. All the words regarding the mulberry cultivation group in the cluster "cultivation": for example, words pertaining to the agronomic techniques and to the plant biology. The cluster "silkworm" collects all the words referring to the biology of silkworms (e.g., strains or different stages of their life). In the cluster named "rearing" the authors pooled all the words regarding the activity of silkworm rearing, including diseases or environmental factors (e.g., air temperature) that are fundamental for the insect development. The cluster "process" includes all terms that are linked to the industrial processing of products deriving from both silkworm rearing and mulberry cultivation. The cluster "product" groups all the words that describe the production obtainable by both mulberry (e.g., important metabolites) and silkworm. Although these clusters and rules were defined to avoid ambiguities and overlapping, the authors used the combination matrix and the specific research of documents on SCOPUS to define which words best fit to different cluster. Table 2 shows the cluster composition.

Table 2. Name of the considered clusters and words that compose them.

Cluster	Words
Cultivation	Crop, cultivar, fruit, germplasm, leaf, *Morus* sp. Pl, M. alba, *M. nigra*, mulberry, plant, root, seed, shoot, soil, species, variety, tree, water
Silkworm	Body, *Bombyx*, *B. mori*, breed, cocoon, diapause, diet, disease, DNA, egg, embryonic, expression, female, gene, gland, hemolymph, hormone, instar, larva, male, hybrid, immune, infection, instar, juvenile, larva, Lepidoptera, male, midgut, molting, moth, prothoracic, pupa, sex, shell, silkmoth, silkworm, spinning, stage, strain
Rearing	*B. bassiana*, feeding, heat, humidity, nutrition, *N. bombycis*, NucleoPolyhedroVirus, rearing, sericulture, stress, survival, temperature
Process	Drying, fermentation, industry, purification
Product	Acid, amino acid, anthocyanin, antioxidant, composition, diabetic, edible, enzyme, extract, fiber, fibroin, filament, Product flavonoid, food, juice, lipid, oil, peptide, phenol, polyphenol, powder, product, protein, quality, sericin, serine, silk, yield

The authors classified the most frequent words by assigning to them a score calculated as weighed average of the word in the considered years. The authors calculated, per word and per year, the ratio of the number of occurrences to the total number of publications in the sericultural field; in this way the authors could objectively characterize the relative importance of the selected word in a certain year and study its trend over the considered interval of time. The authors assigned the highest weight to the most recent years to pay more attention to the last trend of research in sericulture. The score for each word was calculated as proposed by Ferrari [41] (see Equation (1)).

$$W_i = \frac{\sum_{i=1}^{21} w_i \cdot \frac{o_i}{S_i}}{\sum_{i=1}^{21} w_i}, \qquad (1)$$

In Equation (1), w_i is the weight of the *i*-th year (from 2000 to 2020), o_i is the absolute number of occurrences in the *i*th year and S_i is the total number of publications in the ith year.

The authors analyzed the impact of the five most frequent terms of each cluster, according to Equation (1). To this end, the impact of the papers related to such terms was quantified in terms of Hirsch's h-index. Thus, for a given term, the h-index has been calculated here by counting the number *h* of publications including that term and cited at least that same number *h* of times. The average was then computed for the five terms of each cluster.

The conceptual flux of the analysis is represented in Figure 1.

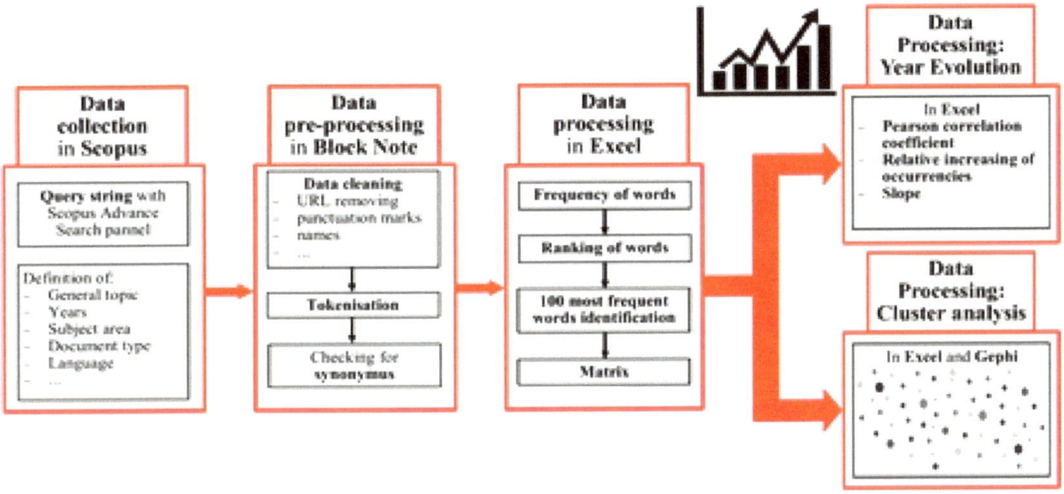

Figure 1. Scheme of the whole analysis process of the text data.

3. RESULTS

3.1 Analysis of the Trends

As first analysis, the authors considered the number of articles published per year in the cumulative research on mulberry and silkworm and the ratio to the total number of publications in the agricultural and biological science field as an indicator of the general interest for the topic. As can be seen in Figure 2, the number of publications recorded a constant and quite rapid increase between 2000 and 2011, with an average increase of 20 articles per years. Between 2012 and 2015 a drop in the number of publications per year can be detected, reaching a relative minimum in 2015. Then, the number of publications per year increased until 2020.

On the basis of Figure 2, in the time span from 2000 to 2020 two main peaks of interest for the silkworm and mulberry topics can be detected, in 2005 and 2011, also considering the ratio of the number of publications regarding the mulberry and silkworm topics to the total number of publications in the agricultural and biological science field. From 2012 to 2018, the ratio of the selected topics to the total number of publications decreased, reaching its minimum in 2015, and then remained stable for almost five years, although the absolute number of publications in the silkworm and mulberry topics increased.

To characterize the geographic distribution of the research in the sericultural topic, an analogue analysis was performed considering the number of publications per country. Figure 3 shows the top five countries for number of publications in the 2000–2020 period.

Figure 2. Publications per year (bars) and ratio between publications on the mulberry and silkworm topic in the sector "Agri" and total publications in Agricultural and Biological Science field (line).

The three most important countries for the research in the sericultural topic are China, where the number of publications increased from 3 to 180 per years, India and Japan, where the interest increased more or less until 2010 and then decreased.

A more specific analysis about the geographical extent of the research is the one fo- cused on collaborations among Universities and Research Institutes from different countries that aims to determine how the chosen topic can generate collaborations in the scientific community. Figure 4 shows in a graphic way how the web of collaborations for the seri- cultural topic is well distributed in the World among different countries. More detailed information about collaborations among different countries are presented in Table 3, where data about the first five countries for the total number of collaborations that led to articles publication between 2000 and 2020 are summarized; data of collaborations were derived from the affiliation metadata of the SCOPUS database.

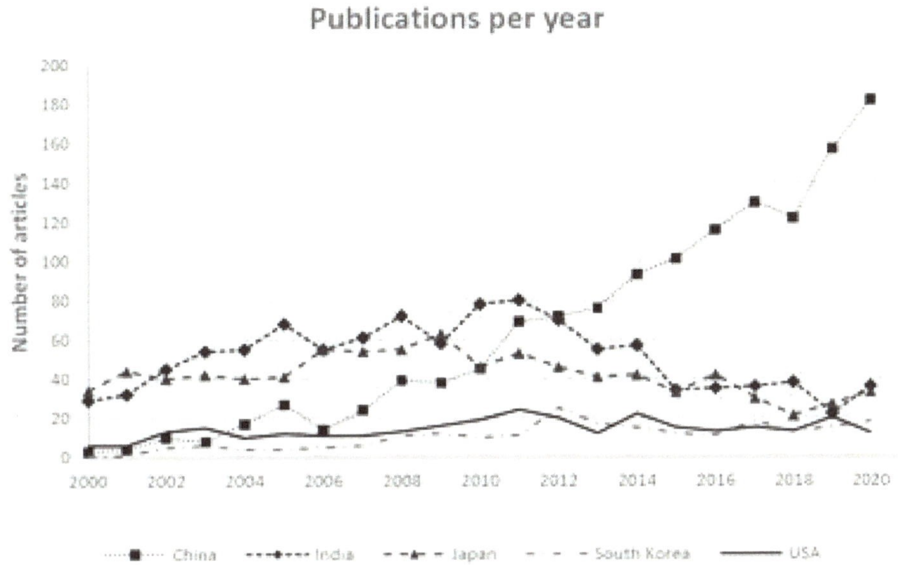

Figure 3. Geographic distribution of interest for the sericulture topic.

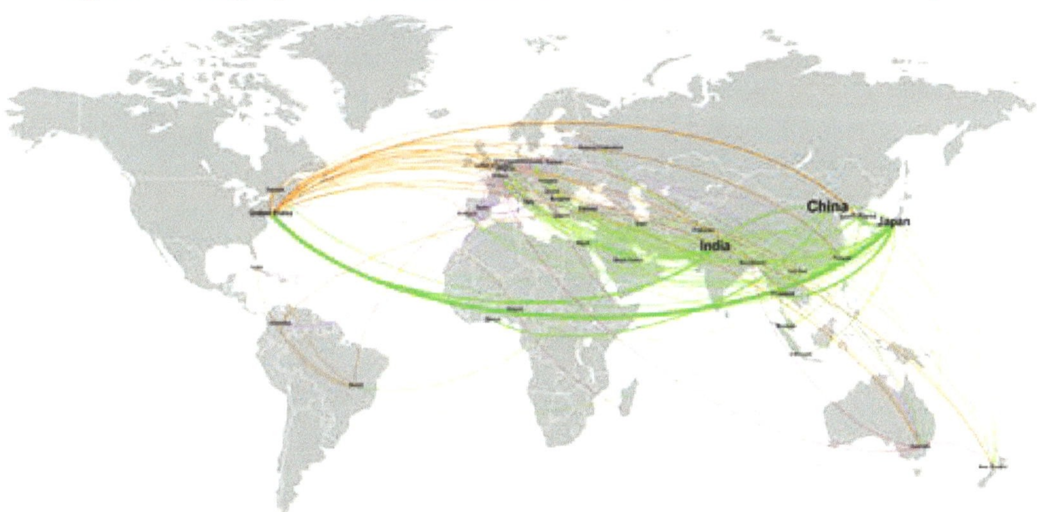

Figure 4. Geographic extent of collaborations for research in sericulture-related topics.

Country	Total Number of Collaborations	Number of Partner Countries	Average Number of Collaborations per Country
Japan	142	37	3.9
USA	102	41	2.6
India	74	24	3.1
China	52	25	2.1
Italy	27	22	1.6

As it can be noticed, the first country for number of collaborations is Japan with a total of 142 articles derived from cooperation. Japan collaborated with 37 different countries, with an average of 3.9 articles derived from the collaboration with each one. The USA are the first country for number of partners of research, but with a lower total amount of published articles and average number of collaborations per country.

3.2 Analysis of the Most Important Words

The authors then analyzed the collection of the most recurrent words previously derived as described in Section 2.2. The authors started the analysis on the list of the most recurrent words both in a cumulative way and considering their evolution over time too; this study was performed to characterize in a better way the research outcomes during the last two decades.

The authors choose three different parameters to characterize the evolution over time of the selected 100 words objectively. For the computation of those parameters, the authors used Microsoft Excel. The first parameter is the mean slope of the regression line that fits the number of publications per year; this parameter is calculated by Microsoft Excel dividing the variation on *y*-axis by the variation on *x*-axis. This parameter allows the authors to understand the rate of increment or decrement of the selected word during the investigated span of time. The second parameter is the *Pearson correlation coefficient*. This parameter is adimensional and ranges between +1 and −1. It expresses the linearity of the trend (i.e., evaluates whether the increment or decrement of the occurrences, for the selected word, are constant over the years). The third parameter is the last *five years relative change*, calculated as ratio between the average number of occurrences per word from 2016 to 2020 and the average number of occurrences from 2000 to 2015. Table 4 resumes some of the most cited nouns and their associated parameters.

Table 4. The most important words considering their slope, Pearson correlation coefficient and their relative increment in the past five years.

Word	Slope	Pearson Correlation Coefficient	Five Years Relative Changing [%]	Graph
Gene	18.2	0.85	+63%	
Protein	12.6	0.90	+68%	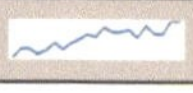
Acid	8.7	0.93	+172%	
Fruit	7.3	0.92	+134%	
Extract	9.6	0.93	+159%	
Morus alba	6.4	0.92	+118%	
Morus nigra	3.3	0.91	+161%	

Antioxidant	8.6	0.97	+218%	
Silk	4.9	0.81	+47%	
Expression	14.3	0.93	+106%	
Anthocyanin	4.5	0.93	+270%	
Phenolic	3.4	0.87	+278%	
Polyphenol	1.9	0.91	+276%	
Flavonoid	3.0	0.90	+274%	
Immune	2.4	0.88	+307%	
Juice	2.3	0.78	+234%	

Then, the same parameters used for characterizing words in the previous analysis were calculated out of the normalized number of occurrences per year; the normalized number of occurrences was calculated as ratio of the absolute value of the occurrences of the selected word and the absolute number of publications in the corresponding year. Results for the same nouns of Table 4 are reported in Table 5.

Table 5. Most important words normalized per topic considering their slope, Pearson regression coefficient and their relative increment in the past five years.

Word	Slope	Pearson Correlation Coefficient	Five Years Relative Changing [%]	Graph
Gene	2.30	0.54	+15%	
Protein	1.32	0.47	+16%	
Acid	2.24	0.87	+101%	
Fruit	1.99	0.85	+77%	
Extract	2.64	0.87	+96%	
Morus alba	1.24	0.76	+50%	
Morus nigra	0.83	0.76	+94%	
Antioxidant	2.73	0.96	+155%	
Silk	0.42	0.26	+3%	
Expression	3.45	0.84	+53%	
Anthocyanin	1.41	0.88	+194%	
Phenol	1.50	0.85	+215%	
Polyphenol	0.64	0.89	+217%	
Flavonoid	0.94	0.88	+200%	
Immune	0.76	0.86	+212%	
Juice	0.71	0.66	+166%	

3.3 Cluster Analysis

The 100 top frequent words have been grouped into six clusters. With reference to Equation (1), the authors used the weight assigned to each term to define the relative weight of the cluster to which that specific word belongs. Therefore, the relative weight of each cluster was calculated on the basis of the total sum of the relative weights of all the *n* words contained in the cluster itself. Table 6 shows the constitution of each cluster, with the relative weights of all the considered words.

Table 6. Clusters composition with the relative weight of each word and the overall relative weight of each cluster.

Cluster	Words	Cluster Relative Weight
Cultivation	Crop (1.6%), cultivar (1.4%), fruit (6.5%), germplasm (0.6%), leaf (15.2%), *Morus* sp.pl. (3.4%), *Morus alba* (6.5%), *Morus nigra* (2.9%), mulberry (32.2%), plant (7.7%), root (2.4%), seed (1.0%), shoot (1.3%), soil (2.0%), species (6.5%), tree (2.9%), variety (2.6%), water (3.3%)	23.6%
Silkworm	Body (1.8%), *Bombyx* (0.6%), *Bombyx mori* (10.7%), breed (1.6%), cocoon (3.4%), diapause (1.1%), diet (1.3%), disease (1.7%), DNA (1.9%), egg (2.3%), embryonic (0.5%), expression (7.4%), female (1.5%), gene (12.6%), gland (2.0%), hemolymph (1.1%), hormone (1.3%), hybrid (1.4%), immune (0.8%), infection (2.3%), instar (1.5%), juvenile (0.5%), larva (5.8%), Lepidoptera (1.5%), male (0.9%), metabolism (0.9%), midgut (1.2%), moth (0.8%), prothoracic (0.3%), pupa (2.7%), sex (0.8%), shell (0.9%), silkmoth (0.3%), silkworm (15.3%), spinning (2.9%) stage (2.4%), strain (2.9%), voltinism (1.0%)	46.4%
Rearing	Beauveria bassiana (5.1%), feeding (23.9%), heat (4.5%), humidity (2.7%), Nosema bombycis (5.1%), NucleoPolyhedroVirus (13.1%), rearing (8.0%), sericulture (7.4%), stress (9.5%), survival (7.2%), temperature (13.5%)	6.1%
Process	Drying (47.6%), fermentation (17.2%), industry (14.0%), purification (12.1%)	1.6%
Product	Acid (8.6%), aminoacid (2.4%) anthocyanin (3.4%), antioxidant (6.9%), composition (7.7%), diabetic (2.5%), edible (0.8%), enzyme (2.9%), extract (8.7%), lipid (4.3%), fiber (0.7%), fibroin (1.5%), filament (0.8%), flavonoid (2.3%), food (3.3%), juice (1.9%), oil (1.4%), peptides (1.9%), phenol (3.5%), polyphenol (1.4%), powder (1.4%), product (2.1%), protein (18.3%), quality (3.2%), sericin (1.5%), serine (1.1%), silk (8.2%), yield (3.5%),	22.2%

With reference to Table 4, the most important clusters are, in descending order, "silk-worm", "cultivation", and "product". The "silkworm" cluster represents about one half (46.4%) of the total occurrences considered for this analysis. The authors grouped in this cluster all the nouns that refer to the biology of the silkworm *B. mori*. In this cluster, about one forth (26.0%) of occurrences are in general about the silkworm itself, respectively with the words "silkworm" (15.3%) and "*B. mori*" (10.7%). In absolute terms the occurrences of the word "gene" (12.6%) saw a rapid increase (slope = 18.2) that was constant and quite linear during the considered period (Pearson coefficient = 0.85); this trend is maintained considering the ratio of occurrences over the number of publications per year in sericulture. In relation to "gene", the word "expression" (7.4%) shows a significant increasing both in absolute (slope = 14.3) and relative terms (slope = 3.45) Another important theme in the biology of silkworm is represented by the sex identification, and in this analysis this topic represents the 3.3% of the weight of the cluster, respectively with words "female" (1.5%), "male" (0.9%), and "sex" (0.8%). Other important nouns belonging to this cluster are "larva", that represents 5.8% of the total weight of the cluster, "cocoon", with 3.4%, "shell", with 0.9% and "spinning", with 2.9%. The cluster named "cultivation" groups all the aspects linked to the mulberry cultivation and mulberry field management; its overall relative weight is 23.6%. The most recurrent word is "mulberry", with about one third of weight of the cluster (32.2%). In particular, the most present one is the white mulberry, *M. alba* (L.), whose weight in the cluster is 6.5%; then the second most studied species is the black mulberry, *M. nigra* (L.), with 2.9% of the total weight of the cluster. For both the species, the increment of interest in the last two decades in absolute terms showed a constant and quite linear increase, with a slope = 6.4 for *M. alba* and 3.3 for *M. nigra* and a Pearson coefficient = 0.92 and 0.91 respectively. *M. nigra* also shows the highest increase of interest in the last five years, trend confirmed both in absolute terms (+161% vs. +118%) and in relative terms (+94% vs. +50%). Other two important words of this cluster are "leaf" (15.2%) and "fruit" (6.5%). The noun "leaf" shows a good growth but not so constant in time in absolute terms (slope = 10.2 and Pearson coefficient = 0.73). However, in relation to the increasing number of publications in the sericulture field the interest for it decreased (slope and Pearson coefficient near to zero but negative and relative increment in the last five years ± 0%). At the opposite, the interest of researchers for mulberry fruit increased both in absolute terms (slope = 7.3, Pearson coefficient = 0.92 and relative increment in the past five years = +134%) and in relative terms (slope = 1.99, Pearson coefficient = 0.85 and relative increment in the past five years = +77%). The third most relevant cluster is "Product", which is dedicated to the products obtainable from both mulberry cultivation and silkworm rearing. The relative weight of this cluster is 22.2%. In this cluster, the most important word is "protein". The importance and interest for the word "protein" arose in a rapid and linear way during the last two decades (slope = 12.6 and Pearson coefficient = 0.90) in absolute terms; in relative terms this increase is not so high, with a peak of interest during the beginning of 2000s (see graph in Table 5). An analogue consideration could be made for the word "silk" that shows a quite high interest in absolute terms, but this increase in interest is biased by the increasing number of publications per year in the sericultural field. In fact, when we perform the normalization of the trend on the number of publications per year, we can see that this interest has been constant for researchers. Linked to the word "silk" there are nouns referring to its principal components, "fibroin" (1.5%) and "sericin" (1.5%). In this cluster, another important category of words is represented by beneficial compounds that could be extracted from mulberry leaves. These words are "antioxidant" (6.9%), "anthocyanin" (3.4%), "phenol" (3.5%), "flavonoid" (2.3%) and "polyphenol" (1.4%) and their sum weighs about 17% of the whole cluster. In particular, the interest for this group of words rapidly increased in the past five years, both in absolute and relative terms, as attested by the parameter of relative increase reported in Tables 4 and 5.

As last analysis, the authors evaluated the temporal trend of the aggregated number of occurrences per cluster. Furthermore, the trend of the normalized total occurrences as a ratio to the number of publications was analyzed. The results are reported in Tables 7 and 8.

Table 7. Analysis of cluster considering their slope, Pearson regression coefficient, and their relative increment in the past five years.

Word	Slope	Pearson Correlation Coefficient	Five Years Relative Changing [%]	Graph
Cultivation	83.4	0.93	+80%	
Process	7.4	0.89	+144%	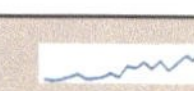
Product	90.0	0.98	+117%	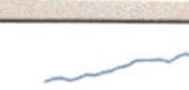
Rearing	18.9	0.87	+71%	
Silkworm	105.7	0.79	+33%	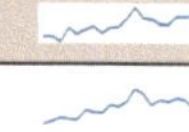

Table 8. Analysis of clusters normalized per cluster by considering their slope, Pearson regression coefficient, and their relative increment in the past five years.

Word	Slope	Pearson Correlation Coefficient	Five Years Relative Changing [%]	Graph
Cultivation	11.57	0.79	+25%	
Process	1.64	0.82	+73%	
Product	18.53	0.93	+55%	
Rearing	1.55	0.40	+16%	
Silkworm	-6.67	-0.35	-10%	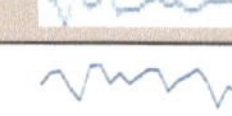

From the analysis of Tables 5 and 6 some interesting considerations can be derived. In particular, the "silkworm" cluster, in absolute terms, has the highest value of slope, followed by "product" and "cultivation"; the cluster "process" is the one with the major relative increase of publications in the last five years. On the other hand, when we normalize the number of publications per year, the silkworm cluster evidences a negative slope, which, although linked to the relative increment of the last five years, testifies a slow decreasing interest in this topic. "Cultivation", "process" and "product" are the cluster that maintain a high relative interest in the field of sericultural research.

3.4 Impact Factor Analysis

The objective of this analysis was to derive sound information about the impact that the most frequent terms generated on scientific research. The proposed impact factors of considered word are reported in Table 9.

Table 9. Calculated impact factor of the most frequent words for the five clusters. The words are sorted in descending order considering their weights according to Equation (1); h-indices in brackets.

Cluster	Words	Cluster Mean
Cultivation	Mulberry (89), leaf (77), plant (94), *Morus alba* (75), species (84)	84
Silkworm	Silkworm (78), gene (95), *Bombyx mori* (77), expression (73), larva (63)	77
Rearing	Feeding (62), temperature (55), nucleopolyhedroviral (35), stress (60), rearing (61)	51
Process	Drying (38), fermentation (39), industry (42), purified (66)	46
Product	Protein (82), extract (82), acid (97), silk (56),antioxidant (77)	79

As reported in Table 9, the "cultivation" and "product" clusters are the ones exhibiting the highest impact, highlighting a vivid interest by the scientific community on these topics. Also, the "silkworm" cluster gives evidence of a significant impact: in this case, the high h-index should also be related to the highest number of papers published on this topic since 2000. A lower interest is apparently arising from research related to the process or to entomological aspects. On the other hand, the specific terms, "acid" (97) and "gene" (95) gained the highest impact, having collected more than 35,000 citing documents each.

3.5 Interrelationships among Words

The objective of this analysis was to derive sound information about the interrelation- ships among the analyzed words. To achieve this aim, each of the 100 most frequent words previously mentioned was coupled with each of the remaining 99 words, thus generating a table of 4950 possible combinations (avoiding repetitions). Each noun of this collection of paired words was studied in terms of total number of occurrences for the last considered 20 years.

As reported in Table 10, "silkworm" is the cluster with the highest number of cooccurrences, both inside the cluster itself and with other clusters; the maximum number of occurrences was reached by the interaction of "silkworm" with itself (61,864), with the cluster "cultivation", (20,626) and with the cluster "product" (8112), attesting the great attention of the scientific community to themes focused on the biology and life cycle of silkworms and to the production that could be derived from them. The cluster "cultivation", as expected, is the second one for number of co-occurrences; as for the cluster "silkworm", the co-interactions are mainly concentrated on the cluster "cultivation" itself and on the cluster "product". In this case, the interaction with the cluster "process" is also high.

Table 10. Number of interrelationships among clusters

	Cultivation	Process	Product	Rearing	Silkworm
Cultivation	15,203	-	-	-	-
Process	17,010	25	-	-	-
Product	19,756	2156	13,028	-	-
Rearing	5415	520	6186	924	-
Silkworm	20,626	2396	50,422	12,135	61,864

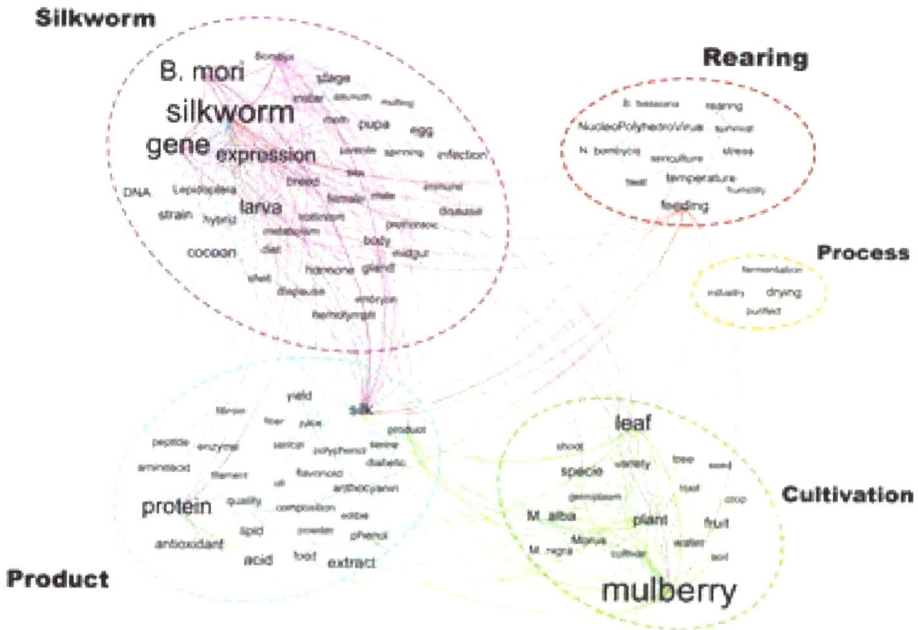

Figure 5. Interrelations among the top 100 most frequent words. In this picture, the curved lines represent relevant interrelations (number of co-occurrences is greater than 100).

Then, the authors proposed a brief analysis of the most recurrent couples of nouns, independently on the original cluster, to stress the importance of the most relevant topics. The couple of words with the highest number of co-occurrences is silk-silkworm (2435) *Bombyx-B. mori* (2097), and *B. mori*-Silk (1953). The first couple for co-occurrences related to the theme of mulberry cultivation is the couple leaf-mulberry (924) followed by *Morus-M. alba* (666); several couples of terms regard the area of genetic studies, as Silk-Gene (1366), Silkworm-Gene (1297), *Bombyx*-Gene (1245) and *B. mori*-Gene (1243).

Taking advantages of the Open-Source software Gephi, the authors generated the map of the interrelationship between the 100 most frequent terms showed in Figure 5.

4. DISCUSSION

The main core of this research is represented by a text-mining analysis focused on the words in the title, abstract, and keywords of articles related to silk production. Different considerations can be drawn by the analysis of the most frequent nouns, which were previously identified and evaluated. Taking advantage of a procedure algorithm already tested by Cogato et al. and Ferrari et al. [38,39], the most significant relationships among nouns were recognized. Additionally, by including time parametrization, it was also possible to characterize the evolution of the research interests in sericulture. Thanks to the large number of documents considered in this study, a statistically robust analysis of research was achieved, supporting hypotheses on influences and trends in sericulture

From the characterization of the number of released publications in sericulture (see Figure 2), it is worth noting that the absolute number of articles published per year has been increasing since 2000; such trend reflects the great interest in the topic and the interdisciplinarity of the matter. By considering this trend, two main periods can be identified: from 2000 to 2011 the trend was constantly increasing, both in absolute number and in relative percentage (considering the total number of publications indexed by the SCOPUS database), reaching a peak in 2011; the second period began in 2012 and exhibited a decrease followed by an increase until 2017. Currently, the interest is still apparently growing in absolute terms. In relative terms, the interest appears quite stable; this is due to a rather great relative increase in the total number of publications per year (from about 62 k in 2000 to more than 200 k in 2020, documents indexed by the SCOPUS database). The next decades should be monitored to determine whether these new trends are stable, or they are not. From a geopolitical and temporal point of view, such a trend should be explained by two main reasons: the decreasing number of publications per year from India, and the increasing number of publications per year from China, which is now the leading country for the research in the sericultural field.

From the analysis of the trends of single words and clusters, interesting and more specific information have been extracted. From an aggregate point of view, considering the clusters, the main arguments of investigation of the research in the past twenty years were the study of the *Morus* cultivation, industrial processes, and products that could derive from sericulture, from both *Morus* cultivation and *B. mori* rearing.

From the analysis of the impact of the most important words of each cluster, it is interesting to note how those words are able to generate a considerable impact in the frame of sericulture, with an overall average impact factor of about 67.

The cluster "silkworm", related to the silkworm biology, is the largest one in terms of number of included words (according to the data of Table 6, 46.4%) but during the last decade, and in particular in the last five years, the interest of the scientific research decreased relatively to this sericultural main topic (Table 8). In particular, the most important words of the cluster are "silkworm" (15.3%) and "*B. mori*" (10.7%), also characterized by considerable high values of the impact factor (respectively 78 and 77); this is easily understandable since the work on the living organism is the main focus of the research on this topic. The silkworm has always been an important laboratory tool for genetics and physiology of insects; in particular the independent genome sequencing of 2004 by a Chinese [42] and a Japanese team [43] and the integrated silkworm database of the end of 2009 [44], gave probably place to the peak of publication on sericulture recorded in 2011. However, from 2012 (the year of FAO experts' consultation in Rome about entomophagy "Insects to feed the world") new insects started to be studied in relationship with the new trend of edible insects as perspective feed and food for the future; for example, *Tenebrio molitor* genome was sequenced in 2020 [45]. Exploitation of edible insects gave rise to a new impulse to research on the physiology of new and less known insects as *T. molitor* and *Hermetia illucens*, probably resulting in a diminished interest for traditional insects used as laboratory tools like silkworms.

However, silkworm genetics, still accounts for about 22% of the weight of this cluster. The Nucleopolyhedrosis virus (NPV) has also been used for genetic studies, and in particular as a vector for the recombinant protein technology [46–48]. NPV research could, therefore, be considered for its importance in determining both methods for biotechnologies and silkworm physiology studies or in establishing correct rearing techniques for disinfection, disease prevention and rearing facility planning. For this reason, NPV research and other studies on silkworm diseases like pebrine (*Nosema bombycis*) or white muscardine (*Beauveria bassiana*) could be regarded as horizontal literature common to the cluster "Silkworm" and "Rearing".

Another important theme of the silkworm biology is represented by the sex identifi- cation. Indeed, in order to obtain a highly-productive polyhybrid, the best male and best female individuals of different strains are reared and bred. As stated by Raj et al. [23], different solutions for non-destructive sex-sorting methods for silkworms have been proposed [23,41,49]. In fact, an efficient sexing method has not been found yet. Three other important terms of the cluster "Silkworm", counting about 10% of the weight of the cluster, are "larva", "cocoon" and "shell", the raw materials for farmers who rear the larval stage of silkworm and obtain, at the end of their fifth larval instar, fresh silk cocoons.

The cluster "cultivation", related to mulberry cultivation, is the second most important one in terms of the relative weight of the included words (according to the data of Table 6, about one fourth). Since mulberry is the general word for the identification of *Morus* sp. pl. plants, the most interesting and reliable results are about specific species. As stated before, silkworms feed mainly on *M. alba* leaves [27,28], and the term "*Morus alba*" showed a high increases in the past twenty years. The term "*Morus nigra*" recorded a lower increase, located in particular in the 2016–2020 period. The term "leaf" is the second most important word in this cluster, and this is related to the fact that mulberry leaves represent the only source of food for silkworms [27,28]. It is worth noting how "leaf" is a word that has not exhibited an increase, as for others considered terms (Table 5). The explanation of this trend could be related to the fact that *Morus* is currently studied much more for fruits and active ingredients for pharmaceuticals than as feed for the silkworm, for which artificial diets are an innovative and attractive matter of research.

In the "product" cluster, the most recurrent word is "protein", which represents the most important class of nutrients in mulberry leaves, the only source of nitrogen for silkworms. Thus, silkworms use the proteins obtained through dietary ingestion in order to extract the amino acids necessary to synthesize other required proteins both for their growth and for spinning cocoons [50]. Proteins are, therefore, also important products obtained from cocoons (fibroin and sericin) and from silkworm pupae. Recombinant proteins achieved from engineered larvae also represent a "hot spot" of sericulture [46–48,51]. It is worth noticing how the nouns "acid" (97), "extract" (82) and "protein" (82) are characterized by very high values of the impact factor here proposed, attesting a great interest of the scientific research about the topics related to these nouns.

The "process" cluster is the one with the lowest number of occurrences and thus with the lowest weight (1.6%). Nonetheless, an actual interest on the topic can be recognized, especially in the last five years (2016–2020), when the cluster gave evidence of the high- est increment. New industrial processes are currently required for recombinant protein purification, for fibroin and sericin dissolution and reconstitution, to build 3D scaffolds, biomedical products, biocompatible implantable devices and biofilms. Furthermore, to purify mulberry organic compounds, fat and proteins from pupae, new methods and processing technologies are necessary and this explains the increase of the last years linked to the emerging fields of sericultural biotechnologies, pharmaceuticals from mulberry, and edible insects [52–54].

In the "rearing" cluster the two most relevant themes are those related to the words "feeding" (already discussed above) and "temperature". The word "temperature" is the one with highest weight in the cluster (13.5%) and this trend is somehow confirmed also by the high attention paid by farmers as well as by researchers to the temperature control for the sericultural practice [26,55,56]; in fact, some authors proposed sensor-based systems that allow the automatic control of temperature (concurrently with other environmental parameters) [57]. For the same reason, a clear interest is also focused on other two strictly related nouns namely "heat" (4.5%) and "humidity" (2.7%). We should also highlight that climatic changes have impacted on all the agricultural activities and sericulture was also much affected by the extreme environmental conditions recorded in the last years.

5. CONCISE SUMMARY

In light of the renewed interest in natural fibers, as seen by the increase in publications from 367 in 2000 to 1,768 in 2020 according to the SCOPUS database, this study presents a comprehensive overview of the findings from scientific research on silk. The analytical results indicated a significant quantity of materials pertaining to traditional issues of silk production, including silk quality, the biology of Bombyx mori, and the cultivation of Morus alba.

The investigation has identified knowledge gaps in sericulture, particularly a paucity of research concerning process automation themes. The automation of labor-intensive tasks such as feeding and environmental regulation in rearing rooms can be effectively achieved through the advancements provided by Machine Learning and Artificial Intelligence, representing a significant area of research for the future of sericulture.

REFERENCES

1. Goldsmith, M.R.; Shimada, T.; Abe, H. The Genetics and Genomics of the Silkworm, Bombyx Mori. *Annu. Rev. Entomol.* **2005**, *50*, 71–100. [CrossRef] [PubMed]

2. Kurin, R.; Borden, C.M. The silk road: Connecting cultures, creating trust. *Talk Story* **2002**, *21*, 1–11.

3. Datta, R.K.; Nanavaty, M. *Global Silk Industry: A Complete Source Book*; Universal-Publishers: Irvine, CA, USA, 2005; ISBN 1581124937.

4. Lopez, R.S. China silk in Europe in the Yuan period. *J. Am. Orient. Soc.* **1952**, *72*, 72–76. [CrossRef]

5. International Sericultural Commission. Statistics about Global Silk Producton. Available online: https://www.inserco.org/en/ statistics (accessed on 24 May 2022).

6. Meinel, L.; Betz, O.; Fajardo, R.; Hofmann, S.; Nazarian, A.; Cory, E.; Hilbe, M.; McCool, J.; Langer, R.; Vunjak-Novakovic, G.; et al. Silk based biomaterials to heal critical sized femur defects. *Bone* **2006**, *39*, 922–931. [CrossRef]

7. Zhou, Q.Z.; Fu, P.; Li, S.S.; Zhang, C.J.; Yu, Q.Y.; Qiu, C.Z.; Zhang, H.B.; Zhang, Z. A Comparison of Co-expression Networks in Silk Gland Reveals the Causes of Silk Yield Increase During Silkworm Domestication. *Front. Genet.* **2020**, *11*, 225. [CrossRef]

8. Porter, D.; Guan, J.; Vollrath, F. Spider silk: Super material or thin fibre? *Adv. Mater.* **2013**, *25*, 1275–1279. [CrossRef]

9. Mortimer, B.; Gordon, S.D.; Holland, C.; Siviour, C.R.; Vollrath, F.; Windmill, J.F.C. The speed of sound in silk: Linking material performance to biological function. *Adv. Mater.* **2014**, *26*, 5179–5183. [CrossRef]

10. Altman, G.H.; Diaz, F.; Jakuba, C.; Calabro, T.; Horan, R.L.; Chen, J.; Lu, H.; Richmond, J.; Kaplan, D.L. Silk-based biomaterials. *Biomaterials* **2003**, *24*, 401–416. [CrossRef]

11. Jewell, M.; Daunch, W.; Bengtson, B.; Mortarino, E. The development of SERI® Surgical Scaffold, an engineered biological scaffold. *Ann. N. Y. Acad. Sci.* **2015**, *1358*, 44–55. [CrossRef]

12. van Turnhout, A.A.W.M.; Franke, C.J.J.; Vriens-Nieuwenhuis, E.J.C.; van der Sluis, W.B. The use of SERITM Surgical Scaffolds in direct-to-implant reconstruction after skin-sparing mastectomy: A retrospective study on surgical outcomes and a systematic review of current literature. *J. Plast. Reconstr. Aesthetic Surg.* **2018,** 71, 644–650. [CrossRef]

13. Holland, C.; Numata, K.; Rnjak-Kovacina, J.; Seib, F.P. The Biomedical Use of Silk: Past, Present, Future. *Adv. Healthc. Mater.* **2019**, *8*, 1800465. [CrossRef] [PubMed]

14. Melke, J.; Midha, S.; Ghosh, S.; Ito, K.; Hofmann, S. Silk fibroin as biomaterial for bone tissue engineering. *Acta Biomater.* **2016**, *31*, 1–16. [CrossRef] [PubMed]

15. Kasoju, N.; Bora, U.; Kasoju, N.; Bora, U. Silk Fibroin in Tissue Engineering. *Adv. Healthc. Mater.* **2012**, *1*, 393–412. [CrossRef] [PubMed]

16. Gupta, D.; Agrawal, A.; Chaudhary, H.; Gulrajani, M.; Gupta, C. Cleaner process

for extraction of sericin using infrared. *J. Clean. Prod.* **2013**, *52*, 488–494. [CrossRef]

17. Chlapanidas, T.; Faragò, S.; Lucconi, G.; Perteghella, S.; Galuzzi, M.; Mantelli, M.; Avanzini, M.A.; Tosca, M.C.; Marazzi, M.; Vigo, D.; et al. Sericins exhibit ROS-scavenging, anti-tyrosinase, anti-elastase, and in vitro immunomodulatory activities. *Int. J. Biol. Macromol.* **2013**, *58*, 47–56. [CrossRef]
18. Dong, X.; Zhao, S.X.; Yin, X.L.; Wang, H.Y.; Wei, Z.G.; Zhang, Y.Q. Silk sericin has significantly hypoglycaemic effect in type 2 diabetic mice via anti-oxidation and anti-inflammation. *Int. J. Biol. Macromol.* **2020**, *150*, 1061–1071. [CrossRef]
19. Aramwit, P.; Siritientong, T.; Srichana, T. Potential applications of silk sericin, a natural protein from textile industry by-products. *Waste Manag. Res.* **2012**, *30*, 217–224. [CrossRef]
20. Zhang, Y.Q.; Tao, M.L.; Shen, W.D.; Zhou, Y.Z.; Ding, Y.; Ma, Y.; Zhou, W.L. Immobilization of L-asparaginase on the microparticles of the natural silk sericin protein and its characters. *Biomaterials* **2004**, *25*, 3751–3759. [CrossRef]
21. Rocha, L.K.H.; Favaro, L.I.L.; Rios, A.C.; Silva, E.C.; Silva, W.F.; Stigliani, T.P.; Guilger, M.; Lima, R.; Oliveira, J.M.; Aranha, N.; et al. Sericin from *Bombyx mori* cocoons. Part I: Extraction and physicochemical-biological characterization for biopharmaceutical applications. *Process Biochem.* **2017**, *61*, 163–177. [CrossRef]
22. Zhang, Y.Q.; Ma, Y.; Xia, Y.Y.; Shen, W.D.; Mao, J.P.; Xue, R.Y. Silk sericin-insulin bioconjugates: Synthesis, characterization and biological activity. *J. Control. Release* **2006**, *115*, 307–315. [CrossRef]
23. Raj, A.N.J.; Sundaram, R.; Mahesh, V.G.V.; Zhuang, Z.; Simeone, A. A Multi-Sensor System for Silkworm Cocoon Gender Classification via Image Processing and Support Vector Machine. *Sensors* **2019**, *19*, 2656. [CrossRef]
24. Grekov, D.; Kipriotis, E.; Tzenov, P. *Sericulture Training Manual*; National Agricultural Research Foundation: Komotini, Greece, 2005.
25. Cappellozza, S.; Saviane, A. Aspetti attuali dell'allevamento del baco da seta. In *Atti Seminario Insetti Utili*; MULSA (Museo Lombardo di Storia dell'Agricoltura: Milano, Italy, 2014; pp. 71–78.
26. Rahmathulla, V.K. Management of climatic factors for successful silkworm (*Bombyx mori* L.) crop and higher silk production: A review. *Psyche* **2012**, *2012*, 121234. [CrossRef]
27. Lim, S.-H.; Kim, Y.-T.; Lee, S.-P.; Rhee, I.-J.; Lim, J.-S.; Lim, B.-H. *Sericulture*

Training Manual; Food & Agriculture Org.: Rome, Italy, 1990; ISBN 9251029040.

28. ITO, T. Nutritional requirements of the silkworm, *Bombyx mori* L. *Proc. Jpn. Acad.* **1967**, *43*, 57–61. [CrossRef]

29. Ayandokun, A.E.; Alamu, O.T. Cocoon production efficiency of silkworm (*Bombyx mori* L.) in response to host shift between two selected mulberry varieties. *Int. J. Trop. Insect Sci.* **2020**, *40*, 49–52. [CrossRef]

30. Zhang, M.; Chen, M.; Zhang, H.Q.; Sun, S.; Xia, B.; Wu, F.H. In vivo hypoglycemic effects of phenolics from the root bark of *Morus alba*. *Fitoterapia* **2009**, *80*, 475–477. [CrossRef] [PubMed]

31. Imran, M.; Khan, H.; Shah, M.; Khan, R.; Khan, F. Chemical composition and antioxidant activity of certain *Morus* species. *J. Zhejiang Univ. Sci. B* **2010**, *11*, 973–980. [CrossRef]

32. Lin, J.Y.; Tang, C.Y. Determination of total phenolic and flavonoid contents in selected fruits and vegetables, as well as their stimulatory effects on mouse splenocyte proliferation. *Food Chem.* **2007**, *101*, 140–147. [CrossRef]

33. Kim, S.Y.; Gao, J.J.; Lee, W.C.; Ryu, K.S.; Lee, K.R.; Kim, Y.C. Antioxidative flavonoids from the leaves of *Morus alba*. *Arch. Pharmacal Res.* **1999**, *22*, 81–85. [CrossRef]

34. Enkhmaa, B.; Shiwaku, K.; Katsube, T.; Kitajima, K.; Anuurad, E.; Yamasaki, M.; Yamane, Y. Mulberry (*Morus alba* L.) Leaves and Their Major Flavonol Quercetin 3-(6-Malonylglucoside) Attenuate Atherosclerotic Lesion Development in LDL Receptor-Deficient Mice. *J. Nutr.* **2005**, *135*, 729–734. [CrossRef]

35. Chen, Z.; Zhu, C.; Han, Z. Effects of aqueous chlorine dioxide treatment on nutritional components and shelf-life of mulberry fruit (*Morus alba* L.). *J. Biosci. Bioeng.* **2011**, *111*, 675–681. [CrossRef]

36. Kimura, T.; Nakagawa, K.; Kubota, H.; Kojima, Y.; Goto, Y.; Yamagishi, K.; Oita, S.; Oikawa, S.; Miyazawa, T. Food-Grade Mulberry Powder Enriched with 1-Deoxynojirimycin Suppresses the Elevation of Postprandial Blood Glucose in Humans. *J. Agric. Food Chem.* **2007**, *55*, 5869–5874. [CrossRef] [PubMed]

37. Awad, N.E.; Seida, A.A.; Hamed, M.A.; Mahmoud, A.H.; Elbatanony, M.M. Phytochemical and in vitro screening of some *Ficus* and *Morus* spp. for hypolipidaemic and antioxidant activities and in vivo assessment of *Ficus mysorensis* (Roth). *Nat. Prod. Res.* **2012**, *26*, 1101–1111. [CrossRef] [PubMed]

38. Cogato, A.; Meggio, F.; Migliorati, M.D.A.; Marinello, F. Extreme Weather Events in Agriculture: A Systematic Review. *Sustain- ability* **2019**, *11*, 2547. [CrossRef]

39. Ferrari, G.; Pezzuolo, A.; Nizami, A.S.; Marinello, F. Bibliometric Analysis of Trends in Biomass for Bioenergy Research. *Energies* **2020**, *13*, 3714. [CrossRef]

40. Bastian, M.; Heymann, S.; Jacomy, M. Gephi: An Open Source Software for Exploring and Manipulating Networks. *Proc. Int. AAAI Conf. Web Soc. Media* **2009**, *3*, 361–362.

41. Mahesh, V.G.V.; Raj, A.N.J.; Celik, T. Silkworm cocoon classification using fusion of Zernike moments-based shape descriptors and physical parameters for quality egg production. *Int. J. Intell. Syst. Technol. Appl.* **2017**, *16*, 246–268. [CrossRef]

42. Xia, Q.; Zhou, Z.; Lu, C.; Cheng, D.; Dai, F.; Li, B.; Zhao, P.; Zha, X.; Cheng, T.; Chai, C.; et al. A Draft Sequence for the Genome of the Domesticated Silkworm (*Bombyx mori*). *Science* **2004**, *306*, 1937–1940. [CrossRef]

43. Mita, K.; Kasahara, M.; Sasaki, S.; Nagayasu, Y.; Yamada, T.; Kanamori, H.; Namiki, N.; Kitagawa, M.; Yamashita, H.; Yasukochi, Y. The genome sequence of silkworm, *Bombyx mori*. *DNA Res.* **2004**, *11*, 27–35. [CrossRef]

44. Shimomura, M.; Minami, H.; Suetsugu, Y.; Ohyanagi, H.; Satoh, C.; Antonio, B.; Nagamura, Y.; Kadono-Okuda, K.; Kajiwara, H.; Sezutsu, H.; et al. KAIKObase: An integrated silkworm genome database and data mining tool. *BMC Genom.* **2009**, *10*, 486. [CrossRef]

45. Eriksson, T.; Andere, A.A.; Kelstrup, H.; Emery, V.J.; Picard, C.J. The yellow mealworm (*Tenebrio molitor*) genome: A resource for the emerging insects as food and feed industry. *J. Insects Food Feed.* **2020**, *6*, 445–455. [CrossRef]

46. Maeda, S.; Kawai, T.; Obinata, M.; Fujiwara, H.; Horiuchi, T.; Saeki, Y.; Sato, Y.; Furusawa, M. Production of human α-interferon in silkworm using a baculovirus vecto. *Nature* **1985**, *315*, 592–594. [CrossRef] [PubMed]

47. Ogawa, S.; Tomita, M.; Shimizu, K.; Yoshizato, K. Generation of a transgenic silkworm that secretes recombinant proteins in the sericin layer of cocoon: Production of recombinant human serum albumin. *J. Biotechnol.* **2007**, *128*, 531–544. [CrossRef] [PubMed]

48. Wang, F.; Xu, H.; Yuan, L.; Ma, S.; Wang, Y.; Duan, X.; Duan, J.; Xiang, Z.; Xia, Q. An optimized sericin-1 expression system for mass-producing recombinant proteins in the middle silk glands of transgenic silkworms. *Transgenic Res.* **2013**, *22*, 925–938. [CrossRef] [PubMed]

49. Liu, C.; Ren, Z.H.; Wang, H.Z.; Yang, P.Q.; Zhang, X.L. Analysis on gender of silkworms by MRI technology. In Proceedings of the BioMedical Engineering and Informatics: New Development and the Future—Proceedings of the 1st

International Conference on BioMedical Engineering and Informatics, BMEI 2008, Sanya, China, 27–30 May 2008; Volume 2, pp. 8–12. [CrossRef]

50. Yogananda Murthy, V.N.; Ramesh, H.L.; Lokesh, G.; Munirajappa, D.Y.B.R.; Yadav, B.R. Leaf quality evaluation of ten mulberry (*Morus*) germplasm varieties through phytochemical analysis. *Int. J. Pharm. Sci. Rev. Res.* **2013**, *21*, 182–189.

51. Tomital, M.; Munetsuna, H.; Sato, T.; Adachi, T.; Hino, R.; Hayashi, M.; Shimizu, K.; Nakamura, N.; Tamura, T.; Yoshizato, K. Transgenic silkworms produce recombinant human type III procollagen in cocoons. *Nat. Biotechnol.* **2003**, *21*, 52–56. [CrossRef] [PubMed]

52. Nongonierma, A.B.; FitzGerald, R.J. Unlocking the biological potential of proteins from edible insects through enzymatic hydrolysis: A review. *Innov. Food Sci. Emerg. Technol.* **2017**, *43*, 239–252. [CrossRef]

53. Kim, H.W.; Setyabrata, D.; Lee, Y.J.; Jones, O.G.; Kim, Y.H.B. Pre-treated mealworm larvae and *silkworm pupae* as a novel protein ingredient in emulsion sausages. *Innov. Food Sci. Emerg. Technol.* **2016**, *38*, 116–123. [CrossRef]

54. Grabowski, N.T.; Klein, G. Microbiology of processed edible insect products—Results of a preliminary survey. *Int. J. Food Microbiol.* **2017**, *243*, 103–107. [CrossRef]

55. Hsieh, F.K.; Yu, S.J.; Su, S.Y.; Peng, S.J. Studies on the thermotolerance of the *silkworm, Bombyx mori*. *Chin. J. Entomol.* **1995**, *15*, 91–101.

56. Ueda, S.; Lizuka, H. Studies on the effects of rearing temperature affecting the health of silkworm larvae and upon the quality of cocoons-1 Effect of temperature in each instar. *Acta Sericologia Jpn.* **1962**, *41*, 6–21.

57. Dixit, M.A.; Kulkarni, A.; Raste, N.; Bhandari, G. Intelligent control system for sericulture. In Proceedings of the 2015 International Conference on Pervasive Computing (ICPC), Pune, India, 8–10 January 2015. [CrossRef]

58. Yang, J.; Guo, X.; Li, Y.; Marinello, F.; Ercisli, S.; Zhang, Z. A survey of few-shot learning in smart agriculture: Developments, applications, and challenges. *Plant Methods* **2022**, *18*, 28. [CrossRef] [PubMed]

4 Chapter

THE ROLE OF SILK FIBROIN IN BIOMEDICAL ENGINEERING.

1. INTRODUCTION: AN OVERVIEW

In recent years, research in the biomaterials domain has significantly advanced due to remarkable developments and the introduction of numerous new materials [1]. Consequently, biomaterials are extensively utilized in medicine owing to their unique amalgamation of durability and biological compatibility. Their capacity to stimulate tissue and organ regeneration has rendered it an exceptional choice for therapeutic interventions [2,3]. Furthermore, contemporary medicine has progressively employed many synthetic or natural substances, together referred to as "biomaterials," to enhance human quality of life and lifespan. A biomaterial is designed to engage with biological systems to assess, treat, augment, or substitute any tissue, organ, or bodily function.

Numerous biomaterials are used in the human body, including metals, ceramics, and synthetic and natural polymers [4]. Of these, natural polymers (biopolymers) form the largest class of biomaterials used in medicine, possessing a number of unique properties covering a wide variety of clinical applications, especially in regenerative medicine and tissue engineering [5]. At the same time, they are able to mimic the structure of human tissue due to their physical and chemical resemblance, and most importantly, they are less toxic to native healthy tissue and more biocompatible compared to most synthetic materials [6,7].

Among biopolymers, silk fibroin (SF) has attracted more and more attention from life scientists all over the world from various fields such as chemistry, physics, engineering, biology, and last but not least, medicine. This protein is regarded as a remarkable bioactive material for tissue engineering applications due to its biocompatibility, biodegradability, and tunability [8], and it is studied in the production of 3D scaffolds mixed with various materials (natural or synthetic polymers, drugs, nanoparticles, growth factors, bioactive molecules, etc.) that together promote wound healing [9–14]. In addition, it can be processed in a wide range of formats, including sponges, wafers, gauze, particles, hydrogels, fibers, and films [15], being feasible for many other applications. SF as a biomaterial dis- plays remarkable biological and mechanical properties such as controlled biodegradability, high biocompatibility, optical transparency, flexibility, mechanical resistance and processability. It is one of the main natural fibers produced by the *Bombyx mori* L. silkworm, and the purification and further processing of this protein as regenerated silk fibroin (RSF) can be obtained by eco-friendly methods that reduce or remove the need for chemical sol- vents and energy-intensive equipment. These eco-properties combined with a water-based extraction and purification process make SF a promising material for sustainable manu- facturing, enabling it to partially replace synthetic, plastic-based and non-biodegradable material [16,17].

In recent years, various research have examined the extensive application of SF as a functional biomaterial in the biomedical domain. This review aims to elucidate the exceptional mechanical and biological properties of silk fibroin (SF), the eco-friendly processing techniques, recent advancements in tissue engineering where SF is pivotal in replicating cellular microenvironments, and the significance of SF-based biomaterials in drug delivery and cancer treatment. Figure 1 illustrates a schematic representation of the primary SF application.

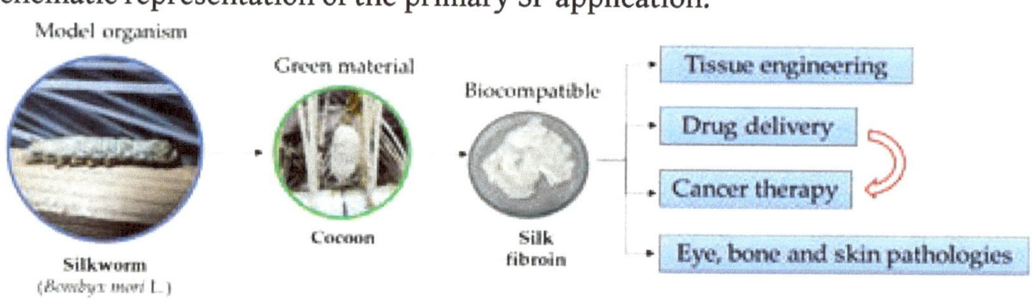

Figure 1. Schematic representation of SF main applications.

2. SF—OVERVIEW

Silks, as natural polymers, play a pivotal role for insects, being involved in the processes of survival and reproduction. Owning remarkable properties, silk has been the subject of various studies for more than a century [18,19], and it is appreciated by the scientific community in the field of biology, biotechnology and materials [20]. The main sources for this outstanding natural protein biopolymer are the spider (*Nephila clavipes*) and the silkworm (*B. mori*). *N. clavipes* exhibits certain behaviors that do not allow for the spider's large-scale harvesting [21]. Conversely, for more than 5000 years, silkworms have been domesticated and are currently being reared in numerous countries, such as China, India, and Japan, etc. As anticipated, the silk obtained from *B. mori* has been the most explored. Keeping this in mind, silkworms are the most feasible biological system in the context of silk production [22–24]. In the past, the silkworms were reared in order to obtain silk for textile applications; however, currently, silk receives considerable attention for its various purposes in life science areas [25].

The silk produced by *B. mori* is composed of two proteins, fibroin and sericin. Of the two proteins, fibroin (~400 kDa) is the core peptide. The silk fibers involve two fibroin filaments banded together by sericin, which has a glue-like activity [26,27]. Every fibroin molecule involves two chains, the fibroin heavy chain (FibH) and the fibroin light chain (FibL). Between the two chains, there is a disulfide bond through which the two proteins are linked. Besides FibH and FibL, SF also includes one more protein, namely fibrohexamerin/P25. The FibH chain contains multiple repetitive sequences that are involved in the development of anti-parallel structures. However, the FibL chain contains non-repetitive sequences that are responsible for the chain's hydrophilicity and fibroin's elasticity [28]. The molar ratio of FibH, FibL and P25 in fibroin molecules is 6:6:1 [29]. In a recent study, it was shown that the genes that encode the three fibroin proteins are directly regulated by the promoter-interacting proteins. Particularly, their data revealed that there are numerous proteins of this type that impact the ribosome and metabolism pathways [30]. The architecture and mechanical strength of SF assembled by *B. mori* are defined by the protein's amino acid structure and sequence. With regard to the fibroin's primary structure, three main amino acids are described, namely glycine, alanine and serine [31]. However, tyrosine, valine, histidine and tryptophan are also involved in fibroin's structure, but in smaller amounts. SF possesses crystalline domains, but it also contains amorphous domains [32]. This outstanding protein, SF, displays three distinct polymorphs (I, II, III) [26,33]. Silk I and silk II are the crystalline forms of fibroin. The first one represents the structure of fibroin before spinning and belongs to the orthorhombic crystal system, and the last one is the protein's solid state obtained after the spinning process, as part of the monoclinic crystal system. However, by treating silk I with potassium phosphate, it will be converted to the second polymorph [28].

The silk gland is the silk producing unit, with fibroin being assembled in the posterior silk gland (PSG) and sericin synthesized in the middle silk gland. After being synthetized, the proteins form an aqueous solution and are deposited into the silk gland lumen. When the process of spinning occurs, the proteins proceed through the anterior silk gland (ASG) duct [34].

3. SF KEY PROPERTIES

SF exhibits numerous properties that convert this outstanding polymer into one of the most appreciated biomaterials (Figure 2). Currently, there are various processing forms for SF: scaffolds, sponges, or films, etc. [35–40]. It has been shown that the biocompatibility feature is influenced by the purification method and the purification process. Numerous studies have reported that SF's biocompatibility is maximized after the removal of the sericin [41,42]. However, by using sericin as a biomaterial, minimal inflammatory responses have been reported, indicating that the immune system is active by the combination of the two proteins [43].

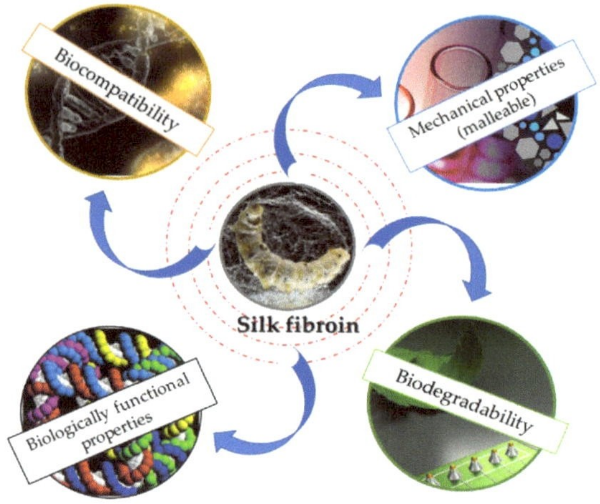

Figure 2. Schematic representation of SF properties.

3.1 Biocompatibility

The applicability of SF in the medical field dates to the 19th century. It has been intensively used as sutures due to its high biocompatibility with the human organism. There are numerous reports that have confirmed this extraordinary property of SF [44–51]. Burn wounds represent one of the most severe categories of injuries; they cause massive loss of life worldwide. When a burn injury occurs, the patient exhibits great vulnerability to numerous photogenic bacteria [52]. Through reoccurring bacterial infections, the healing process can be severely delayed. In order to prevent bacterial infections, the scientific community is currently focused on developing advanced dressings that can exhibit great protection against certain pathogens [53]. Yin et al. (2022) developed SF-based hydrogels and incorporated rhein into its structure, due to the fact that rhein possesses certain antibacterial and anti-inflammatory properties. After developing the target biomaterial, the authors determined the hydrogel biocompatibility by evaluating the hemolysis ratio observed after the blood was exposed to the hydrogel for 1 h. Their results showed favorable biocompatibility. Furthermore, the authors used mice as an animal model to demonstrate in vivo the protective activity against infections, thus highlighting the feasibility of using SF-based dressings for accelerated tissue healing [54].

One of the most common neurodegenerative disorders is Alzheimer's disease. In Alzheimer's disease, the hippocampus is the most vulnerable are of the brain

[55,56]. Tang et al. (2008) [57] performed an in vitro study in order to investigate the biocompatibility of hippocampal neurons with SF biomaterials. The authors removed the sericin from the silk fibers and developed an SF-based substrate in order to use it to culture the hippocampal neurons. With regard to central nervous system therapy, one of the most used biomaterials is polyglycolic acid (PGA). Their findings showed that by using SF as a biomaterial, similar results with PGA fibers were observed, indicating that, in vitro, fibroin displays the essential conditions for a biomaterial [57]. Another in vitro study evaluated the biocompatibility of SF fibers and extract fluid, respectively, with rat dorsal root ganglia and Schwann cells derived from sciatic nerves. The data indicated that SF did not exhibit any cytotoxicity against either biological specimens. This progress is extremely important for the development of artificial nerve grafts [46].

Skin represents the largest and one of the most important organs of humans. It has numerous functions; thus, the wound healing process plays a pivotal role in human health [58]. In this direction, Zhang et al. (2017) [59] explored the effectiveness of using SF-based films in the skin repair process by using rabbits and porcine as animal models. This study highlighted the high biocompatibility and the effectiveness of SF as a biomaterial [59]. Macrophages play a key role in the regulation of the host's immune response. An ideal biomaterial should exhibit great biocompatibility; therefore, it should not induce appreciable macrophage response. The macrophage response had to be evaluated in order to confirm the biocompatibility of each biomaterial. In this direction, a research group investigated the macrophage response by adhering SF films to L929 murine fibroblasts. By performing this experiment, extremely low macrophage activation was observed, suggesting the high biocompatibility of SF as a biomaterial [60].

Moreover, in a recent study, SF produced by *B. mori* was used to develop bioink for three-dimensional (3D) bioprinting. Bioink is in the spotlight due to its applicability in the area of tissue and organ engineering. In this study, the authors used glycidyl methacrylate to chemically modify the SF in order to obtain the bioink. They acquired SF-based hydrogels by employing 3D bioprinting and demonstrated its great biocompatibility by using NIH/3T3 fibroblasts. Furthermore, in order to investigate the long-term biocompatibility of this specific biomaterial, the SF hydrogels were used for printing various complex biological structures, such as cartilaginous trachea or lungs. This research highlighted the feasibility of SF-based hydrogels for 3D bioprinting [61].

3.2 Biodegradability

Biodegradation is the process of disruption of natural polymers into numerous small-scale compounds. Regarding the applicability of biomaterials in the medical area, one of the most key requirements is biodegradability. Due to SF's biodegradable behavior, this natural biopolymer, together with chitosan or collagen, represents one of the most used biomaterials for tissue engineering applications and beyond. There are three main proteolytic enzymes that degrade SF, namely chymotrypsin, carboxylase and actinase. After the process of biodegradation occurs, both the structure and the molecular mass of SF suffer notable changes [62].

SF's degradation behavior is impacted by several factors. The most important factor that influences the degradation rate of SF-based biomaterials is pore size [63–65]. By comparing the degradation rate of SF-based scaffolds with large and small pore sizes, it has been demonstrated that the biomaterials that had large pores degraded faster than the ones with smaller pores. Conversely, scaffolds with higher pore density deteriorated slowly compared to the ones with lower pore density [66].

In a recent study, Sun et al. (2022) [67] developed SF sponges by using the ice-coldinduced phase separation-based method. Moreover, the cryo-sponges contained exosomes in order to obtain controlled delivery of the specific extracellular vesicles. Their data showed that the encapsulated exosomes maintained their characteristics, and for two months, the vesicles remained undigested. Furthermore, in vivo assessments showed that the sustained release promoted cell migration activity, but also stimulated the development of novel blood vessels. Moreover, their results showed that the SF sponges' biodegradability rate was correlated with the fibroin concentrations [67]. This study highlights the great potential of SF as a biodegradable biomaterial for sustained drug delivery.

Catto et al. (2015) [68] successfully designed SF-based tubular vascular grafts. They tested, in vitro, the degradation behavior of SF tubes and confirmed the fact that the enzymatic activity impacted the SF's crystallinity [68]. In another study, SF was associated with chitin in order to develop scaffolds that harbor silver nanoparticles that exhibit antimicrobial activity for wound dressings. The authors demonstrated the antimicrobial effect of nanocomposite scaffolds against several bacteria. The biodegradability of the obtained biomaterial was tested in vitro by using lysozyme. Their data showed that scaffolds exhibit great biodegradable behavior; moreover, the degradation products did not exhibit cytotoxicity [69].

Fan et al. (2009) [70] used pigs as an experimental model in order to evaluate the potential of employing SF scaffolds for clinical applications. By using an SF scaffold in pigs, the authors aimed to regenerate one of the key players for the strength of the knee joint, namely the anterior cruciate ligament. Furthermore, they developed mesenchymal stem cells-seeded SF-based scaffolds and implanted them in the experimental animal model. After 24 weeks, the mesenchymal stem cells displayed fibroblast morphology; moreover, great scaffold degradation was perceived. Conversely, after 24 weeks, the ultimate tensile load of the repaired anterior cruciate ligament was sustained [70].

In another study, 3D porous SF scaffolds were obtained by using two different pro- cessing methods, aqueous or organic solvent-based techniques. Additionally, the impact of several processing variables on SF scaffold behavior was investigated; specifically, the authors explored how the SF concentration and scaffolds' pore size influenced short- and long-term in vivo behavior of this biomaterial. As an animal experimental model, two types of rats were used, nude and Lewis. The aqueous-derived scaffolds compared with the organic solvent (hexafluoroisopropanol)-derived, exhibited more homogeneous degradability. However, the hexafluoroisopropanol-derived scaffolds were more impacted by the SF concentration and pore size; more specifically, a slower degradation rate was observed by using a higher SF concentration and smaller pore size [71].

3.3 Mechanical Properties

Silkworm SF has a toughness of 80–78 MJ m-3 and exhibits an ultimate tensile strength (UTS) of 300–740 MPa. The breaking strain of this natural polymer is 4–26%, and the Young's modulus of SF is 10–17 GPa. The presence of β-sheet crystallites determines the strength and stiffness of SF; thus, these structures impact the stability of this polymer [33]. Conversely, by removing the sericin, the UTS is impacted. In the biomaterial area, the mechanical strength and the elasticity of SF are remarkable properties. However, the properties of SF are impacted by several parameters, for instance the processing techniques and β-sheet patterns [72]. Nevertheless, it was reported that by influencing the molecular weight of SF, and its solvent composition, the stiffness of this polymer is impacted [73]. There are several studies that explored the impact of temperature on SF properties. By applying heat treatment on this natural polymer, its mechanical features changed, specifically the β - sheet patterns became predominant [32]. Conversely, numerous studies described complex strategies to improve the properties of SF by blending it with other polymers [74–76].

The extraordinary mechanical strength, elasticity and resistance of SF produced by *B. mori* are essential, especially for regenerative medicine and tissue regeneration applications [77]. Keeping this in mind, Chen et al. (2018) [78] developed 3D SF-based scaffolds in order to explore the potential of this biomaterial for tracheal defect repairing. They tested in vitro and in vivo the performance of SF scaffolds by analyzing the regeneration process of tracheal epithelium. Their data showed that SF scaffolds sustained tracheal mucosa regeneration within four weeks [78]. Another research group developed SF scaffolds by using high concentrations of SF aqueous solutions. Two different methods were combined for preparing the scaffolds, namely the salt-leaching and freeze-drying techniques. It was observed that the concentration of SF directly impacted the mechanical properties of the scaffolds. The authors revealed that after 30 days the biomaterials maintained their initial structure, suggesting that SF scaffolds are an appropriate choice for cartilage regeneration [79]. Table 1 points out the mechanical differences between several polymers.

Table 1. Comparison between the mechanical properties of several polymers.

Polymer	Source	UTS (MPa)	Modulus (GPa)	Breaking Strain (%)	References
SF	*B. mori*	300–740	10–17	4–26	[33]
Silk	*B. mori*	740	10	20	[80]
Silk	N. clavipes	875	10.9	16.7	[33]
Polylactide	Corn	69.8 ± 3.2	1777 ± 42	5.7 ± 0.3	[81]
Polyethylene-terephthalate	Synthetic	56	2.2	7300	[82]
Polypropylene	Synthetic	34.5	1.7	400	[82]
Cellulose	Bacteria	11.6 ± 0.8	180.3 ± 10.6	8.2 ± 0.6	[83,84]

3.4 Biologically Functional Properties

One of the most complex processes in the human body is wound healing. This process consists of four different phases, specifically hemostasis, inflammation, cell proliferation and resolution. However, there are numerous factors that affect this process [85]. In consequence, major efforts are being made for the development of biomaterials that promote wound healing, such as oxygenation, infections, or age and gender [86]. SF's extraordinary properties are continually being studied for numerous applications in the medical area, including the process of wound healing. Mrowiec et al. (2012) [87] investigated the impact of fibroin and sericin in the wound healing process. Furthermore, they explored the molecular basis of biological functional properties of the two proteins. For this purpose, the authors used breast cancer (MDA-MB-231) and epithelial (Mv1Lu) cells and performed the wound healing scratch assessment. Their results confirmed that the silk proteins induced cell migration. Moreover, they determined that fibroin and sericin are involved in two signaling pathways, namely ERK and JNK, which are members of the family of mitogen- activated protein kinases. By phosphorylating the c-Jun, the two kinases activate and upregulate it [87].

Besides the fact that SF promotes cell migration, it has been shown that this protein improves cell adhesion. Nikam et al. (2020) [88] investigated the impact of SF nanofibers on fibroblast-like cells. After culturing the target cells on silk nanofibers for one week, their data showed that SF significantly improved cell adhesion [88].

Due to SF's great characteristics, it has been shown that it is a promising biomaterial as a coating agent in terms of drug delivery [89]. In a recent study, SF was used by Kwon et al. (2021) [90] for coating probiotic strains. The aim of this study was to improve the stability of probiotics in the human body. By performing this experiment, the authors observed that the survival rate of the target probiotics that were SF coated was improved. Furthermore, by using SF as a biomaterial, cell adhesion was significantly increased [90].

3.5 Enhanced SF by Genetic Engineering

Aiming to meet the high demand of the medical field of biomaterials, numerous efforts have been made by the scientific community. For this purpose, in the last decades, genetic engineering has made extraordinary progress. In this direction, various studies have reported the development of enhanced SF fibers (Table 2) [91–93].

Spider silk exhibits superior mechanical properties that could be extraordinary in the biomaterials branch. It has been shown that the great toughness and elasticity of spider silk are determined by the presence of major ampullate silk protein (MaSp). The MaSp protein has a large and highly repetitive sequence, and due to this fact, it could not be obtained by using bacteria, yeast or plants [94]. Conversely, *B. mori* possesses the ability of spinning highly repetitive proteins. Xu et al. (2018) [24] obtained SF with enhanced mechanical properties by genetically manipulating the *B. mori* genome. The authors simultaneously knocked-out the *FibH* gene and knocked-in the major ampullate spidroin-1 gene from N. clavipes. On the same page, Kuwana et al. (2014) [94] obtained transgenic *B. mori* that expressed an altered dragline spider protein from *Araneus ventricosus*. By performing this, they obtained a tougher silk; furthermore, this mechanical property was improved by 53% [94].

Table 2. Advances in genetically manipulated B. mori for obtaining enhanced SF.

Exogenous Gene	Enhanced SF	Reference
Insulin-like Growth factor-1	Displays improved strength, elongation and tenacity	[95]
Human acidic fibroblast growth factor	Promotes cell proliferation	[96]
Human basic fibroblast growth factor and transforming growth factor-b1	Promotes cell proliferation and exhibits anti-inflammatory activity	[97]
Cecropin B and moricin	Antimicrobial activity	[98]
Green fluorescent protein and cecropin	Antibacterial activity and fluorescence	[99]

Enhanced green fluorescent protein, DsRed monomer fluorescent protein and monomeric Kusabira oange	Exhibits fluorescence	[100]
Laminin and fibronectin peptide adhesive fragments	Exhibits increased adhesive activity	[101]
Polyalanine motifs	Displays improved mechanical	[102]

4. SF AS A GREEN MATERIAL

SF is a protein produced by the silkworm *B. mori*, and it is an extensively researched material with applications in various fields, including the cosmetics industry, biomedicine, and biomaterials [104,105]. Despite having a natural origin and being a biocompatible, mechanically superior, biodegradable and functionalizable material [33], the processing methods of SF oftentimes include highly complex processes that can have a damaging impact on the environment. This disadvantage comes in contradiction with one of the main motivators of researching SF's potential as a biomaterial, which is its eco-friendly origin and properties (Figure 3). Moreover, the way that processing techniques can affect its microstructure and properties is still not fully understood.

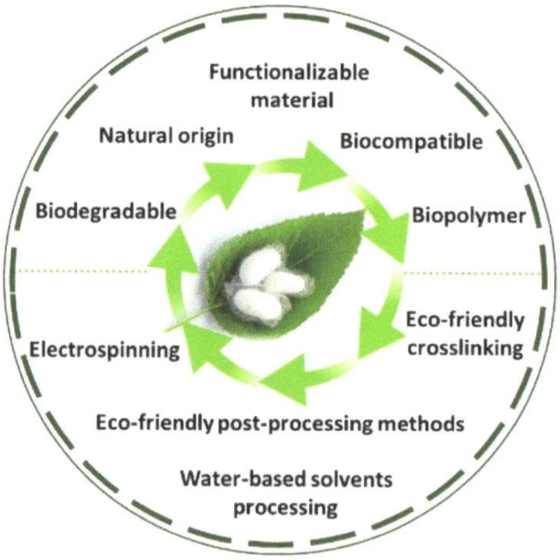

Figure 3. The main properties and methods of processing SF as a green material.

The classical regeneration of SF largely includes the use of toxic organic solvents, including calcium chloride/formic acid [106–109], hexafluoroisopropanol (HFIP) [110–112], lithium salt solutions such as lithium thiocyanate (LiSCN) [113] and lithium bromide (9.3 M $LiBr\text{-}H_2O$) [114] and calcium nitrate/methanol ($Ca(NO_3)_2$)/CH_3OH) mixtures [115]. Unfortunately, these solvents are hazardous and tend to be environmentally unfriendly [116]. Furthermore, they can directly affect the characteristics of SF and particularly impact the mechanical properties of the resulting silk biomaterials in subsequent processing [117,118].

Along with legislation and evolving attitudes toward environmental issues, new (ecological) alternative solvents that are used for SF processing have appeared. These include 0.02 M sodium carbonate (Na_2CO_3) [114], N-methyl morpholine-N-oxide (NMMO) [119], ionic liquids [120–122] and the so-called Ajisawa's reagent consisting of calcium chlo- ride/water/ethanol ($CaCl_2/H_2O/C_2H_5OH$) [123].

Water-based solvents have a lower environmental impact, and interest in their potential use in the processing of SF has gradually increased. Reizabal et al. (2021) [124] analyzed the effect of two dissolving techniques ($CaCl_2/H_2O/EtOH$ ternary and $LiBr/H_2O$ binary solutions), three regeneration approaches (gas foaming, lyophilization, gelation), and one post-processing method (ethanol—EtOH) on porosity, macro- and microstructure, molecular and structural conformation, thermal behavior, water uptake and stability, and last but not least, heavy metal adsorption. Their results were promising, showing that it is possible to control SF properties through processing and that the green processing methods have the potential of expanding, even further, the applicability of SF-based biomaterials [124].

Bae et al. (2020) [125] prepared a water-soluble SF derivative by the methacrylation (introduction of glycidyl methacrylate—GMA) of SF's reactive side chains (—NH_2, —OH, —COOH). Through a green, eco-friendly electrospinning process using distilled water as a spinning solvent, GMA-modified SF water-soluble ultrathin fibers were obtained. Next, a structural change from random coils to b-sheets was facilitated by the treatment with an aqueous ethanol solution in order to insolubilize the fibers, and chemical crosslinking improved the water resistance. Electrospinning is a suitable alternative for the green processing of SF without the use of toxic substances [125]. Another study by Yang et al. (2017) [10] reported the feasibility of developing natural green composite matrices based on SF and MN (Manuka honey), used as an antimicrobial dressing in wound healing (ulcer). The SF was dissolved in a ternary solvent system of a $CaCl_2/H_2O/CH_3CH_2OH$ solution (1:8:2 in molar ratio) and was mixed with various amounts of MH over 12 h to obtain stable spinning solutions. Subsequently, the mixtures loading with different amounts of MH were manufactured by a process of green electrospinning, where the ecological damage associated with solvents used is often overlooked [126]. In this regard, SF played a key role in indicating permeability of these membranes and in biocompatibility.

Freezing-induced silk I crystallization is another entirely green alternative for obtaining a water-insoluble SF material with the silk I crystalline structure from an SF aqueous solution. Crystallization induction allows for the obtainment of a water-insoluble silk I structure without any highly processing approaches or the use of organic solvents. Through a freezing-generated concentration and thermodynamically driven assembly, the silk I crys- tallization process can be controlled and can be environmentally friendly [127].

Yang et al. (2016) [128] improved the strain performance, up to 120%, of a green electro-spun SF nanofiber through an eco-friendly 24 h process of hyaluronic acid (HA)/1-ethyl-3-(3-dimethylaminopropyl) carbodiimide (EDC)/N-hydroxysuccinimide (NHS)-crosslinking. Not only did their SF nanofibrous matrices have better tensile performance than the matrices treated conventionally (ethanol soaking or fumigating), but the hydrophilicity was higher, and the cytocompatibility was good. Thus, HA/EDC/NHS-crosslinking might be an eco-friendly option to improve the mechanical properties of SF nanofibers, which are the main restriction in their use as tissue engineering scaffolds [128].

Fei et al. (2013) [129] used SF as a biotemplate to produce silver nanoparticles in situ under light (both incandescent light and sunlight) at room temperature. The composite solution (RSF-AgNPs) was prepared by an ecological method where the $AgNO_3$ powder was added to RSF solution and mixed until a transparent RSF-$AgNO_3$ solution formed. The mixture solution obtained was then exposed under the light with an incandescent bulb and was incubated at room temperature for 24 h to produce an RSF-AgNPs composite solution. In this case, SF serves as the reducing agent of silver and the dispersing and stabilizing agent of the resulting silver nanoparticles. As the reaction does not need any other chemicals and only uses light as a power source, the synthetic route of silver nanoparticles reported here is environmentally friendly and energy saving.

Raho et al. (2020) [130] used a green synthesis approach to create a composite hydrogel (CoHy) of carboxymethylcellulose-Na (CMC-Na) stabilized and loaded with Ag- NPs. An RSF solution was obtained by dissolving the fibroin with a solution of $CaCl_2$ ($CaCl_2/CH_3CH_2OH/H_2O$) instead of LiBr and mixing it with a CMC-Na aqueous solution. The silver nitrate ($AgNO_3$) was weighed and slowly added to the solution. For a great crosslinking, the composite solution was sonicated and exposed to UV radiation (for green synthesis of AgNPs). In this work, the synthesis of the hydrogel (CoHy) was based on a green process with ecological methods and natural components such as SF that served as a reducing agent in the green synthesis of AgNPs [129,131].

Switching toward eco-friendlier processing techniques of SF has gained popularity in materials science research and has raised the awareness of the negative impact of more environmentally toxic approaches. More research is necessary in order to find the ideal methods for maximizing SF's potential as a biomaterial, with the lowest impact on the environment.

5. SF AND TISSUE ENGINEERING

The purpose of tissue engineering is to combine cells with scaffold materials and growth factors in order to help the regeneration of or even replace damaged tissue and/or organs. The need for a biomaterial matrix that could aid in the development of viable and biologically active tissue, either in vivo or in vitro, has increased in the last few years [132]. It has been proven that SF stimulates the attachment and growth of human cells [133]; thus, it is not surprising that it became a Food and Drug Administration (FDA)-approved biomaterial that can successfully be used as a scaffold in tissue engineering due to its unique properties and complex structure [2].

In order for it to be successfully used in tissue engineering, SF has to undertake a dissolution process, with different morphologies of the regenerated SF observed according to the different solvent systems used [134]. SF has a tightly packed structure; thus, it is insoluble in most of the solvents used for the dissolution of polymers for biotechnology applications, including water. Obtaining an SF solution requires the use of near-saturated bromides or thiocyanates and dialysis, after which the solution is unstable and hard to store for a longer time [135].

Additionally, multiple techniques have been used to process SF, including chemical (surface modifications only, entire material modification, and sonochemical methods) and enzymatic approaches (surface/bulk material modifications) [135]. From gas plasma treatments [136] and UV irradiation to more complex systems where SF films were first am- inated by ammonia, NH_3, and then covalently sulfonated by sulfur dioxide [137], chemical modifications of SF are still of interest, despite their disadvantages. Biochemical enzy- matic treatments of SF that have been studied include protease XIV and a-chymotrypsin treatment of SF for corneal regeneration [138], proteinase K for a porous network by a co-precipitation process with hydroxyapatite solution [139], and collagenase IA [140]. Fur- thermore, the enzymatic grafting of biomolecules such as lactoferrin [141] and chitosan was also described [142], and they are beginning to reveal their potential in tissue engineering.

Among the multiple and complex applications of SF in tissue engineering, SF dressings are a viable and promising option for the improvement of wound healing strategies through tissue engineering techniques [14]. Through the use of SF in developing sponges [143], hydrogels [144,145], SF micro- and nanoparticles [146], nanofibrous matrices [147], scaffolds, and composite films [148], SF and its potential applications in tissue engineering have been intensely studied in the last few years.

SF dressings can also improve the management of chronic wounds, as shown by Li et al. (2017) [149]. They prepared a functionalized SF dressing with topical bioactive insulin release by coaxial electrospraying of aqueous SF solution under mild processing conditions. This was based on the ability of insulin to stimulate cell migration and certain wound recovery methods [150]. The encapsulated insulin in the inner layer of the SF microparticles could thus be gradually released up to 28 days. The therapeutic effects of these microparticles loaded into an SF sponge (obtained by multilayer loading and freeze-drying method) was evaluated in vivo on dorsal full-thickness wounds of diabetic Sprague–Dawley rats. The degree of wound healing by measuring the area of wound closure showed the promotion of wound healing through accelerated wound closure and stimulation of vascularization and collagen deposition [149].

Moreover, processed SF after extraction of silk–sericin was also studied as a potential material for tissue engineering of the anterior cruciate ligament (ACL). A six-cord silk wire-rope matrix was designed that, based on ultimate tensile strength, linear stiffness, yield point, and percent elongation at break, had similar mechanical properties with the human ACL cells. This silk matrix supported both the attachment, as well as the expansion and differentiation, of adult human progenitor bone marrow stromal cells [132].

A highly porous 3D silk scaffold was designed by Wang et al. (2005) [151] and was combined with autologous adult mesenchymal stem cells (MSCs) in an attempt to perform in vitro cartilage tissue engineering. After 3 weeks of cell cultivation, RT-PCR analysis for cartilage-specific ECM gene markers and histological and immunohistochemical evalua- tions of cartilage-specific ECM components showed that the MSCs took the chondrogenic pathway with zonal architecture, including spatial cell arrangement and type II collagen distribution similar to native cartilage tissue [151]. Since cartilage has a limited self-repair capacity, and the current approaches for treating cartilage damage are not good enough for restoring the lost functionality, the potential of silk scaffolds in cartilage tissue engineering is of great importance and should be further evaluated in vivo. More recently, a 4D bio- printing system including SF was designed by Kim et al. (2020) [152]. Their system was based on digital light processing 3D bioprinting technology and photopolymerizable SF bioink. After obtaining a 4D-bioprinted silk hydrogel and using it to make trachea mimetic tissue, they successfully implanted it in the damaged trachea of an 8-week-old rabbit and proved the applicability of this new bioprinting system in vivo [152].

In another study, Mallepally et al. (2015) [153] used co_2 assisted acidification for the synthesis of SF hydrogels that were subsequently converted to SF aerogels using the salt-leaching method [154]. Moreover, these SF aerogels were structurally, mechanically and in vitro biologically characterized using human foreskin fibroblast cells. The results showed that the aqueous SF concentration plays an important role in tuning the morphology and textural properties of the SF aerogels. These presented a significantly higher surface area and pore volume compared to the freeze-dried SF scaffolds. Furthermore, the SF-aerogels presented are highly porous with an interconnected network of nanofibrils, contributing to the success of cytocompatibility with human foreskin fibroblasts cells and their propagation. The SF aerogel scaffolds obtained by supercritical co_2 have the potential for applications in tissue engineering, and the method of their synthesis can be used as an alternative method for 3D scaffold preparation.

Pacheco et al. (2020) [155] reported the use of several biopolymers in the development of SF/CS/SA multilayer membranes as a system for controlled drug release in wound healing. In this regard, the aim of the paper was to combine, in a single-composite material, the mechanical and biocompatibility properties of SF, the antimicrobial action of chitosan (CS) and the ideal exuded absorption of alginate (AS), thus achieving a synergistic effect. The membranes were prepared by pouring, where diclofenac sodium was incorporated as a model drug, into the chitosan solution before the solvent evaporated, which was stored in the middle layer of the membrane. The results showed that the incorporation of the drug did not affect the mechanical, thermal or barrier properties. Drug release was evaluated in vitro using a simulated solution of a wound exudate at $37^\circ C$, where Fickian diffusion behavior was shown to be the dominant release mechanism. The results supported the idea that these multilayer membranes (SF/CS/SA multilayer membranes) could serve in biomedical applications as high-performance wound dressings.

Among the varieties of materials tested in the field of tissue engineering, SF continues to be of increasing interest in the medical field, this being a promising material in scaffold (3D) fabrication, mimicking the natural extracellular matrix (Figure 4). Furthermore, the ease of processing, the excellent biocompatibility, the remarkable mechanical properties and the personalized degradability of SF make it a competitive material in tissue engineering and regenerative medicine.

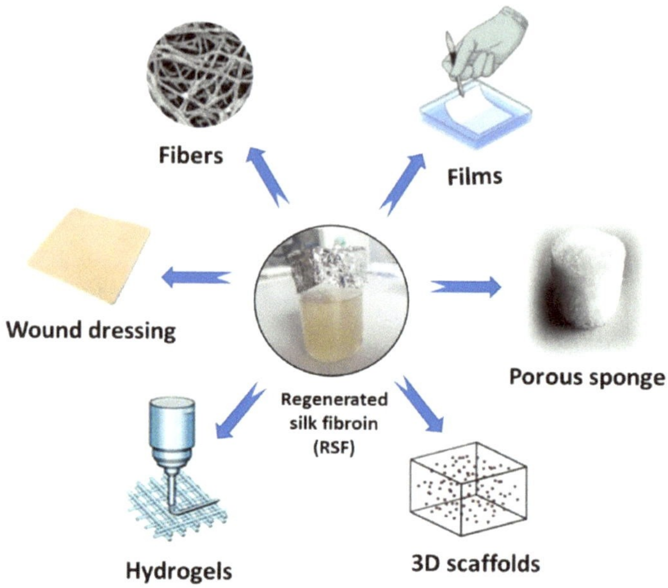

Figure 4. SF as a functional biomaterial for the tissue engineering field.

6. SF INVOLVED IN THE TREATMENT OF THE EYE, BONE AND SKIN REGENERATION

6.1 SF Involved in the Treatment of the Eye

The cornea is a half-millimeter thick structure with three types of tissues: corneal epithelium (anterior to an acellular area—Bowmann's layer), avascular corneal stroma with mesenchymal cell types (keratocyte), and an innermost single layer of corneal endothelial cells (posterior to an acellular zone—Descemet's membrane).

Harkin et al. (2011) [156] hypothesized that, among the complex eye structures that SF could be used as for a bioengineered tissue replacement would be the corneoscleral limbus, the corneal stroma, the corneal endothelium, and Ruysch's complex (the outer blood-retinal barrier). For the corneoscleral limbus, they believe that a composite material such as a prosthetic basement membrane could be used to grow epithelium on, which would support the exchange of nutrients and regulatory substances that must be present for the maintenance of the stem cells. Overlying this prosthetic basement membrane, they proposed a 3D scaffold that is more biomimetic for the stromal cells [156]. Although, for the corneoscleral limbus, transplant from a deceased donor has been associated with a high rejection rate [157], this type of procedure seems to be highly successful for corneal stroma reparation. The risk of transmission of different diseases through a transplant makes it desirable that viable alternatives from bioengineered tissues should be available. In order for a biomaterial to contribute to corneal stroma regeneration, it should support the growth of an epithelial layer and provide a vehicle for the transplantation of keratocytes. However, they should also maintain a high level of transparency, unlike corneoscleral limbus reconstruction.

Two of the methods currently used for obtaining SF films for corneal applications are the centrifugal casting technique and the dry casting method. Lee et al. (2016) [158] used both of these methods and compared the results. Their results showed that the centrifugal method offers more benefits, with less surface roughness of the films, increased growth of corneal fibroblasts (primary human corneal keratocytes) and better tensile strength and transparency [158]. Thus, the centrifugal casting technique is a promising method in tissue engineering of SF films for corneal regeneration.

The potential of SF as a biomaterial for corneal epithelial cell sheet generation has further been evaluated by Liu et al. (2012) [159], who compared the biological cell behavior of human and rabbit primary and immortalized corneal epithelial cells on SF and denuded human amniotic membrane (AM). The cells adhered to and proliferated on both SF and AM, and expression of DNp63a, a progenitor cell marker [160], and keratin 3/12, a differentiation marker [161], existed on both cell culture types. Moreover, after being cultured at the air– liquid interface for 7 days, cells on the SF formed a stratified graft with a 2- to 3-cell layering with compact columnar cells on the basal layer and squamous cells present on the apical layers [159]. Since no cytotoxic response or inhibition of cell growth was observed in the SF cultures, their study strongly supports the high biocompatibility of SF biomaterial to corneal epithelial cells and their potential for corneal regeneration.

Lawrence et al. (2009) [162] designed 2-micromillimeter thick, surface-patterned silk films with 0.5–5.0-micromillimeter pores and demonstrated human and rabbit corneal fibroblast proliferation, alignment and corneal extracellular matrix expression in both 2D and 3D cultures on these films. The thickness was carefully chosen to imitate corneal collagen lamellae dimensions, whereas the surface pattern was meant to guide cell align- ment. Pores were used for the enhancement of translamellar diffusion of nutrients and promotion of cell–cell interactions [162]. Their results further strengthen the potential of SF in bioengineered tissues for corneal injuries and pathologies.

The eye is an extremely complex organ, and its functionality in normal parameters highly influences the quality of life. Multiple research branches are currently performing joint work with the sole purpose of improving the regeneration of eye components. In this context, SF has proven to be a suitable candidate for a biomaterial that can be used for tissue engineering of eye components, especially for the cornea. Despite the enthusiasm that SF has seeded, more research is needed in order to better understand the interaction of its properties as a biomaterial and the human living organism, especially within the eye.

6.2 SF Used in the Treatment of the Bone

Bone is a rigid tissue with the essential propriety of providing structural support and protection of certain internal organs while supporting muscular contraction and serving as a deposit for certain minerals [163]. While it heals and remodels without leaving gross scars, efficient therapeutic strategies are needed for the efficient management of bone injuries. Bone grafts were the main option for a while, but their multiple disadvantages [164,165] have encouraged further research for viable alternatives.

For this purpose, in a recent study, Zhao et al. (2022) [166] developed SF-based scaffolds. One of the key players when it comes to bone repair and healing processes is hydroxyapatite. Being the foremost mineral element in bones, the authors combined SF with hydroxyapatite and obtained porous scaffolds. Moreover, they developed poly(lactic-co-glycolic acid) (PLGA)-based microspheres for controlled release of naringin, which is known for its ability to promote stem cell differentiation into osteoblasts. The microspheres were attached to the developed scaffold. First, the authors used human bone mesenchymal stem cell cultures in order to investigate the impact of the developed biomaterial on stem cell differentiation potential. Their data showed that the microspheres inlaid into the developed scaffolds have great potential in promoting osteogenic differentiation of target cells. Second, they used rabbits as a model organism to evaluate the bone repair process; more specifically, they surgically induced significant lesions of distal femoral epiphysis. The results indicated that the obtained biomaterial has great potential in the bone regeneration area by promoting bone defect healing [166].

A study conducted by Meinel et al. (2005) [167] used SF to form porous scaffolds on which human mesenchymal stem cells (hMSC) were cultured for 5 weeks. Afterward, they used the tissue as an implant for mouse critical calvarial defects. After another 5 weeks, multiple analyses including gene expression analysis, biochemical assays, and X- ray diffractometry demonstrated that SF successfully induced advanced bone formation [167].

The properties of SF of providing an osteogenic environment for bone-related out- comes was also proven by Zhao et al. (2009) [168]. They used apatite-coated SF scaffolds seeded with osteogenically induced autologous bone marrow stromal cells (bMSCs). The graft was used to repair inferior mandibular border defects in a canine model in comparison with bMSCs and SF scaffolds (no apatite-coating), apatite-coated SF scaffolds alone, SF scaffolds alone, autologous mandibular grafts, and untreated blank defects. Four weeks after the operation, new bone formation was detected through histological and radio- graphic examinations, with the defects being completely repaired for the bMSCs together with the apatite-coated SF group [168]. Their favorable results suggest that SF scaffolds could successfully be used with an apatite coating in combination with osteogenically in- duced autologous bone marrow stromal cells for the management of bone defects through tissue engineering.

Hydroxyapatite-coated SF composites were proven through fluorescence microscopy and scanning electron microscopy (SEM) to sustain cell adhesion and proliferation of mesenchymal cells derived from EGFP-expressing transgenic rat bone marrow in 10-day cultures. The results were comparable to the adhesion and growth of the cells seeded on tissue culture polystyrene dishes, which are standard scaffolds for the culture of cells, and which strongly support the use of SF scaffolds for regenerative medicine with applications in bone-related pathologies [169].

Another novel nanocomposite containing SF that might be used as tissue engineering scaffold or bone replacement was obtained by Cui et al. (2007) [170]. During its synthesis in an aqueous solution containing SF, hydroxyapatite (HA) nanocrystallites nucleated and grew preferentially along the plane (002) of the HA crystal, suggesting that SF might have an essential role in the mineralization of HA nanocrystallites. Moreover, the resulting nanocomposite had a compression strength of 97.6 MPa, which is higher than that of woven bone [170]. This compression strength was much higher than the one found in the case of the SF-chitosan/nano-hydroxyapatite porous scaffolds by Wen et al. (2007) [171]. Although their scaffold had good bioactivity, proven in the simulated body fluid experiments, when a layer of bone-like apatite crystals that were randomly oriented formed on the scaffold surface, their compressive strength varied only from 0.26 to 1.96 MPa. Both the porosity and the compressive strength seemed to change in relation to organic phase concentration and pre-freezing temperatures, and the authors concluded that a pre-freezing temperature of 80°C and 5% organic concentration would be the best choices for bone tissue engineering [171].

The multiple properties and functions that bones have in the body make it necessary to have more than one viable option for bone regeneration, especially in the context of the high frequency of accidents involving bony structures. Considering the disadvantages of bone grafts and the impact that post-traumatic bone lesions can have on the quality of life through the impairment it causes, SF research as a potential tissue engineering biomaterial for bone-replacement structure synthesis has recently gained popularity. Although this potential still has to be tested in vivo, the properties of SF in current research, as well as the in vitro experimental work that has been performed, suggest that this is the proper direction to go.

6.3 SF Involved in the Treatment of Skin Regeneration

Skin has a complex function, including protection, thermo-regulation and a barrier be- tween the outside of the body and the underlying tissues. Because skin lesions are frequent, over time, multiple approaches have been suggested for the proper treatment of injuries and pathologies developed at this level. Skin grafts have multiple classifications (based on thickness, donor origin, etc.), and each has different advantages and disadvantages. Their limitations of size, donor site availability and tolerability make skin grafting an imperfect option, and they have encouraged the search for a bioengineered skin equivalent [172].

A complex SF-based formic acid-crosslinked 3D nonwoven scaffold was studied by Dal Pra et al. (2006) [173] and was proven to be a great candidate for human tissue engineering. Among the methods they used was scanning electron microscopy (SEM) and differential scanning calorimetry, as well as thermogravimetric and tensile strength studies. These showed that the scaffolds are a composite material with anisotropic SF fibers enclosed within an isotropic matrix of SF in film form. They also obtained fibers, which as well as films, were firmly crosslinked and water insoluble. The crosslinking was hypothesized to have been induced by formic acid treatment and the lack of water solubility to their ß-sheet crystalline structure. Both human embryonic kidney cells and human dermal fibroblasts were attached to the scaffolds within 3 h and started proliferating and colonizing the scaffolds after 24 h. The cultures kept growing, with an active metabolism and no sign of protein catabolism, for at least 15 weeks [173].

A chitosan/SF blend film was evaluated as a biomaterial for skin tissue engineering by Luangbudnark et al. (2012) [174]. Chitosan appears to have enhanced β-sheet conformation of fibroin as well as increased flexibility, enzyme degradation, and swelling index. Fourier transform infrared spectroscopy and differential scanning calorimetry analysis showed intermolecular interactions between chitosan and SF. Furthermore, XTT assay showed no toxicity of the chitosan/SF blend film on fibroblast cell cultures, which showed proliferation. The blend films had a percentage of elongation at the break between 6.4% and 14.4%, which seemed to increase proportionally with the chitosan content, while the tensile strength was between 52.8 and 58.3 MPa, independent of the chitosan proportion. Conversely, increasing the fibroin content of the CS/SF blend films led to lower flexibility [174].

Another type of recently studied blended scaffold of SF was in combination with human hair keratin with a potential use as a dermal substitute [175]. Three-dimensional (3D) blended scaffolds were fabricated by freeze-drying and were tested on the L929 mouse-fibroblastic cell line. Functional fibroblasts showed good biocompatibility and proliferation on the 3D scaffolds, with a greater expression of collagen type I proven by immunohistochemical staining methods. Cellular compatibility evaluated through MTT assay showed good cell viability for 14 days of culture. Further, these composite scaffolds supported cell attachment and proliferation along with intact extracellular matrix (ECM) deposition, specifically for Col I. These results demonstrate that SF–keratin blended scaffolds are promising substrates for dermal reconstruction, wound healing, and other biomedical applications [175].

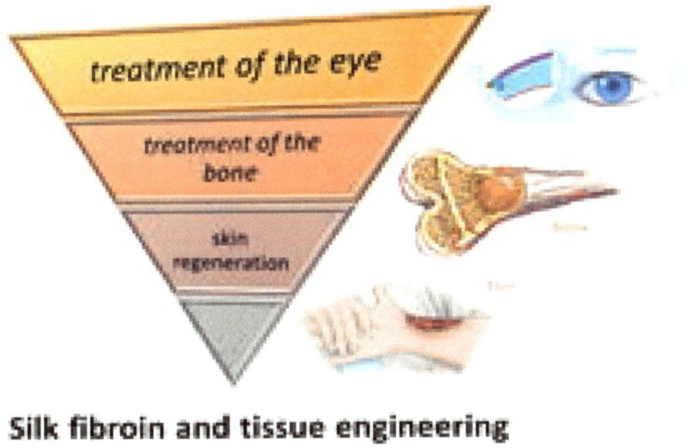

Figure 5. SF involved in the treatment of the eye, bone, and skin regeneration.

Although, for the treatment of skin injuries, skin grafting has been studied and used for a long time, they have a limitation of size, are dependent on donor size availability, and have the potential to be rejected by the host organism. These limitations cannot be overlooked, especially in complex and extended burn-cases that require additional support and special care. The biocompatibility of SF, as well as its natural origin and abundance, give it the potential to overcome certain issues of skin grafts (Figure 5). Besides its potential use for bioactive wound dressings, SF might become an important resource for in vivo regeneration of the skin.

7. SF IN CANCER THERAPY

Globally, cancer is the leading cause of mortality [176]. Currently, chemotherapy is the standard therapy for numerous types of cancer. However, there are numerous disadvantages when it comes to chemotherapy, the most important being its high toxicity [177].

Keeping these aspects in mind, there is a major need for the development of drugs that exhibit reduced toxicity. Due to its extraordinary properties, SF has great potential as a drug delivery system. Another great advantage of using SF is its great feasibility of modification and the ease of processing [178,179]. Being extremely versatile, SF has been used to de- liver numerous compounds, such as vaccines, proteins, therapeutic compounds and so on. Sun et al. (2019) [180] encapsulated doxorubicin in SF-based nanoparticles for exploring its feasibility in target cancer therapy. In addition, on the nanoparticles' surface, the authors covalently grafted folic acid in order to target the tumor cells. Their results showed that the prepared SF nanoparticles did not affect cell viability or proliferation. This approach displayed promising results, as the particles were able to target the tumor cells. Moreover, the doxorubicin was controllably released [180].

In a recent study, Saqr et al. (2021) [181] developed SF nanoparticles that were loaded with docetaxel to target breast cancer cells. Their data revealed that the developed nanoparticles exhibited an enhanced cytotoxic effect against the targeted cancer cells [181]. Mottaghitalab et al. (2017) [182] encapsulated gemcitabine in SF nanoparticles in order to target lung cancer cells. As an experimental model animal, the authors used mice. The results of this approach showed that SF nanoparticles were able to reduce the side effect of gemc- itabine. Furthermore, the therapeutic activity resulting from this formulation was higher than other approaches [182]. In the same year, Moin et al. (2021) [183] developed both nano- and micro-fibroin-based particles as drug delivery systems to encapsulate doxorubicin. The authors used two types of cancer cell lines, specifically MCF-7 and SAOS2, and one normal human cell line, namely HFF. By using the MTT assessment, they evaluated the toxicity exhibited by micro- and nanoparticles on cell lines. Their data showed that cell growth was inhibited. Furthermore, the impact of SF-based particles on *p53* expression was evaluated. Numerous studies have highlighted that *p53* overexpression plays a key role when administering doxorubicin to cancer patients; specifically, the overexpression if this gene is linked to higher sensitivity of malignant cell sensitivity to this chemotherapy agent. By using the real-time PCR method to evaluate *p53* expression, the doxorubicin-loaded SF nanoparticles induced higher *p53* expression in the malignant cell lines and decreased its expression in the normal cell line. Conversely, the doxorubicin-loaded SF microparticles decreased *p53* expression in the SAOS cell line and significantly increased it in the MCF-7 and HFF cell lines. Moreover, the authors investigated the reactive oxygen species levels after treating the cell lines with the developed biomaterials and observed that the reactive oxygen species levels were lower in two cell lines, namely MCF-7 and HFF. These findings highlight the great contribution of SF-based biomaterials in cancer therapy [183].

In 2022, a group of researchers successfully developed SF-based double-layer micronee- dles. The authors developed the delivery devices in order to encapsulate and administer triptorelin in a controlled released manner. Triptorelin is widely used in the treatment of prostate gland carcinomas due to its inhibitory effect on testosterone; however, it is an injectable suspension that exhibits certain disadvantages such as cold-chain storage or the administration method. After developing the SF-based microneedles and successfully encapsulating triptorelin in the microneedles' tip, the authors subcutaneously implanted the devices in rats in order to evaluate its therapeutic effects. The authors showed that by encapsulating triptorelin into SF microneedles, its half-life was extended compared with the half-time obtained after directly subcutaneously injecting the drug. In order to confirm the inhibitory effect of encapsulated triptorelin on testosterone concentration, as controls, castrated and healthy rats were used. Their results showed that by using the developed microneedles to administer triptorelin, the testosterone levels were the same as the levels observed in the castrated rats, and the inhibitory effect was maintained for more than seven days. Conversely, by subcutaneously injecting the target drug, the repressive action of triptorelin was maintained for less than one day. These findings suggest that SF-based microneedles are a promising approach for delivering triptorelin [184].

Furthermore, SF is currently being used as a tumor model to investigate cancer biology. By using a tumor model to mimic the tumor microenvironment, valuable insights could be observed. Dondajewska et al. (2018) [185] used an SF-based scaffold to construct a breast cancer model. The model provided insights into the interactions between the stromal fibroblasts and cancer cells. This complex model represents a promising approach for expanding the understanding of tumor biology [185]. In another study, scaffolds were developed that contained both SF and chitosan to mimic a tumor's microenvironment. These scaffolds allow for the investigation of tumor drug resistance mechanisms and provide insights for the evaluation of therapeutic agents [186]. Table 3 shows numerous studies that successfully used SF in cancer therapy, and Figure 6 illustrates the use of SF in cancer therapy.

Table 3. SF-based biomaterials used in cancer therapy.

Biomaterial	Type of Cancer	Reference
SF hydrogels	Hepatocellular carcinoma	[187]
SF–Thelebolan matrix	Soft tissue carcinoma	[188]
Doxorubicin loaded SF nanoparticles	Brain cancer	[189]
SF–Sodium alginate nanocarriers	Colorectal cancer	[190]
SF-based metastasis model	Breast cancer	[191]
Triptolide–Celastrol-loaded SF nanoparticles	Pancreatic cancer	[192]
Alpha-mangostin loaded SF nanoparticles	Colon cancer; Breast cancer	[193]
SF rods	Breast cancer	[194]
Quercetin loaded SF nanoparticles	Breast cancer; Lung metastasis	[195]
Biliverdin–SF hydrogel	Glioma	[196]
Floxuridine-loaded SF nanospheres	Digestive tract cancer; Lung	[197]
Curcumin-loaded SF nanoparticles	Breast cancer	[198]

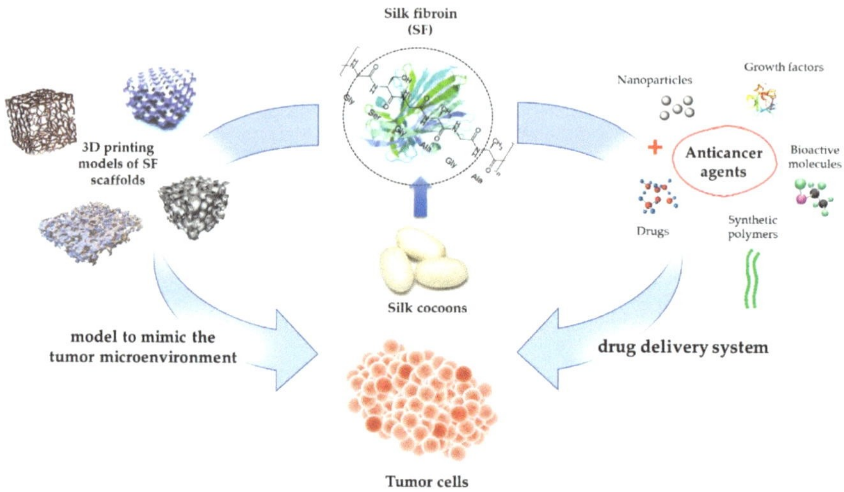

Figure 6. SF-based drug delivery and the production of 3D SF scaffolds for the growth of cancer cells.

8. CONCISE SUMMARY

SF constitutes the principal structural component of the silk generated by B. mori. A multitude of studies has demonstrated the distinctive features of this protein. A key characteristic is the high degree of biocompatibility with human organisms. Furthermore, SF has excellent mechanical characteristics and biodegradability. This protein possesses unique features that confer exceptional potential in the medical profession and beyond. It has been historically utilized as a suture in surgical procedures, it is now the primary constituent of several biomaterials. Significant progress has been achieved in the medical field owing to SF's capacity to produce various biomaterials, including scaffolds and nanoparticles. Furthermore, many strategies can be employed to manipulate and reshape the characteristics of SF. In recent years, SF has garnered attention as it may be classified as an eco-friendly material.

SF is presently utilized in many fields, and an array of approaches has been implemented to acquire distinct sorts of SF-based biomaterials. Various forms of scaffolds have been created from this natural polymer, including hydrogels, films, nanospheres, sponges, and coatings. Current advancements in understanding the structure, characteristics, and processing methods of SF present new potential for its application. Furthermore, SF has been linked with several chemicals to enhance its mechanical and biological functionalities. Numerous studies have utilized SF for the cornea, bone, and its uses. In addition to SF's remarkable role in tissue engineering, this biomaterial significantly contributes to cancer therapy. It is being utilized as a coating agent in the treatment of several cancer kinds, such as lung and breast cancer. Moreover, molecular engineering has been utilized to get improved SF for biomedical applications..

REFERENCES

1. Ozkale, B.; Selman, M.; Mooney, D.J. Active Biomaterials for Mechanobiology. *Biomaterials* **2021**, *267*, 120497. [CrossRef] [PubMed]
2. Sun, W.; Gregory, D.A.; Tomeh, M.A.; Zhao, X. Silk Fibroin as a Functional Biomaterial for Tissue Engineering. *Int. J. Mol. Sci.* **2021**, *22*, 1499. [CrossRef] [PubMed]
3. Holland, C.; Numata, K.; Rnjak-Kovacina, J.; Seib, F.P. The Biomedical Use of Silk: Past, Present, Future. *Adv. Healthc. Mater.* **2019**, *8*, 800465. [CrossRef] [PubMed]
4. Sunija, A.J. Biomaterials and Biotechnological Schemes Utilizing TiO_2 Nanotube Arrays—A Review. In *Fundamental Biomaterials: Metals*; Elsevier Ltd: Amsterdam, The Netherlands, 2018; pp. 175–196. [CrossRef]
5. Shanmugam, K.; Sahadevan, R. *Bioceramics—An Introductory Overview*; Elsevier: Amsterdam, The Netherlands, 2018. [CrossRef]
6. Sionkowska, A. Current Research on the Blends of Natural and Synthetic Polymers as New Biomaterials: Review. *Prog. Polym. Sci.* **2011**, *36*, 1254–1276. [CrossRef]
7. Ullah, S.; Chen, X. Fabrication, Applications and Challenges of Natural Biomaterials in Tissue Engineering. *Appl. Mater. Today* **2020**, *20*, 100656. [CrossRef]
8. Wu, R.; Li, H.; Yang, Y.; Zheng, Q.; Li, S.; Chen, Y. Bioactive Silk Fibroin-Based Hybrid Biomaterials for Musculoskeletal Engineering: Recent Progress and Perspectives. *ACS Appl. Bio Mater.* **2021**, *4*, 6630–6646. [CrossRef]
9. Kostag, M.; Jedvert, K.; El, O.A. Engineering of Sustainable Biomaterial Composites from Cellulose and Silk Fibroin: Fundamentals and Applications. *Int. J. Biol. Macromol.* **2021**, *167*, 687–718. [CrossRef]
10. Yang, X.; Fan, L.; Ma, L.; Wang, Y.; Lin, S.; Yu, F.; Pan, X.; Luo, G.; Zhang, D.; Wang, H. Green Electrospun Manuka Honey/Silk Fibroin Fibrous Matrices as Potential Wound Dressing. *Mater. Des.* **2017**, *119*, 76–84. [CrossRef]
11. Tao, G.; Cai, R.; Wang, Y.; Liu, L.; Zuo, H.; Zhao, P.; Umar, A.; Mao, C.; Xia, Q.; He, H. Bioinspired Design of AgNPs Embedded Silk Sericin-Based Sponges for Efficiently Combating Bacteria and Promoting Wound Healing. *Mater. Des.* **2019**, *180*, 107940. [CrossRef]
12. Bakhsheshi-Rad, H.R.; Fauzi, A.; Aziz, M.; Akbari, M.; Hadisi, Z.; Omidi, M.; Chen, X. Development of the PVA/CS Nano Fibers Containing Silk Protein Sericin as a Wound Dressing: In Vitro and in Vivo Assessment. *Int. J. Biol. Macromol.* **2020**, *149*, 513–521. [CrossRef]

13. Zhou, L.; Yu, K.; Lu, F.; Lan, G.; Dai, F.; Shang, S.; Hu, E. Minimizing Antibiotic Dosage through in Situ Formation of Gold Nanoparticles across Antibacterial Wound Dressings: A Facile Approach Using Silk Fabric as the Base Substrate. *J. Clean. Prod.* **2020**, *243*, 118604. [CrossRef]

14. Patil, P.P.; Reagan, M.R.; Bohara, R.A. Silk Fibroin and Silk-Based Biomaterial Derivatives for Ideal Wound Dressings. *Int. J. Biol. Macromol.* **2020**, *164*, 4613–4627. [CrossRef]

15. Zakeri-Siavashani, A.; Chamanara, M.; Nassireslami, E.; Shiri, M.; Hoseini-Ahmadabadi, M.; Paknejad, B. Three Dimensional Spongy Fibroin Scaffolds Containing Keratin/Vanillin Particles as an Antibacterial Skin Tissue Engineering Scaffold. *Int. J. Polym. Mater. Polym. Biomater.* **2022**, *71*, 220–231. [CrossRef]

16. DeBari, M.L.; King, C.I.; Altgold, T.A.; Abbott, R.D. Silk Fibroin as a Green Material. *ACS Biomater. Sci. Eng.* **2021**, *7*, 3530–3544. [CrossRef]

17. Benfenati, V.; Toffanin, S.; Chieco, C.; Sagnella, A.; Virgilio, N. Di; Posati, T.; Varchi, G.; Natali, M.; Ruani, G.; Muccini, M.; et al. Silk Fibroin Based Technology for Industrial Biomanufacturing. In *Factories of the Future*; Springer International Publishing: Berlin/Heidelberg, Germany, 2019; pp. 409–430. [CrossRef]

18. Hu, W.; Lu, W.; Wei, L.; Zhang, Y.; Xia, Q. Molecular Nature of Dominant Naked Pupa Mutation Reveals Novel Insights into Silk Production in *Bombyx mori*. *Insect Biochem. Mol. Biol.* **2019**, *109*, 52–62. [CrossRef]

19. Song, J.; Chen, Z.; Liu, Z.; Yi, Y.; Tsigkou, O.; Li, J.; Li, Y. Controllable Release of Vascular Endothelial Growth Factor (VEGF) Bywheel Spinning Alginate/Silk Fibroin Fibers for Wound Healing. *Mater. Des.* **2021**, *212*, 110231. [CrossRef]

20. Pérez-Rigueiro, J.; Ruiz, V.; Cenis, J.L.; Elices, M.; Pugno, N.M. Lessons From Spider and Silkworm Silk Guts. *Front. Mater.* **2020**, *7*, 1–8. [CrossRef]

21. Xu, J.; Dong, Q.; Yu, Y.; Niu, B.; Ji, D.; Li, M.; Huang, Y.; Chen, X. Mass Spider Silk Production through Targeted Gene Replacement in *Bombyx mori*. *Proc. Natl. Acad. Sci. USA* **2018**, *115*, 8757–8762. [CrossRef]

22. Etebari, K.; Mirhoseini, S.Z.; Matindoost, L. A Study on Interaspecific Biodiversity of Eight Groups of Silkworm (*Bombyx mori*) by Biochemical Markers. *Insect Sci.* **2005**, *12*, 87–94. [CrossRef]

23. Abdelli, N.; Peng, L.; Keping, C. Silkworm, *Bombyx mori*, as an Alternative Model Organism in Toxicological Research. *Environ. Sci. Pollut. Res.* **2018**, *25*, 35048–35054. [CrossRef]

24. Xu, H.; O'Brochta, D.A. Advanced technologies for genetically manipulating the silkworm Bombyx mori, a model lepidopteran insect. *Proc. R. Soc. B Biol. Sci.* **2015**, *282*, 20150487. [CrossRef]

25. Zhu, Z.; Imada, T.; Asakura, T. Preparation and Characterization of Regenerated Fiber from the Aqueous Solution of *Bombyx mori* Cocoon Silk Fibroin. *Mater. Chem. Phys.* **2009**, *117*, 430–433. [CrossRef]

26. Marín, C.B.; Fitzpatrick, V.; Kaplan, D.L.; Landoulsi, J.; Guénin, E.; Egles, C. Silk Polymers and Nanoparticles: A Powerful Combination for the Design of Versatile Biomaterials. *Front. Chem.* **2020**, *8*, 1–22. [CrossRef]

27. Wang, F.; Guo, C.; Yang, Q.; Li, C.; Zhao, P.; Xia, Q.; Kaplan, D.L. Protein Composites from Silkworm Cocoons as Versatile Biomaterials. *Acta Biomater.* **2021**, *121*, 180–192. [CrossRef]

28. Qi, Y.; Wang, H.; Wei, K.; Yang, Y.; Zheng, R.Y.; Kim, I.S.; Zhang, K.Q. A Review of Structure Construction of Silk Fibroin Biomaterials from Single Structures to Multi-Level Structures. *Int. J. Mol. Sci.* **2017**, *18*, 237. [CrossRef]

29. Wang, J.; Yan, S.; Lu, C.; Bai, L. Biosynthesis and Characterization of Typical Fibroin Crystalline Polypeptides of Silkworm *Bombyx mori*. *Mater. Sci. Eng. C* **2009**, *29*, 1321–1325. [CrossRef]

30. Ma, Y.; Luo, Q.; Ou, Y.; Tang, Y.; Zeng, W.; Wang, H.; Hu, J. New Insights into the Proteins Interacting with the Promoters of Silkworm Fibroin Genes. *Sci. Rep.* **2021**, *11*, 1–12. [CrossRef]

31. Lotz, B.; Cesari, F.C. The Chemical Structure and the Crystalline Structures of *Bombyx mori* Silk Fibroin. *Biochimie* **1979**, *61*, 205–214. [CrossRef]

32. Vidya, M.; Rajagopal, S. Silk Fibroin: A Promising Tool for Wound Healing and Skin Regeneration. *Int. J. Polym. Sci.* **2021**, *2021*, 9069924. [CrossRef]

33. Koh, L.D.; Cheng, Y.; Teng, C.P.; Khin, Y.W.; Loh, X.J.; Tee, S.Y.; Low, M.; Ye, E.; Yu, H.D.; Zhang, Y.W.; et al. Structures, Mechanical Properties and Applications of Silk Fibroin Materials. *Prog. Polym. Sci.* **2015**, *46*, 86–110. [CrossRef]

34. Dong, Z.; Zhao, P.; Zhang, Y.; Song, Q.; Zhang, X.; Guo, P. Analysis of Proteome Dynamics inside the Silk Gland Lumen of *Bombyx mori*. *Sci. Rep.* **2016**, *6*, 1–10. [CrossRef] [PubMed]

35. Leem, J.W.; Fraser, M.J.; Kim, Y.L. Transgenic and Diet-Enhanced Silk Production for Reinforced Biomaterials: A Metamaterial Perspective. *Annu. Rev. Biomed. Eng.* **2020**, *22*, 79–102. [CrossRef] [PubMed]

36. Panda, D.; Konar, S.; Bajpai, S.K.; Arockiarajan, A. Thermodynamically-Consistent Constitutive Modeling of Aligned Silk Fibroin Sponges: Theory and Application to Uniaxial Compression. *Int. J. Solids Struct.* **2018**, *138*, 144–154. [CrossRef]

37. Michelle, G.; Agostini, M.; Moraes, D.; Cecília, A.; Rodas, D.; Zazuco, O.; Masumi, M. Hydrogels from Silk Fibroin Metastable Solution: Formation and Characterization from a Biomaterial Perspective. *Mater. Sci. Eng. C* **2011**, *31*, 997–1001. [CrossRef]

38. Bassani, G.A.; Vincoli, V.; Biagiotti, M.; Valsecchi, E.; Zucca, M.V.; Clavelli, C.; Alessandrino, A.; Freddi, G. A Route to Translate a Silk-Based Medical Device from Lab to Clinic: The Silk Biomaterials Srl Experience. *Insects* **2022**, *13*, 212. [CrossRef]

39. Ghalei, S.; Handa, H. A Review on Antibacterial Silk Fibroin-Based Biomaterials: Current State and Prospects. *Mater. Today Chem.* **2022**, *23*, 100673. [CrossRef]

40. Khademolqorani, S.; Tavanai, H.; Chronakis, I.S.; Boisen, A. The Determinant Role of Fabrication Technique in Final Characteristics of Scaffolds for Tissue Engineering Applications: A Focus on Silk Fibroin-Based Scaffolds. *Mater. Sci. Eng. C* **2021**, *122*, 111867. [CrossRef]

41. Jo, Y.; Kweon, H.; Kim, D.; Baek, K.; Chae, W.; Kang, Y.; Oh, J.; Kim, S.; Garagiola, U. Silk Sericin Application Increases Bone Morphogenic Protein-2/4 Expression via a Toll-like Receptor-Mediated Pathway. *Int. J. Biol. Macromol.* **2021**, *190*, 607–617. [CrossRef]

42. Teuschl, A.; Griensven, M. Van; Redl, H.; Teuschl, A.H.; Bioreactors, E. Sericin Removal from Raw *Bombyx mori* Silk Scaffolds of High Hierarchical Order. *Tissue Eng. Part C Methods* **2014**, *20*, 1–25. [CrossRef]

43. Gholipourmalekabadi, M.; Sapruc, S.; Samadikuchaksaraei, A.; Reis, R.L.; Kaplan, D.L.; Subhas, C.K. Silk Fibroin for Skin Injury Repair: Where Do Things Stand? *Adv. Drug Deliv. Rev.* **2019**, *153*, 28–53. [CrossRef]

44. Crakes, K.R.; Herrera, C.; Morgan, J.L.; Olstad, K.; Hessell, A.J.; Ziprin, P.; Liwang, P.J.; Dandekar, S. Efficacy of Silk Fibroin Biomaterial Vehicle for in Vivo Mucosal Delivery of Griffithsin and Protection against HIV and SHIV Infection Ex Vivo. *J. Int. AIDS Soc.* **2020**, *23*, 1–12. [CrossRef]

45. Madden, P.W.; Klyubin, I.; Ahearne, M.J. Silk Fibroin Safety in the Eye: A Review That Highlights a Concern. *BMJ Open Ophtalmol.* **2020**, *5*, 1–10. [CrossRef]

46. Yang, Y.; Chen, X.; Ding, F.; Zhang, P.; Liu, J.; Gu, X. Biocompatibility Evaluation of Silk Fibroin with Peripheral Nerve Tissues and Cells in Vitro. *Biomaterials* **2007**, *28*, 1643–1652. [CrossRef]

47. Zhang, J.; Huang, H.; Ju, R.; Chen, K.; Li, S.; Wang, W.; Yan, Y. In Vivo Biocompatibility and Hemocompatibility of a Polytetra Fl Uoroethylene Small Diameter Vascular Graft Modified with Sulfonated Silk Fibroin. *Am. J. Surg.* **2017**, *213*, 87–93. [CrossRef]

48. Tian, Y.; Wu, Q.; Li, F.; Zhou, Y.; Huang, D.; Xie, R.; Wang, X.; Zheng, Z.; Li, G. A Flexible and Biocompatible *Bombyx mori* Silk Fibroin/Wool Keratin Composite Scaffold with Interconnective Porous Structure. *Colloids Surf. B Biointerfaces* **2021**, *208*, 112080. [CrossRef]

49. Wang, D.; Wang, L.; Lou, Z.; Zheng, Y.; Wang, K.; Zhao, L.; Han, W.; Jiang, K.; Shen, G. Biomimetic, Biocompatible and Robust Silk Fibroin-MXene Film with STable 3D Cross-Link Structure for Flexible Pressure Sensors. *Nano Energy* **2020**, *78*, 105252. [CrossRef]

50. Wang, S.; Zhang, Y.; Wang, H.; Dong, Z. Preparation, Characterization and Biocompatibility of Electrospinning Heparin-Modified Silk Fibroin Nanofibers. *Int. J. Biol. Macromol.* **2011**, *48*, 345–353. [CrossRef]

51. Choi, Y.; Cho, D.; Lee, H. Development of Silk Fibroin Scaffolds by Using Indirect 3D-Bioprinting Technology. *Micromachines* **2022**, *13*, 43. [CrossRef]

52. Stoica, A.E.; Chircov, C.; Grumezescu, A.M. Hydrogel Dressings for the Treatment of Burn Wounds: An Up-To-Date Overview. *Materials* **2020**, *13*, 2853. [CrossRef]

53. Zheng, L.; Li, S.; Luo, J.; Wang, X. Latest Advances on Bacterial Cellulose-Based Antibacterial Materials as Wound Dressings. *Front. Bioeng. Biotechnol.* **2020**, *8*, 1–15. [CrossRef]

54. Yin, C.; Han, X.; Lu, Q.; Qi, X.; Guo, C.; Wu, X. Rhein Incorporated Silk Fibroin Hydrogels with Antibacterial and Anti- Inflammatory Efficacy to Promote Healing of Bacteria-Infected Burn Wounds. *Int. J. Biol. Macromol.* **2022**, *201*, 14–19. [CrossRef]

55. Mu, Y.; Gage, F.H. Adult Hippocampal Neurogenesis and Its Role in Alzheimer's Disease. *Mol. Neurodegener.* **2011**, *6*, 1–9. [CrossRef]

56. Zhong, S.; Wang, M.; Zhan, Y.; Zhang, J.; Yang, X.; Fu, S.; Bi, D.; Gao, F. Single-Nucleus RNA Sequencing Reveals Transcriptional Changes of Hippocampal Neurons in APP23 Mouse Model of Alzheimer's Disease. *Biosci. Biotechnol. Biochem.* **2020**, *84*, 919–926. [CrossRef]

57. Tang, X.; Ding, F.; Yang, Y.; Hu, N.; Wu, H.; Gu, X. Evaluation on in Vitro Biocompatibility of Silk Fibroin-Based Biomaterials with Primarily Cultured Hippocampal Neurons. *J. Biomed. Mater. Res. A* **2008**, *91*, 166–174. [CrossRef]

58. Reinke, J.; Sorg, H. Wound Repair and Regeneration. *Eur. Surg. Res.* **2012**, *49*, 35–43. [CrossRef]

59. Zhang, W.; Chen, L.; Chen, J.; Wang, L.; Gui, X.; Ran, J. Silk Fibroin Biomaterial Shows Safe and Effective Wound Healing in Animal Models and a Randomized Controlled Clinical Trial. *Adv. Healthc. Mater.* **2017**, *6*, 1–16. [CrossRef]

60. Acharya, C.; Ghosh, S.K.; Kundu, S.C. Silk Fibroin Protein from Mulberry and Non-Mulberry Silkworms: Cytotoxicity, Biocom- patibility and Kinetics of L929 Murine Fibroblast Adhesion. *J. Mater. Sci. Mater. Med.* **2008**, *19*, 2827–2836. [CrossRef]

61. Kim, S.H.; Yeon, Y.K.; Lee, J.M.; Chao, J.R.; Lee, Y.J.; Seo, Y.B.; Sultan, T.; Lee, O.J.; Lee, J.S.; Yoon, S.; et al. Precisely Printable and Biocompatible Silk Fibroin Bioink for Digital Light Processing 3D Printing. *Nat. Commun.* **2018**, *9*, 1–14. [CrossRef]

62. Cao, Y.; Wang, B. Biodegradation of Silk Biomaterials. *Int. J. Mol. Sci.* **2009**, *10*, 1514–1524. [CrossRef]

63. Zhang, L.; Liu, X.; Li, G.; Wang, P.; Yang, Y. Tailoring Degradation Rates of Silk Fibroin Scaffolds for Tissue Engineering. *J. Biomed. Mater.* **2019**, *107*, 104–113. [CrossRef]

64. Wang, H.; Zhang, Y.; Wei, Z. Characterization of Undegraded and Degraded Silk Fibroin and Its Significant Impact on the Properties of the Resulting Silk Biomaterials. *Int. J. Biol. Macromol.* **2021**, *176*, 578–588. [CrossRef] [PubMed]

65. Wang, Y.; Fan, S.; Li, Y.; Niu, C.; Li, X.; Guo, Y.; Zhang, J.; Shi, J.; Wang, X. Silk Fibroin/Sodium Alginate Composite Porous Materials with Controllable Degradation. *Int. J. Biol. Macromol.* **2020**, *150*, 1314–1322. [CrossRef] [PubMed]

66. Luo, Z.; Zhang, Q.; Shi, M.; Zhang, Y.; Tao, W.; Li, M. Effect of Pore Size on the Biodegradation Rate of Silk Fibroin Scaffolds. *Adv. Mater. Sci. Eng.* **2015**, *2015*, 1–8. [CrossRef] [PubMed]

67. Sun, M.; Li, Q.; Yu, H.; Cheng, J.; Wu, N.; Shi, W.; Zhao, F.; Shao, Z.; Meng, Q.; Chen, H.; et al. Cryo-Self-Assembled Silk Fibroin Sponge as a Biodegradable Platform for Enzyme-Responsive Delivery of Exosomes. *Bioact. Mater.* **2022**, *8*, 505–514. [CrossRef]

68. Catto, V.; Farè, S.; Cattaneo, I.; Figliuzzi, M.; Alessandrino, A.; Freddi, G.; Remuzzi, A.; Cristina, M. Small Diameter Electrospun Silk Fibroin Vascular Grafts: Mechanical Properties, in Vitro Biodegradability, and in Vivo Biocompatibility. *Mater. Sci. Eng. C* **2015**, *54*, 101–111. [CrossRef]

69. Mehrabani, M.G.; Karimian, R.; Mehramouz, B.; Rahimi, M.; Kafil, H.S. Preparation of Biocompatible and Biodegradable Silk Fibroin/Chitin/Silver Nanoparticles 3D Scaffolds as a Bandage for Antimicrobial Wound Dressing. *Biol. Macromol.* **2018**, *114*, 961–971. [CrossRef]

70. Fan, H.; Liu, H.; Toh, S.L.; Goh, J.C.H. Anterior Cruciate Ligament Regeneration Using Mesenchymal Stem Cells and Silk Scaffold in Large Animal Model. *Biomaterials* **2009**, *30*, 4967–4977. [CrossRef]

71. Wang, Y.; Rudym, D.D.; Walsh, A.; Abrahamsen, L.; Kim, H.; Kim, H.S.; Kirker-head, C.; Kaplan, D.L. In Vivo Degradation of Three-Dimensional Silk Fibroin Scaffolds. *Biomaterials* **2008**, *29*, 3415–3428. [CrossRef]

72. Melke, J.; Midha, S.; Ghosh, S.; Ito, K.; Hofmann, S. Silk Fibroin as Biomaterial for Bone Tissue Engineering. *Acta Biomater.* **2016**, *31*, 1–16. [CrossRef]

73. Johari, N.; Moroni, L.; Samadikuchaksaraei, A. Tuning the Conformation and Mechanical Properties of Silk Fibroin Hydrogels. *Eur. Polym. J.* **2020**, *134*, 109842. [CrossRef]

74. Grabska-Zielin′ska, S.; Sionkowska, A.; Coelho, C.C.; Grabska-zieli, S.; Monteiro, F.J. Silk Fibroin/Collagen/Chitosan Scaffolds Cross-Linked by a Glyoxal Solution as Biomaterials toward Bone Tissue Regeneration. *Materials* **2020**, *13*, 3433. [CrossRef]

75. Eivazzadeh-keihan, R.; Ahmadpour, F.; Aghamirza, H.; Aliabadi, M.; Radinekiyan, F.; Maleki, A.; Madanchi, H.; Mahdavi, M.; Esmail, A.; Lanceros-m, S. Pectin-Cellulose Hydrogel, Silk Fibroin and Magnesium Hydroxide Nanoparticles Hybrid Nanocomposites for Biomedical Applications. *Int. J. Biol. Macromol.* **2021**, *192*, 7–15. [CrossRef]

76. Chen, T.; Wen, T.; Dai, N.; Hsu, S. Cryogel Hydrogel Biomaterials and Acupuncture Combined to Promote Diabetic Skin Wound Healing through Immunomodulation. *Biomaterials* **2021**, *269*, 120608. [CrossRef]

77. Hoon, D.; Tripathy, N.; Hun, J.; Eun, J.; Geun, J.; Dan, K.; Hum, C.; Khang, G. Enhanced Osteogenesis of B-Tricalcium Phosphate Reinforced Silk Fibroin Scaffold for Bone Tissue Biofabrication. *Int. J. Biol. Macromol.* **2017**, *95*, 14–23. [CrossRef]

78. Chen, Z.; Zhong, N.; Wen, J.; Jia, M.; Guo, Y.; Shao, Z.; Zhao, X.; Accepted, J. Porous Three-Dimensional Silk Fibroin Scaffolds for Tracheal Epithelial Regeneration in Vitro and in Vivo. *ACS Biomater. Sci. Eng.* **2018**, *4*, 2977–2985. [CrossRef]

79. Yan, L.-P.; Oliveira, J.M.; Oliveira, A.L.; Caridade, S.G.; Mano, J.F.; Reis, R.L. Macro/Microporous Silk Fibroin Scaffolds with Potential for Articular Cartilage and Meniscus Tissue Engineering Applications. *Acta Biomater.* **2012**, *8*, 289–301. [CrossRef]

80. Vepari, C.; Kaplan, D.L. Silk as a Biomaterial. *Prog. Polym. Sci.* **2007**, *32*, 991–1007. [CrossRef]

81. Zhang, H.; Fang, J.; Ge, H.; Han, L.; Wang, X.; Hao, Y.; Han, C.; Dong, L. Thermal, Mechanical, and Rheological Properties of Polylactide/Poly (1, 2-Propylene Glycol Adipate). *Polym. Eng. Sci.* **2013**, *53*, 112–118. [CrossRef]

82. Bunster, G.F. Polyhydroxyalkanoates: Production and Use in Medicine. In *Encyclopedia of Biomedical Polymers and Polymeric Biomaterials*; CRC Press: Boca Raton, FL, USA, 2016; pp. 6412–6421. [CrossRef]

83. Guhados, G.; Wan, W.; Hutter, J.L. Measurement of the Elastic Modulus of Single Bacterial Cellulose Fibers Using Atomic Force Microscopy. *Am. Chem. Soc.* **2005**, *21*, 6642–6646. [CrossRef]

84. Pogorelova, N.; Rogachev, E.; Dige, I.; Chernigova, S.; Nardin, D. Bacterial Cellulose Nanocomposites: Morphology. *Materials* **2020**, *13*, 2849. [CrossRef]

85. Guo, S.; Dipietro, L.A. Factors Affecting Wound Healing. *Crit. Rev. Oral Biol. Med.* **2010**, *89*, 219–229. [CrossRef]

86. Chen, Z.; Zhang, Y.; Zheng, L.; Zhang, H.; Shi, H.; Zhang, X.; Liu, B. Mineralized Self-Assembled Silk Fibroin/Cellulose Interpenetrating Network Aerogel for Bone Tissue Engineering. *Mater. Sci. Eng. C* **2021**, 112549. [CrossRef]

87. Martınez-Mora, C.; Mrowiec, A.; Marı, E.; Alcaraz, A.; Jose, F. Fibroin and Sericin from *Bombyx mori* Silk Stimulate Cell Migration through Upregulation and Phosphorylation Of. *PLoS ONE* **2012**, *7*, e42271. [CrossRef]

88. Nikam, V.S.; Punde, D.S.; Bhandari, R.S. Silk Fibroin Nanofibers Enhance Cell Adhesion of Blood-Derived Fibroblast-like Cells— A Potential Application for Wound Healing. *Indian J. Pharmacol.* **2020**, *52*, 306–312. [CrossRef]

89. Gharehnazifam, Z.; Dolatabadi, R.; Baniassadi, M.; Shahsavari, H.; Kajbafzadeh, A.; Abrinia, K.; Baghani, M. Computational Analysis of Vincristine Loaded Silk Fibroin Hydrogel for Sustained Drug Delivery Applications: Multiphysics Modeling and Experiments. *Int. J. Pharm.* **2021**, *609*, 121184. [CrossRef]

90. Kwon, G.; Heo, B.; Kwon, M.J.; Kim, I.; Chu, J.; Kim, B.; Kim, B.; Park, S.S. Effect of Silk Fibroin Biomaterial Coating on Cell Viability and Intestinal Adhesion of Probiotic Bacteria. *J. Microbiol. Biotechnol.* **2021**, *31*, 592–600. [CrossRef]

91. Lee, O.J.; Sultan, M.T.; Hong, H.; Lee, Y.J.; Lee, J.S.; Lee, H.; Kim, S.H.; Park, C.H. Recent Advances in Fluorescent Silk Fibroin. *Front. Mater.* **2020**, *7*, 1–12. [CrossRef]

92. Baci, G.-M.; Cucu, A.-A.; Giurgiu, A.-I.; Muscă, A.-S.; Rat iu, C.A.; Bagameri, L.; Moise, A.R.; Bobis , O.; Dezmirean, D.S. Advances in Editing Silkworms (*Bombyx mori*) Genome by Using the CRISPR-Cas System. *Insects* **2022**, *13*, 28. [CrossRef] [PubMed]

93. Nagano, A.; Tanioka, Y.; Sakurai, N.; Sezutsu, H.; Kuboyama, N.; Kiba, H. Regeneration of the Femoral Epicondyle on Calcium- Binding Silk Scaffolds Developed Using Transgenic Silk Fibroin Produced by Transgenic Silkworm. *Acta Biomater.* **2011**, *7*, 1192–1201. [CrossRef] [PubMed]

94. Kuwana, Y.; Sezutsu, H.; Nakajima, K.I.; Tamada, Y.; Kojima, K. High-Toughness Silk Produced by a Transgenic Silkworm Expressing Spider (*Araneus Ventricosus*) Dragline Silk Protein. *PLoS ONE* **2014**, *9*, e105325. [CrossRef] [PubMed]

95. Fujinaga, D.; Kohmura, Y.; Okamoto, N.; Kataoka, H.; Mizoguchi, A. Insulin-like Growth Factor (IGF)-like Peptide and 20-Hydroxyecdysone Regulate the Growth and Development of the Male Genital Disk through Different Mechanisms in the Silkmoth, *Bombyx mori*. *Insect Biochem. Mol. Biol.* **2017**, *87*, 35–44. [CrossRef]

96. Wang, F.; Xu, H.; Wang, Y.; Wang, R.; Yuan, L.; Ding, H.; Song, C.; Ma, S.; Peng, Z.; Peng, Z.; et al. Advanced Silk Material Spun by a Transgenic Silkworm Promotes Cell Proliferation for Biomedical Application. *Acta Biomater.* **2014**, *10*, 4947–4955. [CrossRef]

97. Wang, M.; Du, Y.; Huang, H.; Zhu, Z.; Du, S.; Chen, S.; Zhao, H. Silk Fibroin Peptide Suppresses Proliferation and Induces Apoptosis and Cell Cycle Arrest in Human Lung Cancer Cells. *Acta Pharmacol. Sin.* **2019**, *40*, 522–529. [CrossRef]

98. Saviane, A.; Romoli, O.; Bozzato, A.; Freddi, G.; Cappelletti, C.; Rosini, E.; Cappellozza, S.; Tettamanti, G. Intrinsic Antimicrobial Properties of Silk Spun by Genetically Modified Silkworm Strains. *Transgenic Res.* **2018**, *27*, 87–101. [CrossRef]

99. Li, Z.; Cao, G.; Xue, R.; Chengliang, G. Construction of Transgenic Silkworm Spinning Antibacterial Silk with Fluorescence Construction of Transgenic Silkworm Spinning Antibacterial Silk with Fluorescence. *Mol. Biol. Rep.* **2014**, *42*, 19–25. [CrossRef]

100. Iizuka, T.; Sezutsu, H.; Tatematsu, K.; Kobayashi, I.; Yonemura, N.; Uchino, K.; Nakajima, K.; Kojima, K.; Takabayashi, C.; Machii, H.; et al. Colored Fluorescent Silk Made by Transgenic Silkworms. *Adv. J. Mater.* **2013**, *23*, 5232–5239. [CrossRef]

101. Asakura, T.; Isozaki, M.; Saotome, T.; Tatematsu, K.; Sezutsu, H.; Kuwabara, N.; Nakazawa, Y. Recombinant Silk Fibroin Incorporated Cell-Adhesive Sequences Produced by Transgenic Silkworm as a Possible Candidate for Use in Vascular Graft. *J. Mater. Chem. B Mater. Biol. Med.* **2014**, *2*, 7375–7383. [CrossRef]

102. Zhao, S.; Ye, X.; Wu, M.; Ruan, J.; Wang, X.; Tang, X.; Zhong, B. Recombinant Silk Proteins with Additional Polyalanine Have Excellent Mechanical Properties. *Int. J. Mol. Sci.* **2021**, *22*, 1513. [CrossRef]

103. Yanagisawa, S.; Zhu, Z.; Kobayashi, I.; Uchino, K.; Tamada, Y.; Tamura, T.; Asakura, T. Improving Cell-Adhesive Properties of Recombinanant *Bombyx*

mori Silk by Incorporation of Collagen or Fibronectin Derived Peptides Produced by Transgenic Silkworms. *Biomacromolecules* **2007**, *8*, 3487–3492. [CrossRef]

104. Nguyen, T.P.; Nguyen, Q.V.; Nguyen, V.; Le, T.; Le, Q.V. Silk Fibroin-Based Biomaterials for Biomedical. *Polymers* **2019**, *11*, 1933. [CrossRef]
105. Santos, M.V.; Paula, K.T.; Andrade, M.B. De; Gomes, E.M.; Marques, L.F.; Ribeiro, S.J.L.; Mendonc, C.R. Direct Femtosecond Laser Printing of Silk Fibroin Microstructures. *Appl. Mater. Interfaces* **2020**, *12*, 50033–50038. [CrossRef]
106. Ho, W.; Jeong, L.; Il, D.; Hudson, S. Effect of Chitosan on Morphology and Conformation of Electrospun Silk Fibroin Nanofibers. *Polymer* **2004**, *45*, 7151–7157. [CrossRef]
107. Çalamak, S.; Erdo, C.; Özalp, M.; Ulubayram, K. Silk Fibroin Based Antibacterial Bionanotextiles as Wound Dressing Materials. *Mater. Sci. Eng. C* **2014**, *43*, 11–20. [CrossRef]
108. Ha, S.; Tonelli, A.E.; Hudson, S.M.; Carolina, N. Structural Studies of *Bombyx mori* Silk Fibroin during Regeneration from Solutions and Wet Fiber Spinning. *Biomacromolecules* **2005**, *6*, 1722–1731. [CrossRef]
109. Zhang, F.; Ming, J.; Dou, H.; Liu, Z. Silk Dissolution and Regeneration at the Nanofibril Scale. *J. Mater. Chem. B* **2014**, *2*, 3879–3885. [CrossRef]
110. Sato, M.; Nakazawa, Y.; Takahashi, R.; Tanaka, K.; Sata, M.; Aytemiz, D.; Asakura, T. Small-Diameter Vascular Grafts of *Bombyx mori* Silk Fibroin Prepared by a Combination of Electrospinning and Sponge Coating. *Mater. Lett.* **2010**, *64*, 1786–1788. [CrossRef]
111. Aznar-cervantes, S.; Roca, M.I.; Martinez, J.G.; Meseguer-olmo, L.; Cenis, J.L.; Moraleda, J.M.; Otero, T.F. Fabrication of Conductive Electrospun Silk Fibroin Scaffolds by Coating with Polypyrrole for Biomedical Applications. *Bioelectrochemistry* **2012**, *85*, 36–43. [CrossRef]
112. Aznar-cervantes, S.D.; Vicente-cervantes, D.; Meseguer-olmo, L.; Cenis, J.L.; Lozano-pérez, A.A. Influence of the Protocol Used for Fibroin Extraction on the Mechanical Properties and Fiber Sizes of Electrospun Silk Mats. *Mater. Sci. Eng. C* **2013**, *33*, 1945–1950. [CrossRef]
113. Yamada, H.; Nakao, H.; Takasu, Y.; Tsubouchi, K. Preparation of Undegraded Native Molecular Fibroin Solution from Silkworm Cocoons. *Mater. Sci. Eng. C* **2001**, *14*, 41–46. [CrossRef]
114. Sofia, S.; Mccarthy, M.B.; Gronowicz, G.; Kaplan, D.L.; Al, S.E.T. Functionalized Silk-Based Biomaterials for Bone Formation. *J. Biomed. Mater. Res.* **2000**, *54*, 139–148. [CrossRef]

115. Mathur, A.B.; Tonelli, A.; Rathke, T.; Hudson, S. The Dissolution and Characterization of *Bombyx mori* Silk Fibroin in Calcium Nitrate-methanol. Pdf. *Fiber Polym. Sci.* **1998**, *42*, 61–74.

116. Kunz, W.; Katharina, H. Some Aspects of Green Solvents. *Comptes Rendus Chim.* **2018**, *21*, 572–580. [CrossRef]

117. Wang, H.; Zhang, Y.; Wei, Z. Dissolution and Processing of Silk Fibroin for Materials Science. *Crit. Rev. Biotechnol.* **2021**, *41*, 406–424. [CrossRef] [PubMed]

118. Wöltje, M.; Kölbel, A.; Aibibu, D.; Cherif, C. A Fast and Reliable Process to Fabricate Regenerated Silk Fibroin Solution from Degummed Silk in 4 Hours. *Int. J. Mol. Sci.* **2021**, *22*, 565. [CrossRef] [PubMed]

119. Freddi, G.; Pessina, G.; Tsukada, M. Swelling and Dissolution of Silk Fibroin (*Bombyx mori*) in N -Methyl Morpholine N -Oxide. *Biol. Macromol.* **1999**, *24*, 251–263. [CrossRef]

120. Carissimi, G.; Baronio, C.M.; Montalbán, M.G.; Víllora, G.; Barth, A. On the Secondary Structure of Silk Fibroin Nanoparticles Obtained Using Ionic Liquids: An Infrared Spectroscopy Study. *Polymers* **2020**, *12*, 1294. [CrossRef] [PubMed]

121. Garc, M.; Aznar-Cervantes, S.D.; Lozano-Perez, A.A.; Cenis, L.; Gloria, V. Production of Silk Fibroin Nanoparticles Using Ionic Liquids and High-Power Ultrasounds. *J. Appl. Polym. Sci.* **2015**, *41702*, 1–8. [CrossRef]

122. Phillips, D.M.; Drummy, L.F.; Conrady, D.G.; Fox, D.M.; Naik, R.R.; Stone, M.O.; Trulove, P.C.; Long, H.C. De; Mantz, R.A. Dissolution and Regeneration of *Bombyx mori* Silk Fibroin Using Ionic Liquids. *J. Am. Chem. Soc.* **2004**, *126*, 14350–14351. [CrossRef]

123. Ajisawa, A. Dissolution Aqueous of Silk Fibroin with Calciumchloride/Ethanol Solution. *J. Seric. Sci. Jpn.* **1997**, *67*, 91–94.

124. Reizabal, A.; Costa, C.M.; Saiz, P.G.; Gonzalez, B.; Perez-Alvarez, L.; Luis, R.F. de; Garcia, A.; Vilas-Vilela, J.; Lanceros-Mendez, S. Processing Strategies to Obtain Highly Porous Silk Fibroin Structures with Tailored Microstructure and Molecular Characteristics and Their Applicability in Water Remediation. *J. Hazard. Mater.* **2021**, *403*, 123675. [CrossRef]

125. Bae, S. Bin; Kim, M.H.; Park, W.H. Electrospinning and Dual Crosslinking of Water-Soluble Silk Fibroin Modified with Glycidyl Methacrylate. *Polym. Degrad. Stab.* **2020**, *179*, 109304. [CrossRef]

126. Mosher, C.Z.; Brudnicki, P.A.P.; Gong, Z.; Childs, H.R.; Lee, S.W.; Antrobus, R.M.; Fang, E.C.; Schiros, T.N.; Lu, H.H. Green Electrospinning for Biomaterials and Biofabrication. *Biofabrication* **2021**, *13*, 035049. [CrossRef]

127. Li, X.; Fan, Q.; Zhang, Q.; Yan, S.; You, R. Freezing-Induced Silk I Crystallization of Silk Fibroin. *R. Soc. Chem.* **2020**, *22*, 3884–3890. [CrossRef]

128. Yang, X.; Wang, X.; Yu, F.; Ma, L.; Pan, X.; Luo, G.; Lin, S.; Mo, X.; Wang, H. Hyaluronic Acid/EDC/NHS-Crosslinked Green Electrospun Silk Fibroin Nanofibrous Scaffolds for Tissue Engineering. *R. Soc. Chem.* **2016**, *6*, 99720–99728. [CrossRef]

129. Fei, X.; Jia, M.; Du, X.; Yang, Y.; Zhang, R.; Shao, Z.; Zhao, X. Green Synthesis of Silk Fibroin-Silver Nanoparticle Composites with Effective Antibacterial and Biofilm-Disrupting Properties. *Biomacromolecules* **2013**, *14*, 4483–4488. [CrossRef]

130. Raho, R.; Nguyen, N.; Zhang, N.; Jiang, W.; Sannino, A.; Liu, H.; Pollini, M.; Paladini, F. Photo-Assisted Green Synthesis of Silver Doped Silk Fibroin/Carboxymethyl Cellulose Nanocomposite Hydrogels for Biomedical Applications. *Mater. Sci. Eng. C* **2020**, *107*, 110219. [CrossRef]

131. El-Sheikh, M.A.; El-Rafie, S.M.; Abdel-Halim, E.S.; El-Rafie, M.H. Green Synthesis of Hydroxyethyl Cellulose-Stabilized Silver Nanoparticles. *J. Polym.* **2013**, *2013*, 1–11. [CrossRef]

132. Altman, G.H.; Diaz, F.; Jakuba, C.; Calabro, T.; Horan, R.L.; Chen, J.; Lu, H.; Richmond, J.; Kaplan, D.L. Silk-Based Biomaterials. *Biomaterials* **2003**, *24*, 401–416. [CrossRef]

133. Kamalathevan, P.; Ooi, P.S.; Loo, Y.L. Silk-Based Biomaterials in Cutaneous Wound Healing: A Systematic Review. *Adv. Ski. Wound Care* **2018**, *31*, 565–573. [CrossRef]

134. Cheng, G.; Wang, X.; Tao, S.; Xia, J.; Xu, S. Differences in Regenerated Silk Fibroin Prepared with Different Solvent Systems: From Structures to Conformational Changes. *J. Appl. Polym. Sci.* **2015**, *132*, 1–8. [CrossRef]

135. Volkov, V.; Ferreira, A.V.; Cavaco-paulo, A. On the Routines of Wild-Type Silk Fibroin Processing Toward Silk-Inspired Materials: A Review. *Macromol. Mater. Eng.* **2015**, *300*, 1199–1216. [CrossRef]

136. Jin, S.C.; Baek, H.S.; Woo, Y.I.; Lee, M.H.; Kim, J.; Park, J.; Park, Y.H.; Lee, S.J. Beneficial Effects of Microwave-Induced Argon Plasma Treatment on Cellular Behaviors of Articular Chondrocytes onto Nanofibrous Silk Fibroin Mesh. *Macromol. Res.* **2009**, *17*, 703–708. [CrossRef]

137. Gu, J.; Yang, X.; Zhu, H. Surface Sulfonation of Silk Fibroin Film by Plasma Treatment and in Vitro Antithrombogenicity Study. *Mater. Sci. Eng. C* **2002**, *20*, 199–202. [CrossRef]

138. Shang, K.; Rnjak-kovacina, J.; Tao, H.; Kaplan, D.L.; Lin, Y.; Hayden, R.S. Accelerated In Vitro Degradation of Optically Clear Low b -Sheet Silk Films by Enzyme-Mediated Pretreatment. *Transl. Vis. Sci. Technol.* **2013**, *2*, 2. [CrossRef]

139. Wang, L.; Nemoto, R.; Senna, M. Three-Dimensional Porous Network Structure Developed in Hydroxyapatite-Based Nanocom- posites Containing Enzyme Pretreated Silk Fibroin. *J. Nanopart. Res.* **2004**, *6*, 91–98. [CrossRef]

140. Li, M.; Ogiso, M.; Minoura, N. Enzymatic Degradation Behavior of Porous Silk Fibroin Sheets. *Biomaterials* **2003**, *24*, 357–365. [CrossRef]

141. Wang, P.; Zhou, Y.; Cui, L.; Yuan, J.; Wang, Q.; Fan, X.; Ding, Y. Enzymatic Grafting of Lactoferrin onto Silk Fibroins for Antibacterial Functionalization. *Fibers Polym.* **2014**, *15*, 2045–2050. [CrossRef]

142. Chi, R.; Cheung, F.; Ng, T.B.; Wong, J.H.; Chan, W.Y. Chitosan: An Update on Potential Biomedical and Pharmaceutical Applications. *Mar. Drugs* **2015**, *13*, 5156–5186. [CrossRef]

143. Li, X.; Li, B.; Wang, X.; Zhang, S. Development of a Silk Fibroin/HTCC/PVA Sponge for Chronic Wound Dressing. *J. Bioact. Compat. Polym. Biomed. Appl.* **2014**, *29*, 398–411. [CrossRef]

144. Karahalilog˘ lu, Z. Curcumin-Loaded Silk Fibroin e-Gel Scaffolds for Wound Healing Applications. *Mater. Technol.* **2018**, *7857*, 1–12. [CrossRef]

145. Thangavel, P.; Ramachandran, B.; Kannan, R.; Muthuvijayan, V. Biomimetic Hydrogel Loaded with Silk and L -Proline for Tissue Engineering and Wound Healing Applications. *Soc. Biomater.* **2016**, *105*, 1401–1408. [CrossRef]

146. Luo, Z.; Jiang, L.; Xu, Y.; Li, H.; Xu, W.; Wu, S.; Wang, Y.; Tang, Z.; Lv, Y.; Yang, L. Biomaterials Mechano Growth Factor (MGF) and Transforming Growth Factor (TGF)—B3 Functionalized Silk Scaffolds Enhance Articular Hyaline Cartilage Regeneration in Rabbit Model. *Biomaterials* **2015**, *52*, 463–475. [CrossRef]

147. Mehrabani, M.G.; Karimian, R.; Rakhshaei, R.; Pakdel, F.; Eslami, H.; Fakhrzadeh, V.; Rahimi, M.; Salehi, R.; Kafil, H.S. Chitin/Silk Fibroin/TiO_2 Bio-Nanocomposite as a Biocompatible Wound Dressing Bandage with Strong Antimicrobial Activity. *Biol. Macromol.* **2018**, *116*, 966–976. [CrossRef]

148. Patil, P.P.; Meshram, J.V.; Bohara, R.A. ZnO Nanoparticle-Embedded Silk Fibroin–Polyvinyl Alcohol Composite Film: A Potential Dressing Material for Infected Wounds. *R. Soc. Chem.* **2018**, *42*, 14620–14629. [CrossRef]

149. Li, X.; Liu, Y.; Zhang, J.; You, R.; Qu, J.; Li, M. Functionalized Silk Fibroin Dressing with Topical Bioactive Insulin Release for Accelerated Chronic Wound Healing. *Mater. Sci. Eng. C* **2017**, *72*, 394–404. [CrossRef]

150. Hrynyk, M.; Neufeld, R.J. Insulin and Wound Healing. *Burns* **2014**, *40*, 1433–1446. [CrossRef]

151. Wang, Y.; Kim, U.; Blasioli, D.J.; Kim, H.; Kaplan, D.L. In Vitro Cartilage Tissue Engineering with 3D Porous Aqueous-Derived Silk Scaffolds and Mesenchymal Stem Cells. *Biomaterials* **2005**, *26*, 7082–7094. [CrossRef]

152. Hee, S.; Been, Y.; Kyu, Y.; Jin, Y.; Sang, H.; Sultan, T.; Min, J.; Seung, J.; Joo, O.; Hong, H.; et al. Biomaterials 4D-Bioprinted Silk Hydrogels for Tissue Engineering. *Biomaterials* **2020**, *260*, 120281. [CrossRef]

153. Mallepally, R.R.; Marin, M.A.; Surampudi, V.; Subia, B.; Rao, R.R. Silk Fibroin Aerogels: Potential Scaffolds for Tissue Engineering Applications. *Biomed. Mater.* **2015**, *10*, 35002. [CrossRef]

154. Nazarov, R.; Jin, H.; Kaplan, D.L. Porous 3-D Scaffolds from Regenerated Silk Fibroin. *Biomacromolecules* **2004**, *5*, 718–726. [CrossRef]

155. Pacheco, S.M.; Eiji, G.; Almeida, L. De; Santos, P.; Agostini, M.; Moraes, D. Silk Fibroin/Chitosan/Alginate Multilayer Membranes as a System for Controlled Drug Release in Wound Healing. *Int. J. Biol. Macromol.* **2020**, *152*, 803–811. [CrossRef]

156. Harkin, D.G.; George, K.A.; Madden, P.W.; Schwab, I.R.; Hutmacher, D.W.; Chirila, T.V. Biomaterials Silk Fibroin in Ocular Tissue Reconstruction. *Biomaterials* **2011**, *32*, 2445–2458. [CrossRef]

157. Calonge, M.; Nieto-Miguel, T.; Mata, A. De; Galindo, S.; Herreras, J.M.; Marina, L. Goals and Challenges of Stem Cell-Based Therapy for Corneal Blindness Due to Limbal Deficiency. *Pharmaceutics* **2021**, *13*, 1483. [CrossRef]

158. Lee, M.C.; Kim, D.; Lee, O.J.; Kim, J.; Ju, H.W.; Lee, J.M.; Moon, B.M.; Park, H.J.; Kim, D.W.; Kim, S.H.; et al. Fabrication of Silk Fibroin Film Using Centrifugal Casting Technique for Corneal Tissue Engineering. *Soc. Biomater.* **2016**, *104*, 508–514. [CrossRef]

159. Liu, J.; Lawrence, B.D.; Liu, A.; Schwab, I.R.; Oliveira, L.A.; Rosenblatt, M.I. Silk Fibroin as a Biomaterial Substrate for Corneal Epithelial Cell Sheet Generation. *Investig. Ophthalmol. Vis. Sci.* **2012**, *53*, 4130–4138. [CrossRef]

160. Pellegrini, G.; Dellambra, E.; Golisano, O.; Martinelli, E.; Fantozzi, I.; Bondanza, S.; Ponzin, D.; Mckeon, F.; Luca, M. De. P63 Identifies Keratinocyte Stem Cells. *Proc. Natl. Acad. Sci. USA* **2001**, *98*, 357–394. [CrossRef]

161. Schermer, A.; Galvin, S. Differentiation-Related Expression of a Major 64K Corneal Keratin In Vivo and In Culture Suggests Limbal Location of Corneal Epithelial Stem Cells. *J. Cell Biol.* **1986**, *103*, 49–62. [CrossRef] [PubMed]

162. Lawrence, B.D.; Marchant, J.K.; Pindrus, M.A.; Omenetto, F.G.; Kaplan, D.L. Biomaterials Silk Film Biomaterials for Cornea Tissue Engineering. *Biomaterials* **2009**, *30*, 1299–1308. [CrossRef] [PubMed]

163. Malliappan, P.; Alp, A.; Burcu, S.; Demir, E.; Cetinel, S. Bone Tissue Engineering: Anionic Polysaccharides as Promising Scaffolds. *Carbohydr. Polym.* **2022**, *283*, 119142. [CrossRef]

164. Sk, S. Fracture Non-Union: A Review of Clinical Challenges and Future Research Needs. *Malays. Orthop. J.* **2019**, *13*, 1–10. [CrossRef] [PubMed]

165. Gillman, C.E.; Jayasuriya, A.C. FDA-Approved Bone Grafts and Bone Graft Substitute Devices in Bone Regeneration. *Mater. Sci. Eng. C* **2021**, *130*, 112466. [CrossRef]

166. Zhao, Z.; Ma, X.; Ma, J.; Kang, J.; Zhang, Y.; Guo, Y. Sustained Release of Naringin from Silk- Fibroin-Nanohydroxyapatite Scaffold for the Enhancement of Bone Regeneration. *Mater. Today Bio* **2022**, *13*, 100206. [CrossRef]

167. Meinel, L.; Fajardo, R.; Hofmann, S.; Langer, R.; Chen, J.; Snyder, B.; Vunjak-novakovic, G.; Kaplan, D. Silk Implants for the Healing of Critical Size Bone Defects. *Bone* **2005**, *37*, 688–698. [CrossRef]

168. Zhao, J.; Zhang, Z.; Wang, S.; Sun, X.; Zhang, X.; Chen, J.; Kaplan, D.L.; Jiang, X. Apatite-Coated Silk Fibroin Scaffolds to Healing Mandibular Border Defects in Canines. *Bone* **2009**, *45*, 517–527. [CrossRef]

169. Hirose, M.; Hamada, K.; Tanaka, T. Nano-Scaled Hydroxyapatite/Silk Fibroin Composites as Mesenchymal Cell Culture Scaffolds. *Key Eng. Mater.* **2006**, *311*, 923–926. [CrossRef]

170. Cui, B.; Liang, L.; Lu, X.; Weng, J. Fabricating HYDROXYAPATITE—SIlk Fibroin Nanocomposite by Bone Bionics. *Key Eng. Mater.* **2007**, *332*, 345–348. [CrossRef]

171. Wen, G.; Wang, J.; Li, M.; Meng, X. Study on Tissue Engineering Scaffolds of Silk Fibroin-Chitosan/Nano-Hydroxyapatite Composite. *Key Eng. Mater.* **2007**, *332*, 971–975. [CrossRef]

172. Sun, B.K.; Siprashvili, Z.; Khavari, P.A. Advances in Skin Grafting and Treatment of Cutaneous Wounds. *Science* **2014**, *346*, 941–945. [CrossRef]

173. Pra, I.D.A.L.; Chiarini, A.; Boschi, A.; Freddi, G.; Armato, U. Novel Dermo-Epidermal Equivalents on Silk Fibroin-Based Formic Acid-Crosslinked Three-Dimensional Nonwoven Devices with Prospective Applications in Human Tissue Engineer- ing/Regeneration/Repair. *Int. J. Mol. Med.* **2006**, *18*, 241–247.

174. Luangbudnark, W.; Viyoch, J.; Laupattarakasem, W.; Surakunprapha, P.; Laupattarakasem, P. Properties and Biocompatibility of Chitosan and Silk

Fibroin Blend Films for Application in Skin Tissue Engineering. *Sci. World J.* **2012**, *2012*, 697201. [CrossRef]

175. Bhardwaj, N.; Sow, W.T.; Devi, D.; Ng, K.W.; Mandal, B.B.; Cho, N.-J. Insight Statement for "Silk Fibroin-Keratin Based 3D Scaffolds as a Dermal Substitute for Skin Repair and Regeneration. *Integr. Biol.* **2014**, *7*, 53–63. [CrossRef]
176. Lyu, J.; Kaur, M.; Dibble, K.E.; Connor, A.E. A National Study of Alcohol Consumption Patterns among Population-Based U. S. Cancer Survivors Compared with Cancer-Free Individuals. *Cancer Epidemiol.* **2022**, *77*, 102101. [CrossRef]
177. Desforges, A.D.; Hebert, C.M.; Spence, A.L.; Reid, B.; Dhaibar, A.; Cruz-topete, D.; Cornett, E.M.; David, A.; Urits, I.; Viswanath, O. Treatment and Diagnosis of Chemotherapy-Induced Peripheral Neuropathy: An Update. *Biomed. Pharmacother.* **2022**, *147*, 112671. [CrossRef]
178. Ma, D.; Wang, Y.; Dai, W. Silk Fibroin-Based Biomaterials for Musculoskeletal Tissue Engineering. *Mater. Sci. Eng. C* **2018**, *89*, 456–469. [CrossRef]
179. Ma, Y.; Canup, B.S.B.; Tong, X.; Dai, F.; Xiao, B.; Wang, J. Multi-Responsive Silk Fibroin-Based Nanoparticles for Drug Delivery. *Front. Chem.* **2020**, *8*, 1–5. [CrossRef]
180. Sun, N.; Lei, R.; Xu, J.; Kundu, S.C. Fabricated Porous Silk Fibroin Particles for PH- Responsive Drug Delivery and Targeting of Tumor Cells. *J. Mater. Sci.* **2019**, *54*, 3319–3330. [CrossRef]
181. Saqr, A. Al; Ud, S.; Wani, D.; Gangadharappa, H.V.; Aldawsari, M.F.; Khafagy, E.; Lila, A.S.A. Enhanced Cytotoxic Activity of Docetaxel-Loaded Silk Fibroin Nanoparticles against Breast Cancer Cells. *Polymers* **2021**, *13*, 1–18.
182. Mottaghitalab, F.; Kiani, M.; Farokhi, M.; Kundu, S.C.; Reis, R.L.; Gholami, M.; Bardania, H.; Dinarvand, R.; Geramifar, P.; Beiki, D.; et al. Targeted Delivery System Based on Gemcitabine Loaded Silk Fibroin Nanoparticles for Lung Cancer Therapy Targeted Delivery System Based on Gemcitabine Loaded Silk Fibroin Nanoparticles for Lung Cancer Therapy. *Appl. Mater. Interfaces* **2017**, *9*, 31600–31611. [CrossRef]
183. Moin, A.; Wani, S.U.D.; Osmadi, R.A.; Lila, A.S.A.; Khafagy, E.-S.; Arab, H.H.; Gangaharappa, H.V.; Allam, A.N. Formulation, Characterization, and Cellular Toxicity Assessment of Tamoxifen-Loaded Silk Fibroin Nanoparticles in Breast Cancer. *Drug Deliv.* **2021**, *28*, 1626–1636. [CrossRef]
184. Lu, X.; Sun, Y.; Han, M.; Chen, D.; Wang, A.; Sun, K. Silk Fibroin Double-Layer Microneedles for the Encapsulation and Controlled Release of Triptorelin. *Int. J. Pharm.* **2022**, *613*, 121433. [CrossRef]

185. Dondajewska, E.; Juzwa, W.; Mackiewicz, A.; Dams-, H. Heterotypic Breast Cancer Model Based on a Silk Fibroin Scaffold to Study the Tumor Microenvironment. *Oncotarget* **2018**, *9*, 4935–4950. [CrossRef] [PubMed]

186. Li, J.; Zhou, Y.; Chen, W.; Yuan, Z.; You, B.; Liu, Y.; Yang, S.; Li, F.; Qu, C.; Zhang, X. A Novel 3D in Vitro Tumor Model Based on Silk Fibroin/Chitosan Sca Ff Olds to Mimic the Tumor Microenvironment. *Appl. Mater. Interfaces* **2018**, *10*, 36641–36651. [CrossRef] [PubMed]

187. Qian, K.; Song, Y.; Yan, X.; Dong, L.; Xue, J.; Xu, Y.; Wang, B.; Cao, B.; Hou, Q.; Peng, W.; et al. Biomaterials Injectable Ferrimagnetic Silk Fibroin Hydrogel for Magnetic Hyperthermia Ablation of Deep Tumor. *Biomaterials* **2020**, *259*, 120299. [CrossRef] [PubMed]

188. Mukhopadhyay, S.K.; Naskar, D.; Bhattacharjee, P.; Mishra, A.; Kundu, S.C.; Dey, S. Silk Fibroin-Thelebolan Matrix: A Promising Chemopreventive Scaffold for Soft Tissue Cancer. *Colloids Surf. B Biointerfaces* **2017**, *155*, 379–389. [CrossRef] [PubMed]

189. Pandey, V.; Haider, T.; Chandak, A.R.; Chakraborty, A.; Banerjee, S.; Soni, V. Technetium Labeled Doxorubicin Loaded Silk Fibroin Nanoparticles: Optimization, Characterization and in Vitro Evaluation. *J. Drug Deliv. Sci. Technol.* **2020**, *56*, 101539. [CrossRef]

190. Anas, M.; Hadianamrei, R.; Sun, W.; Xu, D.; Brown, S.; Zhao, X. Stiffness-Tuneable Nanocarriers for Controlled Delivery of ASC-J9 into Colorectal Cancer Cells. *J. Colloid Interface Sci.* **2021**, *594*, 513–521. [CrossRef]

191. Talukdar, S.; Kundu, S.C. Engineered 3D Silk-Based Metastasis Models: Interactions Between Human Breast Adenocarcinoma, Mesenchymal Stem Cells and Osteoblast-Like Cells. *Adv. Funct. Mater.* **2013**, *9*, 5249–5260. [CrossRef]

192. Ding, B.; Wahid, M.A.; Zhijun, W.; Chen, X.; Arvind, T.; Sunil, P.; Wang, J. Triptolide and Celastrol Loaded Silk Fibroin Nanoparticles Show Synergistic Effect against Human Pancreatic Cancer Cells. *Nanoscale* **2017**, *9*, 11739–11753. [CrossRef]

193. Toan, D.; Saelim, N.; Tiyaboonchai, W. Alpha Mangostin Loaded Crosslinked Silk Fibroin-Based Nanoparticles for Cancer Chemotherapy. *Colloids Surf. B Biointerfaces* **2019**, *181*, 705–713. [CrossRef]

194. Yucel, T.; Lovett, M.L.; Giangregorio, R.; Coonahan, E.; Kaplan, D.L. Silk Fibroin Rods for Sustained Delivery of Breast Cancer Therapeutics. *Biomaterials* **2014**, *35*, 8613–8620. [CrossRef]

195. Zhang, X.; Huang, Y.; Song, H.; Canup, B.S.B.; Gou, S. Inhibition of Growth and Lung Metastasis of Breast Cancer by Tumor- Homing Triple-Bioresponsive Nanotherapeutics. *J. Control. Release* **2020**, *328*, 454–469. [CrossRef]

196. Yao, Q.; Lan, Q.; Jiang, X.; Du, C.; Zhai, Y.; Shen, X.; Xu, H. Bioinspired Biliverdin/Silk Fibroin Hydrogel for Antiglioma Photothermal Therapy and Wound Healing. *Theranostics* **2020**, *10*, 11739–11753. [CrossRef]

197. Yu, S.; Yang, W.; Chen, S.; Chen, M.; Liu, Y.; Shao, Z.; Chen, X. Floxuridine-Loaded Silk Fibroin Nanospheres. *RSC Adv.* **2014**, *4*, 18171–18177. [CrossRef]

198. Mishra, D.; Iyyanki, T.S.; Hubenak, J.R.; Zhang, Q.; Mathur, A.B. Silk Fibroin Nanoparticles and Cancer Therapy. In *Nanotechnology in Cancer*; Elsevier: Amsterdam, The Netherlands, 2017; pp. 19–44. [CrossRef]

5 Chapter

BLACK, CASPIAN SEAS AND CENTRAL ASIA SILK ASSOCIATION (BACSA) FOR THE ADVANCEMENT OF SERICULTURE IN EUROPE AND CENTRAL ASIA.

1. INTRODUCTION: AN OVERVIEW

The Central Asia region of the former Soviet Union comprises five countries: Kazakhstan, Kyrgyzstan, Tajikistan, Turkmenistan, and Uzbekistan, whereas the South Caucasus region consists of three: Armenia, Azerbaijan, and Georgia. These eight nations achieved independence in 1991 following the dissolution of the Soviet Union, and subsequently endured a profound economic and social crises stemming from the shift from a "centrally planned" economy to a "market economy." The rapid and coerced industrialization of the Soviet period was characterized by significant structural distortions and microeconomic inefficiencies. Following the fall of the Soviet Union, numerous industrial businesses in this region lost their former customers and were unable to compete in the new market circumstances. Given that sericulture is an agroindustry, the challenges faced by Central Asian and Caucasian sericulture activities during the 1990s were foreseeable.

Similar problems occurred in the Central and Eastern Europe after 1989 (Berlin Wall's collapse). Economies throughout the region fell into recession in the 1990s. Efforts to privatize and open markets caused unemployment and social inequality. Bulgaria's econ- omy contracted each year from 1989 through 1993, while Romania's GDP (Gross Domestic Product) dropped nearly 13% in 1991 and nearly 9% in 1992 [2].

These political, social, and economic changes greatly impacted sericulture, perhaps more than any other economic activity, in the ex-communist countries. In fact, to manage the sericultural chain it is necessary to coordinate the agricultural production at the farm level, with silkworm eggs sourcing, logistic steps for drying, sorting, preservation, stocking, and marketing of cocoons; furthermore, coordination is also necessary with the subsequent steps performed in the reeling and textile plants.

What happened in the ex-communist countries in Eastern Europe and Central Asia during the transitional period from centralized to market economic system was a sudden stop of the governmental support to sericulture, the breakage of the traditional economic relationships among the countries, thus the destruction of the already established system of integration of the different processes of the sericultural production.

Until the beginning of '90s of the last century the approximate annual fresh cocoon production in Europe, Caucasus and Central Asia was around 50,000 tons, by this occupying the third place in the world after China and India; nearly one million farmer's households were engaged with sericulture [3]. At the Soviet Union breakdown the Central Asian region, including Uzbekistan, Tajikistan, Turkmenistan and Iran was still a remarkable cocoon and silk producer (even now this region is engaging in the industry about 450,000 farmers) [3]; at that time the Eastern part of Europe, which was also under the communist umbrella of the USSR (Union of Soviet Socialist Republics) and also Turkey, were still producing large quantities of silk, while the Western Europe and Greek sericulture activities had already considerably declined.

During the aforementioned timeframe, specifically from 1983 to 1995, global raw silk production had a steady increase, nearly doubling, primarily due to China's significant contribution, which quadrupled its output during the same period. Consequently, in 1995, China held a global monopoly on silk production, accounting for 74% of the total quantity of this fiber. Over time, the persistent influx of Chinese silk into the market resulted in an oversupply of this fiber, leading to a decline in its international price, which plummeted by around 24% in 1997, ultimately reaching USD 22/kg for raw silk. In this scenario, only India and Brazil remained relatively competitive, while all other countries were either excluded from the market or stayed on the periphery (Uzbekistan, Thailand, Vietnam). The concentration of silk production in China was particularly pronounced for countries sourcing silk from the Celestial Empire due to the "double price" policy, which allowed the domestic silk industry to procure raw materials at a cost 20% lower than that of foreign purchasers. The monopoly on silk production, coupled with the dual pricing strategy, enabled China to achieve unequivocal dominance in the pricing of raw silk as well as silk fabrics and garments. [4].

2. ESTABLISHMENT AND PURPOSE OF THE BLACK AND CASPIAN SEAS AND CENTRAL ASIA SILK ASSOCIATION (BACSA)

As mentioned above sericulture is an agro-based industry and agriculture has always been of huge importance for the countries of the Black and Caspian Seas and Central Asia area, supporting the income and providing employment to a large majority of inhabitants of the rural territories, but also protecting the environment through the sustainable use of natural resources. For this reason, the collapse of sericulture following the USSR dissolution was regarded with much concern by NGOs

(Non-Governmental Organizations). The case of Uzbekistan is emblematic: due to the combined effect of the economic turmoil after the independence, and the Chinese monopoly of silk prices, Uzbek cocoon production dropped from 33,000 t in 1990 to 21,000 t in 1997 [5].

In order to counteract this decline in sericulture for the countries of the former USSR or previously enclosed under its political umbrella, in April 2005, the Food and Agriculture Organization (FAO) of the United Nations and the Uzbek Government organized in Tashkent, Uzbekistan, the "International Workshop on Revival and Promotion of Sericultural Industries and Small Enterprise Development in the Black, Caspian Seas and Central Asia Region". Twelve countries including Azerbaijan, Bulgaria, Egypt, Georgia, Greece, Japan, Kazakhstan, Republic of Korea, Tajikistan, Turkey, Ukraine, and Uzbekistan participated in this meeting. During the workshop a regional association, named Black, Caspian Seas and Central Asia Silk Association (BACSA) was established [5].

The main targets of this association were to generate sericulture projects from external resources, including bilateral and multilateral cooperation, sensitize respective governments and prospective donors, promote local and regional joint efforts, which would have allowed the cooperation among the countries of the region of Europe, Caucasus and Central Asia, develop concrete actions to fortify the sustainable development of the sericulture in the region, promote agreements for international scientific-technical cooperation and business relations among the countries involved, encourage market studies, training, and spreading sericultural germplasm, and silkworm eggs. In fact, the BACSA member coun- tries have many problems and issues in their sericulture revival and development, which could be partly or completely solved by the help of regional cooperation among them. Such problems are the access to the EU (European Union) funds for research, specializations and training of students and technical personnel, exchange of sericulture germplasm re- sources and improvement of the mulberry sapling and silkworm egg quality; other issues consist of saving the sericulture germplasm of the BACSA region, increasing silkworm egg production or favoring its revival in some of the member countries, solving criticalities of cocoon and silk marketing, improving the raw silk quality and rising the share of exported or locally processed raw silk, producing small niche textiles from cocoons and developing silk handcrafts [6].

At the time of its establishment, BACSA member states that were producing silk carpets, such as Iran and Turkey, were also comparatively big raw silk importers as the local silk production was not able to satisfy the silk carpet needs of raw material [7]. In that period, as even presently, some BACSA countries did not have any own silkworm egg production and in some of the Central Asian countries the quality of currently locally produced silkworm eggs did not meet the international standards, or their quantity could not satisfy the local needs [8].

Therefore, one of the basic aims of expansion of BACSA inter—regional cooperation was/is to transmit and share sericulture germplasms, silkworm eggs, advanced technolo- gies, training, dry cocoons, raw silk, and silk allied products.

3. THE BACSA ACTIVITY

3.1 The BACSA Composition

When established in 2005 BACSA included 9 countries—Azerbaijan, Bulgaria, Georgia, Greece, Kazakhstan, Tajikistan, Turkey, Ukraine and Uzbekistan. Later on, and gradually, BACSA attracted for membership new countries—Albania, Armenia, Iran, Poland, Roma- nia, Switzerland, Italy, Spain, Germany, Portugal, Slovenia, Russia, UK so that currently the association includes 22 countries and has also 64 individual members and 4 institutional members. The individual members from countries, which are completely out of the enlarged BACSA region are mostly from India (43), but also from China (2), Egypt (2), Korea (1), Ghana (1), Syria (1) and Indonesia (1). Presently BACSA has 26 members of the Executive committee from 22 different countries. The composition of the Executive committee has been totally renewed at the end of 2009 [9] and can be periodically revised according to changes in the membership.

3.2 The BACSA Structure

The management of the association is composed of a President, two Vice-Presidents, one national coordinator for each member country, which are the most representative part of the Executive Committee; all of them are democratically elected. The members of the Executive Committee are the direct coordinators of all the current activities for each country, within the regional context. The Executive Committee is the bridge among each country, the national coordinator and the other countries of the association, to execute the actions defined in the region. The Committee gathers in person at least once every two years; its members maintain contacts among themselves regularly by e-mail/phone, to perform the following functions [9]:

- To evaluate the work made by each national coordinator in the respective countries in relationship to BACSA.
- To recommend people in the association who should receive training abroad.
- To evaluate and to watch the handling of the "Rotatory Funds" and "Research Funds" possibly created and to give recommendations on the best use of these resources.

- To present/display the research proposals that require financing from the "Research Fund" and to approve the necessary resources for this aim.
- To give recommendations and suggestions on all publications and written material that are created within the framework of the BACSA.
- To advise the association's President on the advances and progresses that take place in the development of the activities and give recommendations to her/him on possible modifications and corrections.
- To decide the main BACSA activities, including accepting new members etc.

3.2 The BACSA Conferences

Until the end of 2019 BACSA organized 9 international conferences. Each conference was on a specific subject, connected with the problems of regional sericulture development. The BACSA conferences are shown in Table 1.

Table 1. BACSA conferences (2005–2021).

	BACSA Conference Title	Year	Location
1.	Revival and Promotion of Sericultural Industries and Small Enterprise Development	2005	Taskent, Uzbekistan
2.	Silk Handicrafts Cottage Industries and Silk Enterprises Development	2006	Bursa, Turkey
3.	Sericulture Challenges in the 21st Century	2007	Vratsa, Bulgaria
4.	Possibilities for Using Silkworm and Mulberry for Non-Textile Purposes (First Balkan workshop)	2008	Plovdiv, Bulgaria
5.	Sericulture for multi products—new prospects for development	2011	Bucharest, Romania
6.	Building Value Chains in Sericulture	2013	Padova, Italy
7.	Organic Sericulture—Now and the Future	2015	Sinaia, Romania
8.	Climate changes and chemicals—the new sericulture challenges	2017	Sheki, Azerbaijan
9.	Sericulture preservation and revival—problems and prospects	2019	Batumi, Georgia

The main findings, conclusions and recommendations of these international meetings are listed below:

1. Revival and Promotion of Sericultural Industries and Small Enterprise Development (2005): As recalled before, this conference posed the basis for the development of a short/medium-term strategy for the sericulture revival at a regional level and prepa- ration of projects for the sericultural industry rehabilitation in the region countries, ready for donor search [6].

2. Silk Handicrafts Cottage Industries and Silk Enterprises Development (2006): This conference was particularly focused on the creation of a network collaborating in the different sectors of the supply chain to improve competences in silk reeling and silk industry management specially to establish ecofriendly processes of dyeing and finishing and to promote silk handcraft production and export to foreign countries even through the touristic channel [7].
3. Sericulture Challenges in the 21st Century (2007): The most important theme of this meeting was the preservation and exploitation of germplasm resources (silkworm and mulberry) of the region, which are particularly rich but at risk as their preservation is costly for the concerned countries [7].
4. Possibilities for Using Silkworm and Mulberry for Non-Textile Purposes (2008): The target of the discussion was the proper utilization of secondary and waste products of the sericultural industry, which can generate an extra income in addition to silk, which represents the main output. The by-products of sericulture are sericin, pupae, moths, silkworm frasses, silk waste, mulberry branches, fruits and roots. New commercial products can be obtained from these raw materials with a valuable destination for pharmaceuticals, cosmetics, feed and food, new materials [10].
5. Sericulture for multi products—new prospects for development (2011): This confer- ence enlarged the issues treated in the previous one of 2008, particularly stressing the criticalities of non-textile silk production, from the scientific and regulatory point of view [10–12].
6. Building Value Chains in Sericulture (2013): Stakeholders of different BACSA coun- tries met to plan possible cooperation programmes in the field of science and tech- nology transfer, education and training with the aim of promoting bilateral research projects, financed by each participating country's government through grant competi- tions, announced periodically by the Ministries of education and sciences or bi and/or multilateral projects, financed by the EU, in addition to specialization and train- ing of students and technical personnel in leading research centers and commercial companies, financed by EU/national programmes [13].
7. Organic Sericulture—Now and the Future (2015): The theme of this meeting was about the new trend of organic production in agriculture that well adapts to sericulture, especially in Europe, where mulberry cultivation has always been "naturally organic". How to develop a certification chain for textile and non-textile organic silk production and how to exploit this certification on the market were the key-points treated in the internal discussion [14].

8. Climate changes and chemicals—the new sericulture challenges (2017): In most of the 22 countries currently members of the Black, Caspian Seas and Central Asia Silk Association, wherever they are, Europe, Caucasus and Central Asia, sericulture has been very negatively affected by the use of chemicals in agriculture and by climatic changes. In fact, through fluctuations in temperature, water regimes and the increase in carbon dioxide levels, global climate changes directly influence mulberry, soil, pests, and silkworm rearing. The specific climatic conditions in the BACSA region countries require mulberry to have high cold and drought tolerance and the silkworm strains to possess a good tolerance to adverse rearing conditions like high temperature, daily temperature fluctuations and feeding with coarse mulberry leaves. On the other hand, the wide use of insecticides can easily harm silkworm rearing and even completely destroy the whole sericulture value chains in some regions or countries. A common effort to develop new strategies is required to mitigate the joint effects of the chemical use and climatic changes [15].
9. Sericulture preservation and revival—problems and prospects (2019): During this conference the BACSA members discussed about the world trends of silk production, in particular the decrease of the cocoon quantity annually available in China and the increase of the international price. These phenomena give a prospective of decrease in the silk use by the textile industry and in coming back to consider this fiber as a very luxury good for a niche market. Sericulture may change from an industry for the poorest farmers, to an agribusiness, requiring more investments and production costs, but having high revenues because of the high market price of the products. BACSA countries should be prepared for that [3,16,17].

The BACSA conference foreseen for 2021 was postponed because of the COVID 19 problems.

3.4 The BACSA Project Proposals

The BACSA activity to prepare project proposals is illustrated in Table 2. This activity was very intense between 2006 and 2010, then after many failures, these kinds of attempts ceased in terms of projects studied for the whole area and focused mostly on more limited projects or bilateral agreements between members of BACSA [18].

Table 2. BACSA project proposals to different donors (2006–2021).

Presented Projects	Year of Proposal or Realization	Possible Donor	Financed (Yes/No)
Improvement of Income-Generation Options Based on Revival of Sericultural Industries and Promotion of Small Silk Enterprise Development in Eastern Europe and Central Asia—Concept Note	2006	UNDP, ADA, CIDA, DCI, DEZA, DFID, FAO, GTZ, IFAD, ITC, UNCTAD, JICA, KOICA, NORAD, SIDA, UNIDO, USAID, EBRD, World bank	N
Comparative studies of silkworm hybrids performance for sericultural enterprise development in Black, Caspian seas and Central Asia region	2006–2007	FAO	Y (only partially)
Support for unlocking and developing the research potential of silkworm breeding and innovative management techniques in Bulgaria, Greece and Romania, targeting to the small silk enterprise development	2007	EU FP 7 "Capacities" Program	N
Improvement of Income-Generation Options Based on Revival of Sericultural Industries and Promotion of Small Silk Enterprise Development in Eastern Europe and Central Asia—Full Proposal	2008	FAO	N
Regional sericulture germplasm resources network for Africa, Middle East, Central Asia and Europe	2008	FAO	N
Proposal of a Regional workshop on "Utilization of mulberry and silkworm genetic resources for sericultural enterprises development in Africa, Middle East, Central Asia and Europe".	2008	FAO	N
TCP/ALB 3101 "Revival and Development of Sericulture in Albania"	2008–2009	FAO	Y

SERINNOV- SEE EoI/B/481/1.3/X "Sericulture and silk products in South-East Europe—stepping from the tradition to innovation and strengthening the economic profile of respective regions	2009	EU—SEE program	N
Exploitation of the heavy metal movement and removal pathways and mechanisms in the mulberry- silkworm chain for the establishment of model applications for further practical use in bioremediation of contaminated soils.	2009	FP7—KBBE program	N
TCP/GEO/3201 "Sericulture sector study in Georgia"	2009–2010	FAO	Y

3.5 The BACSA Network for Facilitating Commercial Contacts among Stakeholders

BACSA makes all the efforts to establish connections among the producers, sellers and buyers of different sericultural products such as mulberry saplings, silkworm eggs, dry cocoons, raw silk, silk yarn, fabrics, and garments. These activities are performed through regular updating of the section "sell/buy information" on the BACSA website as well as responding in real time to all the enquiries from possible sellers and buyers, connecting stakeholders together and giving a chance to exhibit sericultural products by organizing international workshops, conferences, etc. [19].

3.6 Results of BACSA Global Activity

BACSA carried out a series of activities that can be better understood when we compare the situation before 2005, when BACSA was established, and at the present time; from this comparison it is evident this association's role in the regional sericulture preservation, revival and development (Table 3).

Table 3. BACSA's activity results (2005–2021).

Item	Situation in 2005	Situation in 2021
Contacts among the key specialists and institutions engaged in the sericultural industries	Very few, well-developed among some of the Ex-Soviet Union and Eastern Europe countries only	Good contacts among all BACSA member countries, key sericulture specialists and most of the institutions

Regional sericulture database	Not available	Available, uploaded on the BACSA website and regularly updated
Number of countries, associated in BACSA	Only 8 Eastern Europe, Caucasus and Central Asia countries	22 countries from Europe, Caucasus and Central Asia; out of them 9 EU member states, which give possibilities to apply for projects financing from the EU funds.
Enquiry system for sericultural products marketing	Not available	Available: enquiries to BACSA → distribution of the enquiries to the national coordinators → distribution of the enquiries to interested stakeholders in each member country for direct contacts; uploading the sell/buy enquiries on the BACSA web site.
Exchange of sericulture germplasm resources among the BACSA member countries	Very few, well developed among some of the Ex Soviet Union and Eastern Europe countries only	At a much higher scale, based on bilateral scientific projects
Export of mulberry saplings and silkworm eggs	Very few, among some of the Ex-Soviet Union states only	There is, but still in a too small scale
Regional conferences and meetings	Very few, well developed among some of the Ex Soviet Union and Eastern Europe countries only	Regular BACSA international conferences, at intervals of 2 years, 9 conferences already organized.
Specializations and training of students and technical personnel in leading research centers and commercial companies in BACSA member countries	No	Present, but still in a too small scale.
Sensitizing the national governments about the regional BACSA executive meetings results, decisions and follow ups	No	Yes, operated by BACSA president and national coordinators
Responding to different enquiries, concerning the sericultural industries in the region	No	Yes, operated by BACSA president and national coordinators

Supporting the sericultural institutions and specialists in the BACSA member countries by providing, when necessary, letters of support, personal recommendations, reviews of scientific monographs and doctoral dissertations etc.	No	Yes, operated by BACSA president and selected experts
Popularization of Europe and Central Asia sericultural industries in the world through participation to international conferences and meetings	Scarce	Yes
Development of regional sericulture projects and looking for donors	No	Yes
Promoting bilateral and multilateral agreements for cooperation among the BACSA member states	No	Yes

4. BACSA BECOMES ATTRACTIVE FOR WESTERN AND CENTRAL EUROPE COUNTRIES

As mentioned before, the situation of sericulture of the countries of Western and Central Europe non-belonging to the Communist block was even worse than that of BACSA countries, because here the decline of sericulture (referred to as agricultural and reeling activity) began much earlier than in the '90s of the last century and precisely after the second World War with the maximum peak between the '60s and '70s. On the other hand, these countries retained a flourishing silk industry (weaving, dyeing, printing, manufacturing), especially in Italy, France, Switzerland and England [9]. The main feature of this silk industry has been its complete dependence for silk from external sourcing, especially from China, for a long time [9]. During the '90s the international price of silk was low and the quantity on the market abundant; this was the reason why the European silk industry was not stimulated to look for other sources than China. However, some signals of market sufferance were already very clear: double pricing of silk and difficulty in finding a constant and high quality of this fiber production in China [4].

In Western Europe some residual cocoon production could be preserved due to the EC (European Communities) subsidies to sericulture (Regulation EEC n. 922/72) establishing a contribution per each reared silkworm box from which a minimum quantity of at least 20 kg of fresh cocoons was obtained. This contribution was completely stopped in Europe in 2014, due to the Common Agricultural Policy

(CAP) reform; the EC governments were allowed to support some strategic sectors only with direct payments; the majority of EC countries decided that sericulture was not a priority, while only the government of Greece continued to pay subsidies to its farmers for fresh cocoon production.

In Italy and France this lack of support from the Government was partly justified by the catastrophic situation of sericulture caused by external factors; in fact, from 1989 to 2010 France and Northern Italy (where some cocoon production had been preserved even in a small amount) were affected for a long time by the pollution due to an Insect Growth Regulator (active ingredient: fenoxycarb, commercial name: Insegar). The chemical was sprayed on the fruit orchards and its drift was transported by the winds onto mulberry leaf. It was so effective to be capable of acting on the silkworm larvae at doses of nanograms (1.0×10^{-9} g) or even lower [10]. Due to this phenomenon the larvae were unable to spin their cocoons ("non-spinning syndrome") [11,12] and French and Italian rearers were completely discouraged from continuing to practice sericulture. For almost 20 years, until the date of the retirement (which was operated by the same producing company) of the chemical from the market and exhaustion of the stocks, it was impossible to rear silkworms in the above-mentioned areas.

In 2009 the INRA (National Research Institute for Agriculture) Sericultural National Unit of Lyon (France) closed and only the Sericulture Laboratory of CREA (Council for Agricultural Research and Economics) in Padua remained responsible for the preservation of all the genetic resources of Western Europe [13]. Furthermore, the International Sericultural Commission based in France, transferred its headquarters to India, which was a signal of disengagement of the European stakeholders from sericulture [14].

In the same period the international price of raw silk began to rise very rapidly, reaching levels of about USD 55–60/kg mostly because of the industrialization of China in the last decades, in particular the Eastern area of the country, which was traditionally more apt to sericulture. Although the Chinese government made a big effort to transfer sericulture to other regions, less industrialized and more devoted to agriculture, the silk quantity and mostly its quality decreased significantly; as a consequence, the price of the best quality of silk in the international market began to increase quite abruptly. The COVID 19 pandemic has been a brief parenthesis in this trend.

Therefore, during the last 10 years a new interest was expressed by the European silk industry (especially from Italy and Switzerland) for countries alternative to China where silk might be produced; on this basis the silk industry may re-consider establishing part of the cocoon production they need in Europe, Caucasus and/ or Central Asia. A sign of this interest was the BACSA conference of 2013, which was held in Italy with the economic support and commitment of the Italian silk industry in collaboration with CREA (the Research organ of the Italian Ministry of Agricultural, Food and Forestry Policies) [15]. A delegation of the executive board was hosted in Como and visited "Ratti", one of the most important silk Italian companies belonging to the Marzotto group. Another sign of interest was the progressive association to BACSA of countries from Western Europe: Italy, Germany, Spain, Portugal, Slovenia, Switzerland, UK which are currently members of BACSA [16].

However, the restoration of the sericulture chain in Europe and BACSA countries is a very large goal that cannot be sustained by the industry alone; a public-private partnership should be envisaged, and the governmental and EC support should be provided to make it a realistic goal, which, however, needs long term investments.

5. PROBLEMS AND PROSPECTS

In this section BACSA will be examined according to the problems and prospects that characterize this organization. An attempt to summarize the criticalities and "plus values" of BACSA is reported in Figure 1.

5.1 Problems Related to the BACSA Structure

BACSA is a quite open and very flexible organization; the membership of the Black, Caspian Seas and Central Asia Silk Association (BACSA) is possible for all individuals/institutes/countries from Europe and Central Asia which are willing to share common goals for an industrial and economic growth through sericultural activities. The expected members may be of two categories: "Internal" from the same geographical region countries, and "External" from other countries, who are possibly interested in technical cooperation or business with Europe and Central Asia region as well as from those countries that are possible donors for sericulture revival projects [20]. Even Chinese, Indian, Japanese, Korean, African scientists/stakeholders applied to this organization for membership. Chinese and Indian delegations are admitted as observers to BACSA meetings. No membership fee is required so far, and this fact permits to participate even to poor countries or individuals. However, what in principle appears to be very democratic and egalitarian might be a prob- lem in the future in case some divisive themes are discussed, or the interest of some single country is concerned. At the moment, the BACSA community reaches a good agreement on any decision basically because of three factors: (1) members are few and many of them have been known each other for years; (2) there are no huge interests at stake because sericulture in the BACSA countries is mostly confined to a niche economy; (3) China's presence in the silk world is so outstanding that any other country should decide its strategy to develop its own silk production/industry reacting to the Chinese monopoly more than fighting against other competitors. Therefore, it is possible that this association should assume another structure in the future; in fact, some problems are the same for most of the countries, others are typical of a more restricted geographical area. The three macro-areas that can be outlined in the BACSA region are Central Asia, the Caucasus and Europe. It might be advisable in the next period to divide BACSA in three different working groups, joining together only on themes of common interest. For example, the income level of farmers in these three macro-areas is completely different, being the minimum in Central Asia, maximum in Western Europe. It appears very clear that Europe should make a big effort to mechanize and specialize its silk production, while countries like Uzbekistan (Central Asia) have the potential of producing high quantity of cocoons at a comparatively low price by exploiting local manpower and can be identified as competitor with China in the middle term. Other principles to form working groups may be assumed: for example, silk industrial converters versus cocoon producing/reeling countries, low quantity silk producing countries versus high quantity silk producing countries, etc.

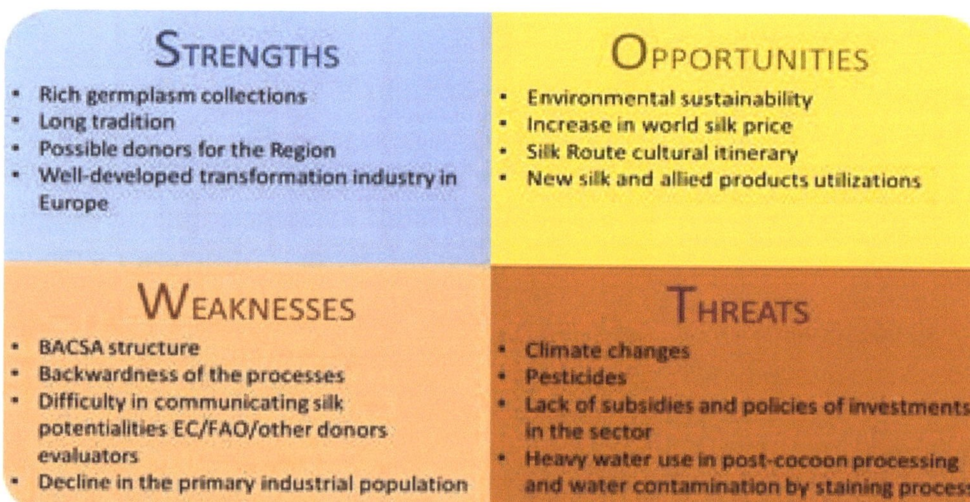

Figure 1. SWOT analysis representing criticalities and qualities of BACSA examined with respect to the possibilities of success in revitalizing the sericultural activity in the area.

5.2 Common Problems

5.2.1 Preservation of Genetic Resources

The global raw silk production was around 91,945 t in 2020, but out of them 53,359 t were produced by China and 33,770 t by India, while all the other countries produced only about 4816 t of raw silk [21] (see Figure 2).

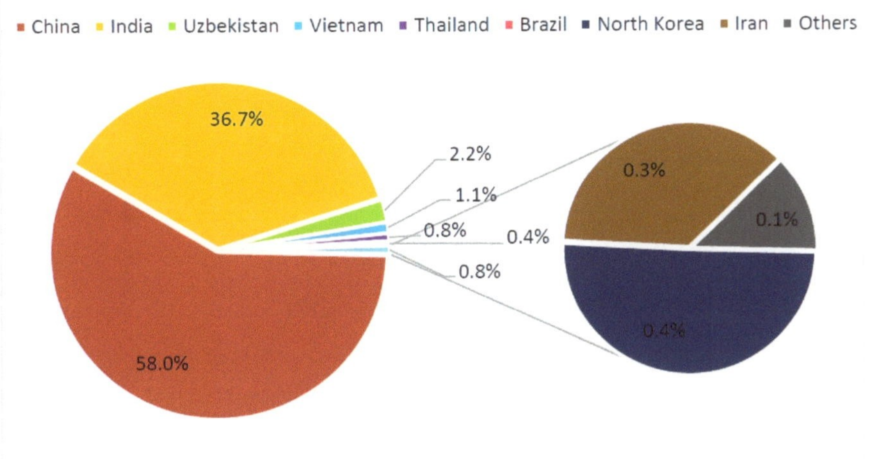

Figure 2. Pie-chart and relative shares of global silk production (data from https://inserco.org/en/ statistics; accessed on 23 November 2021) in 2020; China and India alone cover more than 90% of the worldwide production (94.7%).

That means almost 95% of the total world silk production is from only two countries— China and India. On the other hand, although many countries in the World dealt with sericulture in the past or are still dealing nowadays, the majority of them only make efforts to preserve this activity and only few of them to revive the sericultural industry. Silk production volumes more than doubled from 1990 to 2019 but it saw a decrease over the last five years. Even in those countries that are the biggest world cocoon and silk producers, there are presently entire regions where the sericultural activities have been partly or even completely stopped and the sericultural expertise may be lost. Among these, BACSA associated countries, which still represents the third world producing area, face problems typical of nations with history of long tradition and scarce current production. The first problem for them is how to preserve their mulberry and silkworm germplasms, which usually constitute public genetic resources, mostly located in research institutes [8,22]; the preservation activity is quite expensive, governments, in consideration of the present scarce economical revenues from the sericultural agribusiness, tend to restrict funds destined for conservation. Any hypothesis to concentrate the germplasm in one center only, preserving accessions for all the region, is quite unrealistic and in addition dangerous for the possible losses of genetic material in case of diseases or unforeseen accidents. The best way to preserve genetic resources is their spreading ex-situ; however, this proved to be very difficult because of intellectual properties on the selected strains and varieties and could be done only for the genetic material that does not have any economic significance. Therefore, encouraging the use of these resources and restarting an economically viable sericulture is certainly the best way to guarantee their preservation. A contemporaneous rehabilitation of the activity in the whole BACSA region does not appear to be necessary; a partial restarting in those countries still endowed with well-preserved genetic material might also be useful for those countries that have lost their own sourcing, due to different accidental events, or have germplasm of inferior quality; in fact, even a partial sericultural revival will favor bilateral exchanges and commercial exploitation agreements among different countries, which might be regulated internally within the BACSA framework.

5.2.2 Backwardness of the Agroindustry Chain

Another problem for BACSA countries is the backwardness of the agricultural process (mulberry cultivation and silkworm rearing) and/or the silk industrial transformation. This is partially because the Chinese monopoly on silk has maintained too low the silk price to stimulate innovation and mechanization in countries with capitalistic economies; the only advanced part of the silk chain, in these countries, regards industrial processing after reeling, in which modernization and digitalization play a basic role on the final quality of manufacts for the exclusive fashion industry of the most important world brands. The communist governments highly promoted the mechanization and technical advances in the agriculture. In the communist countries after the '60s of the 20th century a big industrialization occurred; thus, moving of a huge mass of people from the rural areas to the cities resulted in the lack of agricultural workers. For this reason, in the '80s the mechanization in agriculture was a priority. Even in sericulture, where cocoon production, in some communist countries, was in the hands of the big agricultural cooperatives or state enterprises, they managed to organize large scale silkworm rearing, mechanized feeding of young instar larvae, using Japanese machineries, cardboard frame mountages, machines for cocoon deflossing. However, most of these efforts in sericulture appeared to be not economically effective due to the Chinese concurrence and were abandoned. To recover the technological gap of sericulture with respect to other agricultural crops or agro-industrial chains, more competitive in terms of economic revenues for farmers or investors, is not easy and requires many funds to develop innovations. This is one of the main reasons why sericulture is a niche production in several developed countries. This technological gap in sericulture became more evident in the last decades when the effect of climatic changes and environmental pollution began to seriously affect agriculture. For example, dramatic climatic changes began to negatively affect silkworm rearing with serious fluctuations in the average temperatures even in the seasons traditionally favorable to sericulture; the lack of climatic control of silkworm rearing facilities and of digital automatic control of temperature and humidity can cause big problems to larval development; heavy droughts, late frosts, excessive rains compromise mulberry leaf harvests; high temperatures and humidity favor spreading of insect pests, which are also more invasive due to the increased globalization of the transport of goods around the world, which works as an involuntary carrier. Fighting against these new insect pests with insecticides, in turn, affects sericulture. To solve these problems a lot of technology and research related to the environmental management of sericulture would be necessary; however, as previously mentioned, available funds to improve knowledge in the sericultural field are very limited [16].

5.2.3 Funding Access

With regard to structural funding for research, training, dissemination, demonstration and other activities related to the sericulture revival, as mentioned above, some BACSA associated countries are members of the EU; therefore, they should have access to EU funding for research institution and SMEs (Small and Medium Enterprises). However, until now the joint attempts for funding common projects have generally failed [18]. It is likely EC evaluators think that silk-focused projects have a minor impact on EC country development, with other agricultural/ industrial activities considered more important than sericulture. Probably, the correct manner to attract financing is to enclose sericulture as a small part of wide projects focusing on other activities and where sericulture represents a case-study more than the central research or investment attractor. On the other hand, non-EC BACSA countries can have access to FAO funds or to those of other NGOs [18]. However, the lack of experts at the world level in the specific branch of sericulture often results in a minor attention to this theme. For example, two Korean experts gave a great personal contribution (Dr. Hoo Zoo Lea, FAO Senior officer and Dr. Jong Sung Lim, FAO consultant) to the BACSA creation, because of their knowledge and in-depth expertise in this sector [6]. Unfortunately, FAO now is missing these professional officers specialized in sericulture.

5.2.4 Governmental Subsidies

With regard to subsidies of governments to silk production, what occurred was that both in the EC and in the other BACSA countries, where they were applied, they did not prevent sericulture from the decline and, in some cases, they triggered fraud from farmer associations or other organizations. Therefore, it is clear enough that this instrument to guide the market might be useful only if coupled with a general policy of industrial and technological development of this sector.

5.3 Prospects for the Future

China, India, Brazil have probably already reached their maximum level of silk production and are not going to increase further their quote in the world market

[23]. On the other hand, silk consumption, so far, has been a very small quote of total world fiber production (less than 1%) [24], but very stable over time, although with a diminishing trend in the last 5 years [21]. This stability of silk for the textile market is due to the general buyers' identification of silk with a luxury fiber, which is a part of human civilization; in many countries it is intrinsically connected with local culture, and it has a long-standing tradition. Silk is considered as "Queen of textiles"; it has some unique and important characteristics, such as its ability to keep the body warm when it is cool, and cool when it is warm, or being a healthy fiber because it breathes easily and naturally keeps away moisture from the skin or being actually soothing to skin diseases and itches. Therefore, according to the recent trend for a rising demand of natural fibers from the final consumers, who look for comfortable wearing and sustainability of clothes, an increase of the silk price is expected; this phenomenon might give opportunities to BACSA countries to be competitive with China, even considering that the top world quality silk fabrics and garment producing industries are concentrated in Europe (Italy, France, Switzerland, England).

Many BACSA countries might expand their artisanal and handicrafts production, especially by linking it to the tradition of their territories, countryside landscapes, culture. This idea gave birth to a project promoted by the Venice Municipality and the Council of Europe through the creation of a cultural itinerary "The European Silk Route"; it aims to be a local cultural network and infrastructure linking cities, regions, sites, museums and universities in order to enhance knowledge of a shared European cultural heritage, both tangible and intangible, and to promote new relationships within Europe and between Europe and the East through sharing of best practices and cultural tourism activities. The route would ideally be based on Marco Polo's travels eastward and include silk production and trade itineraries in Europe in the following centuries. The narrative to be developed will start from the commercial and religious exchanges that took place along the silk road, and specifically Marco Polo's trips. It will then analyze silk's impact in Europe through four main themes: the textile activity (from artisanal production to industrialization: innovations, technologies, the work world); silkworm rearing and its social, economic, agricultural and environmental consequences; the use of silk in paintings, fashion and design; and research and development in silk production [25]. The work undertaken to reach the certification of the cultural itinerary from the Institute of Luxembourg began in 2018 [25].

In addition to this exciting opportunity, the forecasts for the future represent that the demand of non-textile silk as constitutive proteins will increase at a steady rate, due to the new utilizations of silk as a versatile polymer for different aims (cosmetics, pharmaceuticals, biomedicals). Recently bio-technological sericulture has been developing [26,27]. For the first time in the world, in 2017, the legitimated rearing of genetically modified silkworms in conventional sericulture farms started in Japan. Functional silk is a promising material for medical applications. Using the methods of genetic engineering, absolutely new silks that have unprecedented functions were developed. These are transgenic spider silk [28], hyperfine silk of small diameter, artificial blood vessels, fluorescent silk [29]. Some of the BACSA countries are ready to face this biotechnological challenge. Silk regenerative medical materials like silk sponge, silk hernia mesh, wound dressing, silk surgical tape, hydrogel, films and 3D scaffolds for wound healing and tissue regeneration and reconstruction gels, powders, enzyme immobilization matrices were also created [30–34]. Transgenic sericin is used for several medical reagents, like blood test drugs, biomatrix for tissue engineering and cosmetics [35–38].

Therefore, new kinds of applications are likely to offer new opportunities; the interesting consideration is that, in this case, it is not necessary to produce silk in a huge quantity and it is not necessary to possess big reeling plants or transformation industries. These opportunities might allow BACSA countries to increase their production slowly and steadily. The region possesses some of the richest silkworm and mulberry germplasm collections. Several of the commercial silkworm hybrids, produced in the BACSA area manifest comparatively high productivity, namely single cocoon weight 2.2–2.5 g, shell ratio 23–24%, shell weight 0.500–0.600 g, filament length 1300–1500 m under laboratory conditions [39,40]. This might help in making the member countries attractive for this production. The level of sericultural science and technology in the region countries is comparatively high at a world level. This expertise might be particularly useful because new technological properties might be required for such a kind of silk production for innovative aims.

The EC green deal might also play a great role in promoting the development of sericulture in the BACSA countries: as mentioned before, sericulture might be an organic agricultural practice; the mulberry is environmentally useful to protect soil from erosion, to adsorb carbon dioxide, to prevent desertification in marginal areas; if it is exploited in a non-intensive way it requires limited fertilization and irrigation and no pesticides at all; moriculture can also be practiced in polluted or salty soils to accelerate their recovery to pro- duction [41–43]. The silkworm is an environmental sentinel especially informative on the abuse of pesticides on agricultural crops cultivated nearby the rearing places. Furthermore, sericulture and moriculture can be exploited for circular economies where by-products of some processes can become raw materials for others. Mulberry fruit can be consumed fresh, dry or employed for production of juice, wine, jam and food additives. Pharmaceuticals can be extracted from mulberry branches, roots, leaves (for example, 1-deoxynojirimycin (DNJ) with antidiabetic aims [44]).

6. CONCISE SUMMARY

BACSA, founded merely 16 years ago, is a relatively nascent international organization, particularly when juxtaposed with the International Sericultural Commission, formed in 1960. BACSA is mostly administered on a voluntary basis, relying on the efforts and assistance of individual members, predominantly affiliated with scientific institutions. throughout accordance with its objectives, it has been fortifying connections and disseminating information among sericultural member nations by providing extensive assistance for numerous initiatives aimed at revitalizing sericulture throughout Europe, the Caucasus, and Central Asia. Despite the absence of large-scale resurrection, numerous indications suggest potential future advancements, rendering the support efforts of this Association essential and warranting greater attention from sector players.

REFERENCES

Conflicts of Interest: The authors declare no conflict of interest.

1. Batsaikhan, U.; Dabrowski, M. Central Asia—Twenty-five years after the breakup of the USSR. *Russ. J. Econ.* **2017**, *3*, 296–320. [CrossRef]
2. Soergel, A. How the Fall of Communism in 1989 Reshaped Eastern Europe|Best Countries|US News. Available online: https:// www.usnews.com/news/best-countries/articles/2019-11-08/how-the-fall-of-communism-in-1989-reshaped-eastern-europe (accessed on 23 November 2021).
3. Tzenov, P. Problems and prospects of sericulture preservation and revival in Europe, Causasus and Central Asia. In Proceedings of the SERIVIVAL, Batumi, Georgia, 7–12 April 2019; pp. 14–27.
4. Brunetti, G.; Visconti, F.; Marelli, M. *Euro e Distretti Industriali: Una Ricerca Nella Realtà Lombarda*; Franco Angeli—Fondazione CARIPLO per la Ricerca Scientifica: Milan, Italy, 2000; ISBN 8846419553.
5. Yamada, M.; Kawabata, Y.; Iikubo, M.; Vyacheslav, A.; Onwona-Agyeman, S. Revitalizing Silk-Road Silk Industry: A Case Study in Fergana Region, Uzbekistan. *J. Arid L. Stud.* **2012**, *22*, 179–182.
6. Tzenov, P.; Lea, H.Z. Regional strategies proposed for revival and promotion of sericultural industries and small enterprise development in the countries of Black, Caspian seas and Central Asia region. In Proceedings of the International Workshop on Revival and Promotion of Sericultural Industries and Small Silk Enterprise Development in the Black & Caspian Seas Region, Tashkent, Uzbekistan, 11–15 April 2005.
7. Tzenov, P.; Lea, H.Z. Silk Handcrafts Cottage Industries and Small Enterprises in Africa, Europe, Central Asia and the Near East. In Proceedings of the

International Workshop on Silk Handcrafts Cottage Industries and Silk Enterprises Development in Africa, Europe, Central Asia and the Near East, Bursa, Turkey, 6–10 March 2006; pp. 19–61.

8. Tzenov, P. Present status and utilization of sericulture germplasm and comparative studies of different silkworm hybrids performance for sericultural enterprise development in the Black, Caspian seas and Central Asia (BACSA) region. In Proceedings of the Sericulture Challenges in the 21st Century—SERICHAL, Vratza, Bulgaria, 18–21 September 2007; pp. 13–35.
9. Bacsa Black, Caspian Seas and Central Asia Silk Association (BACSA) Proceedings. Available online: https://www.bacsa-silk. org/en/principles-of-work-organization/ (accessed on 23 November 2021).
10. Ichim, M.; Tanase, D.; Tzenov, P.; Grekov, D. Global trends in mulberry and silkworm use for non-textile purposes. In Proceedings of the Possibilities for Using Silkworm and Mulberry for Non-Textile Purposes, Plovdiv, Bulgaria, 23–26 September 2008; pp. 6–36.
11. Tzenov, P. Black, Caspian Seas and Central Asia Silk Association (BACSA) Activities for Revival and Development of Sericulture in Europe and Central Asia. In Proceedings of the Sericulture for Multi Products-New Prospects for Development—SERIPRODEV, Bucharest, Romania, 11–15 April 2011; pp. 13–18.
12. Tzenov, P.; Ichim, M.; Petkov, Z.; Vasileva, Y.; Arkova-Pantaleva, D. Identification and Possible Utilization of Some Silkworm Rearing Waste Products. In Proceedings of the Exploitation of Agricultural and Food Industry By-Products and Waste material Through Application of Modern Processing Techniques, Bucharest, Romania, 1–3 July 2008; pp. 42–49.
13. Tzenov, P.; Cappellozza, S. The role of regional cooperation between BACSA member countries in building value chains in sericulture. In Proceedings of the Building Value Chains in Sericulture—BISERICA, Padua, Italy, 7–12 April 2013; pp. 21–29.
14. Tzenov, P.; Ichim, M.; Mitova, D. The Organic Sericulture in the Context of Biological Agriculture and Organic Textile. In Proceedings of the Organic Sericulture—Now and the Future—ORGASERI, Sinaia, Romania, 19–24 April 2015; pp. 13–24.
15. Tzenov, P. Climate Changes Effect on Sericulture in Europe, Caucasus and Central Asia. In Proceedings of the Climate Changes and Chemicals—The New Sericulture Challenges—CLISERI, Sheki, Azerbaijan, 2–7 April 2017; pp. 9–16.

16. Ichim, M.; Tzenov, P.; Grekov, D.; Avramova, K. Mulberry Plantation Establishment Methods, Suitable for Europe, Caucasus and Central Asia. In Proceedings of the Sericulture Preservation and Revival—Problems and Prospects—SERIVIVAL, Batumi, Georgia, 7–12 April 2019; pp. 68–80.

17. Moorthy, M.; Sivaprasad, V.; Paramesh, B.; Bindhiya; Hukkeri, S.M.; Kumar, D.; Grekov, D.; Tzenov, P.; Teotia, R.S. Indo- Bulgarian Collaborative Research Project—A Way Forward in Silkworm Breed Improvement Programme in India and Bulgaria. In Proceedings of the Sericulture Preservation and Revival—Problems and Prospects—SERIVIVAL, Batumi, Georgia, 7–12 April 2019; pp. 158–159.

18. Bacsa Black, Caspian Seas and Central Asia Silk Association (BACSA) Proceedings. Available online: https://www.bacsa-silk. org/en/regional-project-proposal/ (accessed on 25 November 2021).

19. Bacsa Black, Caspian Seas and Central Asia Silk Association (BACSA) Proceedings. Available online: https://www.bacsa-silk. org/en/sellbuy-information/ (accessed on 25 November 2021).

20. BACSA Membership: The Black, Caspian Seas and Central Asia Silk Association (BACSA). Available online: https://www.bacsa- silk.org/en/membership/ (accessed on 25 November 2021).

21. International Sericultural Commission INSERCO Statistics. Available online: https://www.inserco.org/en/statistics (accessed on 23 November 2021).

22. Cappellozza, S.; Toso, L.; Saviane, A. La collezione di germoplasma di baco da seta (*Bombyx mori*) e gelso appartenente all'Unità di Apicoltura e Bachicoltura di Bologna, sede di Padova. In *Conservazione Biodiversità, Gestione Banche Dati e Miglioramento Genetico—BIODATI*; D'Andrea, F., Ed.; Edizioni Nuova Cultura: Roma, Italy, 2013; pp. 961–990. ISBN 9788868120986.

23. De Ponti, P. European Silk Forum—Per quanto tempo ancora la Cina fornirà la seta all'Europa? *La Seta* **2011**, *63*, 13–15.

24. Textile Exchange Preferred Fiber & Materials—Market Report 2020. *Prefer. Fiber Mater.—Mark. Rep. 2020* **2020**, 42–43. Available online: https://textileexchange.org/wp-content/uploads/2020/06/Textile-Exchange_Preferred-Fiber-Material-Market-Report_ 2020.pdf (accessed on 30 November 2021).

25. Council of Europe-Venice-European Silk Route: About the Project. Available online: https://www.coe.int/en/web/venice/ about-the-project (accessed on 25 November 2021).

26. Tzenov, P.; Grekov, D. Research Achievements and Future Trends of the

Silkworm *Bombyx mori* L. Breeding Work in Bulgaria. In Proceedings of the Recent Trends in Seribiotechnology—ICTS, Anantapur, India, 27–29 March 2008; pp. 43–54.

27. Tzenov, P.; Kipriotis, E. Product Diversification—An Alternative for Sericulture Development in the Black, Caspian Seas and Central Asia Region Countries. In Proceedings of the 21st Congress of the International Sericultural Commission, Athens, Greece, 3–6 November 2008; pp. 232–240.

28. Xu, J.; Dong, Q.; Yu, Y.; Niu, B.; Ji, D.; Li, M.; Huang, Y.; Chen, X.; Tan, A. Mass spider silk production through targeted gene replacement in *Bombyx mori*. *Proc. Natl. Acad. Sci. USA* **2018**, *115*, 8757–8762. [CrossRef] [PubMed]

29. Iizuka, T.; Sezutsu, H.; Tatematsu, K.I.; Kobayashi, I.; Yonemura, N.; Uchino, K.; Nakajima, K.; Kojima, K.; Takabayashi, C.; Machii, H.; et al. Colored fluorescent silk made by transgenic silkworms. *Adv. Funct. Mater.* **2013**, *23*, 5232–5239. [CrossRef]

30. Kapoor, S.; Kundu, S.C. Silk protein-based hydrogels: Promising advanced materials for biomedical applications. *Acta Biomater.* **2016**, *31*, 17–32. [CrossRef] [PubMed]

31. Huang, W.; Ling, S.; Li, C.; Omenetto, F.G.; Kaplan, D.L. Silkworm silk-based materials and devices generated using bio- nanotechnology. *Chem. Soc. Rev.* **2018**, *47*, 6486–6504. [CrossRef] [PubMed]

32. Chouhan, D.; Mandal, B.B. Silk biomaterials in wound healing and skin regeneration therapeutics: From bench to bedside. *Acta Biomater.* **2020**, *103*, 24–51. [CrossRef] [PubMed]

33. Nguyen, T.P.; Nguyen, Q.V.; Nguyen, V.; Le, T.; Le, Q. Van Silk Fibroin-Based Biomaterials for Biomedical. *Polymers* **2019**, *11*, 1933. [CrossRef] [PubMed]

34. Teramoto, H.; Iga, M.; Tsuboi, H.; Nakajima, K. Characterization and scaled-up production of azido-functionalized silk fiber produced by transgenic silkworms with an expanded genetic code. *Int. J. Mol. Sci.* **2019**, *20*, 616. [CrossRef] [PubMed]

35. Tomita, M. Transgenic silkworms that weave recombinant proteins into silk cocoons. *Biotechnol. Lett.* **2011**, *33*, 645–654. [CrossRef] [PubMed]

36. Arango, M.C.; Montoya, Y.; Peresin, M.S.; Bustamante, J.; Álvarez-López, C. Silk sericin as a biomaterial for tissue engineering: A review. *Int. J. Polym. Mater. Polym. Biomater.* **2021**, *70*, 1115–1129. [CrossRef]

37. Teramoto, H.; Iizuka, T.; Kameda, T.; Okada, E.; Sezutsu, H. Application of high molecular weight intact sericin as a cosmetic material. 2013, 34–35. Available online: http://www.naro.affrc.go.jp/archive/nias/eng/research/h25/nias2013-16.pdf (accessed on 30 November 2021).

38. Teramoto, H.; Nakajima, K.I.; Takabayashi, C. Preparation of elastic silk sericin hydrogel. *Biosci. Biotechnol. Biochem.* **2005**, *69*, 845–847. [CrossRef] [PubMed]

39. Tzenov, P. Silkworm *Bombyx mori* L. Selection and Breeding in Bulgaria Recent Achievements. In Proceedings of the ASEAN Sericulture Conference 2010 and ASEAN Silk Fabric and Fashion Design Contest 2010, Nontaburi, Thailand, 22–27 August 2010.

40. Tzenov, P.; Grekov, D.; Vasileva, Y.; Moorthy, M.; Sivaprasad, V. Prospective hybrids of Bulgarian uni-bivolyine silkworm breeds and Indian polyvoltine breed, C.NICHI. *Sericologia* **2019**, *59*, 9–12.

41. Rohela, G.K.; Shukla, P.; Muttanna; Kumar, R.; Chowdhury, S.R. Mulberry (*Morus* spp.): An ideal plant for sustainable development. *Trees For. People* **2020**, *2*, 100011. [CrossRef]

42. Wan, X.; Lei, M.; Chen, T.; Tan, Y.; Yang, J. Safe utilization of heavy-metal-contaminated farmland by mulberry tree cultivation and silk production. *Sci. Total Environ.* **2017**, *599–600*, 1867–1873. [CrossRef] [PubMed]

43. Alahabadi, A.; Ehrampoush, M.H.; Miri, M.; Ebrahimi Aval, H.; Yousefzadeh, S.; Ghaffari, H.R.; Ahmadi, E.; Talebi, P.; Abaszadeh Fathabadi, Z.; Babai, F.; et al. A comparative study on capability of different tree species in accumulating heavy metals from soil and ambient air. *Chemosphere* **2017**, *172*, 459–467. [CrossRef] [PubMed]

44. Marchetti, L.; Saviane, A.; Montà, A.D.; Paglia, G.; Pellati, F.; Benvenuti, S.; Bertelli, D.; Cappellozza, S. Determination of 1-deoxynojirimycin (1-dnj) in leaves of italian or italy-adapted cultivars of mulberry (*Morus sp.pl.*) by hplc-ms. *Plants* **2021**, *10*, 1553. [CrossRef] [PubMed].

6 Chapter

ENHANCEMENTS IN GENOME EDITING OF SILKWORMS (BOMBYX MORI) UTILIZING THE CRISPR-CAS SYSTEM.

1. INTRODUCTION: AN OVERVIEW

The life sciences research domains were transformed by the remarkable advancement of diverse genome editing technologies. Specific genome editing techniques enable guided modifications to the genomic DNA of all living organisms, including deletions, insertions, and sequence substitutions [1].

In recent years, numerous genome editing technologies have garnered significant attention. Three notable technologies include programmable nucleases: transcription activator-like effector nucleases (TALENs), zinc finger nucleases (ZFNs), and clustered regularly interspaced short palindromic repeat-associated nucleases (CRISPR-Cas) [2,3]. Presently, engineered nucleases are facilitating significant progress in the rectification of genetic mutations, regulation of gene expression, and the creation of therapeutic agents; these methodologies are also employed to enhance the comprehension of gene functions and the mechanisms that contribute to specific genetic disorders or various diseases [4].

In genome editing, it is essential to prevent off-target effects; yet, the CRISPR-Cas system demonstrates consistent results because to its high fidelity [4,5]. Since its identification in bacteria, the CRISPR-Cas system has been persistently utilized, serving as an exceptionally adaptable instrument for the scientific community owing to its reprogrammable characteristic. This technique is presently utilized to modify the genomes of diverse creatures, including bacteria, insects, plants, and human cells [6].

For almost 5000 years, mankind have engaged in the domestication of the economically significant insect, Bombyx mori. B. mori historically earned popularity for silk production, but it is today regarded as one of the most esteemed model animals in the biological sciences. Recent accomplishments include the utilization of B. mori as a bioreactor for the manufacture of recombinant proteins [7–10], and its silk is employed to create exceptional silk-based biomaterials of significant relevance in the medical field [11]. The silk proteome comprises two principal proteins: fibroin and sericin. The silk gland comprises three anatomical regions: the posterior silk gland (PSG), the middle silk gland (MSG), and the anterior silk gland (ASG). Fibroin is produced in the posterior silk gland, while sericin is generated in the middle silk gland.

In 2018, Ma et al. [13] adeptly encapsulated the progress of genome editing technologies in silkworms. This study primarily examines the potential of CRISPR-Cas technology for modifying the genome of B. mori [13]. We examine recent advancements in the application of the exceptional CRISPR-Cas system in B. mori and analyze the latest research employing this methodology to explore gene function, regulate gene expression, and improve resistance to B. mori nucleopolyhedrovirus (BmNPV). The significant importance of B. mori in scientific disciplines will be examined.

2. THE CRISPR-CAS SYSTEM

CRISPR-Cas is one of the key methods employed by many molecular biology scientific laboratories. Since its first description [14], genome editing focused research was implemented by countless research groups [15–17].

2.1 The CRISPR-Cas Complex Role in the Immunity System

When investigating the *iap* gene product in the opportunistic pathogen *Escherichia coli*, Ishino et al. (1987) [18] observed an atypical structure, specifically the repetition of several homologous sequences. Later, this type of structure was observed in various bacterial, as well as archaeal strains [19,20]. Subsequently, these repetitive sequences were linked with exogenous genetic material, and following several years, their assembling mechanism and function were elucidated [21]. This type of sequence can be placed on the chromosomal DNA, but it can also be found on the plasmid DNA [22].

The scientists demonstrated that CRISPR-Cas, which is present in one-third of bacteria and nearly in all archaea, has a key role in host's adaptive immunity. It protects the organism against various intruders, such as viruses, but it also offers protection against other mobile genetic elements, such as transposons or plasmids [23].

The CRISPR-Cas system structure includes three main components, i.e., the CRISPR arrays, the associated Cas proteins, and the leader nucleotide sequence. The first genetic component, the CRISPR locus, is characterized by identical repeats structures (21–37 bp) that are highly conserved and the spacer sequences that are acquired fragments of invader's nucleic acid material. The CRISPR array is located downstream from *cas* genes. The latter encodes for Cas proteins that are crucial to the immune reaction [24].

Initially, only four distinct Cas proteins (1–4) were reported, due to the rapid evolution of biological sciences; currently, there numerous Cas proteins have been described [25,26], Cas1 being the most analyzed [27].

CRISPR-Cas has a great adaptability, with host-related specificities; thus, it exhibits a significant diversity. The varying feature is defined by the CRISPR array and the *cas* gene sequences. The classification of these types of systems is based on the signature Cas proteins. Currently, there are two major classes of CRISPR-Cas systems, each also di-vided in several groups [28]. Regarding the leader nucleotide sequence, it has been shown that it has a key role by carrying the essential promoter sequences for the transcription of CRISPR loci. Besides the promoter, the leader contains specific signals that are crucial for the adaptation stage from the first phase of CRISPR-Cas activation [29].

The adaptation is the first functional stage of the CRISPR-Cas mechanism, during which the foreign nucleic acid is recognized by several Cas proteins [30] and consequently integrated next to a leader sequence. Through this mechanism, in evolution, the spacers are arranged chronologically, and this feature helps bacteria and archaea to enhance their protection against the genetic material of the latest foreign encounter [31]. Each new acquired spacer is accompanied by a repeat sequence; therefore, the CRISPR array expands with every invasion [32].

The CRISPR array is transcribed in the second step, specifically in the biogenesis phase [33]. First of all, it is being transcribed into a precursor CRISPR RNA (crRNA). At the end of this phase, there are numerous mature crRNAs molecules, resulting from the action of RNase III that process the precursor crRNA. Each crRNA includes a spacer and a repeat sequence [31,34,35].

The last step of the system's mechanism is the interference phase. It involves the degradation of the foreign nucleic acid, by targeting and cleaving it [36]. The products of the biogenesis phase, the crRNAs, act like guides for targeting the invader, which is then cleaved following a Cas proteins cascade that act like molecular scissors [37].

2.2 The CRISPR-Cas System as a Genome Editing Tool

When it comes to leading tools in genetic engineering, the CRISPR-Cas system can be considered the foremost instrument. After elucidating its function in various organisms, scientists aimed to exploit its versatility, in order to overcome the disadvantages of other available genome editing tools [38]. Even if scientific studies still report the use of ZFNs and TALENs techniques as editing tools, the CRISPR-Cas system is the most effective genome editing instrument, standing on top in regards to efficiency, cost-effectiveness, and the relative simplicity of use [39] (Table 1). Another considerable advantage of this system is represented by its capacity to simultaneously target multiple genes [40].

Table 1. Comparison between TALEN, ZFN, and CRISPR-Cas gene editing technologies.

Traits	TALEN	ZFN	CRISPR-Cas	References
Origin	Prokaryotic	Eukaryotic	Prokaryotic	[41]
Efficiency (%)	76	12	81	[42]
Specificity	Moderate	Low	High	[43–46]
Target site recognition	Any site	Any site	Pam motif (NGG) required	[43]
Multiplex potential	Low	Low	High	[43,45]
Processing time	Time consuming	Time consuming	Short	[45]
Methylation sensitive	Sensitive	Sensitive	Not sensitive	[42]
Engineering feasibility	Moderate/High	Moderate	Moderate/High	[42,45]
Dimerization required	Yes	Yes	No	[44]
Cost effectiveness	Moderate	No	Yes	[43,44]

Of the numerous CRISPR-Cas systems, CRISPR-Cas9 is currently the most used instrument in laboratories across the world [47]. The Cas9 nuclease is the signature pro- tein of CRISPR-Cas II systems and it is responsible for double strand DNA breaks [27]. Three different methods to deliver the Cas9 endonuclease have been described. It can be directly delivered by microinjection into the embryos, while the other two delivery methods involve a plasmid that expresses the Cas9 enzyme, or a messenger RNA (mRNA) sequence that encodes it. Of the three techniques, in terms of genome engineering, the earliest mentioned is the best option due to its certain advantages. By directly delivering the protein, low immunogen effects were observed. Furthermore, the off-target activity is minimized compared with the other two methods [48]. The use of CRISPR-Cas9 is a simple but power- ful genome editing tool, with various implementations, and their impact on new research trends has been reviewed elsewhere [49].

The CRISPR-Cas9 mechanism relies on the Cas9 nuclease and a guide sequence (gRNA). As the name implies, the gRNA has the role to guide the Cas9 nuclease to a target site in order to cleave the DNA. The key feature of gRNA is the extensive complementarity with the target sequence [50].The protospacer-adjacent motif (PAM) bordering the target complementary sequence has a key role, since in its absence, the CRISPR-Cas systems would degrade their own CRISPR loci. In order to perform a cleavage, the Cas9 protein scans for the PAM sequence. Even if the gRNA is complementary with the target sequence, the Cas9 endonuclease will not cleave it in the absence of PAM [51].

The central factor that influences the success of the gene editing process is the repair path of the double-strand breaks produced by Cas9. There are two main repair path-ways: the homology-directed repair (HDR) and the nonhomologous end joining (NHEJ), respectively [52]. More often, NHEJ is exploited in order to acquire indels mutations, specially to obtain small deletions. These deletions are extremely useful for disclosing gene functions [53]. However, the HDR machinery is used not just to obtain knock-out or knock-down mutations, such as the expected output following NHEJ, but to generate target knock-ins. Therefore, by using HDR, exogenous sequences can be successfully integrated into the host's genome. Currently, major efforts are being made in order to enhance the sequence replacement by using the HDR mechanism [54].

CRISPR-Cas9 is currently used in multiple research fields, such as agriculture (editing of various agricultural plant genomes or pest insect's genome) [55–57], biotechnology, food industry, and medicine (modeling diseases using HeLa cells, deciphering HIV infec- tion mechanisms, using various experimental models, such as *Danio rerio* to tackle cancer and neurological diseases, etc.) [58,59], just to mention a few (Figure 1) [60].

2.3 CRISPR-Cas9 in Entomology

Being the most diverse and numerous category of organisms for decades [61], insects have been intensively studied. Countless studies have been performed due to insects' key roles in ecology, agriculture, and medicine [62,63]. Considering this, numerous research groups aimed to use the CRISPR-Cas9 system to manipulate the insects' genome. The first application of CRISPR-Cas9 was performed in *Drosophila melanogaster* [64] due to its strate- gic importance as arguably the main experimental model organism for life sciences [65]. Besides *D. melanogaster*, the researchers also used the CRISPR-Cas9 applicability on *B. mori*, *Apis mellifera*, *Aedes aegypti*, and *Tribolium castaneum* [66,67].

Gratz et al. (2013) [68] programmed CRISPR-Cas9 to edit *Drosophila's* genome. The authors targeted the *yellow* gene, which is commonly used in various studies. First, they aimed to determine if this genome engineering tool could be efficient and could fulfill its role to induce breaks in the target sequence. By using the CRISPR-Cas9 system in *Drosophila*, not only was the *yellow* gene successfully knocked out, but the genome's alterations were also germline transmitted. Subsequent to the deletion of the target gene, a donor sequence was designed. This sequence provided the template for the HDR repair pathway and its use was to test the accuracy of specific replacement of *yellow* gene with an exogenous sequence. These sequence replacements were transmitted to descendants as well. Their data showed that there was no off-target activity and it highlighted the feasibility of using the CRISPR-Cas9 technology in eukaryotes [68].

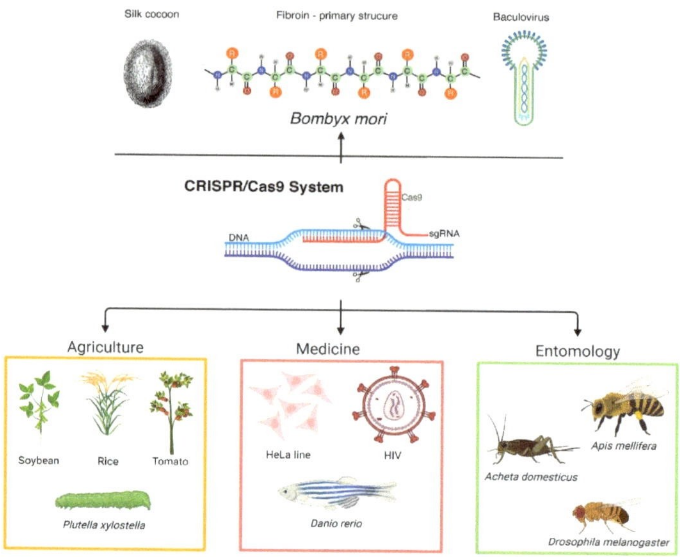

Figure 1. Schematic representation of the most important current applications of CRISPR-Cas9 in entomology, medicine, and agriculture. On top, a simplified description of CRISPR-Cas9 applicability in B. mori that is extensively described in the main text (created with BioRender.com, accessed on 2 December 2021).

Aiming to further highlight the feasibility of choosing this system to perform genome alteration in *Drosophila*, Yu et al. (2013) [69] designed two gRNAs to induce mutations in two regions of the *yellow* gene. In addition, they targeted other six sequences, both euchromatic and heterochromatic loci. Remarkably, a definite mutation in *ms(3)k81* was transmitted to descendants in a proportion of 100%. By successfully targeting heterochromatic loci, their result showed that the CRISPR-Cas9 system is efficient for altering the heterochromatin [69]. *Drosophila* have been used in numerous studies in order to examine the insecticide resistance [70–72]. In this direction, Douris et al. (2020) [56] notably summarized the progress in using CRISPR-Cas9 to explore the genetic basis of this mechanism.

The CRISPR-Cas9 technique was used to perform functional analysis concomitantly on two genes belonging to the cricket (*Gryllus bimaculatus)* [73]. *G. bimaculatus* is an important insect for experimental studies; for example, it plays an important role for evolutionary developmental studies and comparative biology, but it is also a relevant model organism for neurobiology and behavioral sciences [74]. The efficiency of inserting a donor sequence via a homology-independent technique was tested in two *hox* genes, namely *Gb-Ubx* and *Gb-abd-A*. After inserting the donor fragment into essential exons of both genes, their function was lost. Thus, functional investigations of *hox* genes could be carried out by using the knock-in/knock-out approaches [73].

Being one of the most important social insects [75] and as it plays a crucial role as a pollinator, the honeybee (*Apis mellifera*) has been intensively studied. It also plays a pivotal role in various therapeutic areas due to honey production. This natural product has extraordinary benefits for human health, exhibiting antioxidant, antiviral, and antibacterial effects [76]. Due to its special characteristics, the use of honey is not limited to humans, but this natural product is being used to improve certain features of other insects, such as silkworms [77]. There are numerous studies that detail functional analysis of *A. mellifera* genes by exploiting the CRISPR-Cas9 system. For instance, Hu et al. (2019) [67] reported the successful utilization of this system for knocking out the *mrjp 1* gene from the honeybee genome. The CRISPR-Cas9 complex was delivered through microinjection and they tested two specific regions of embryos, for identifying the most convenient structure for delivering the gRNA and the Cas9 endonuclease. By microinjecting the construct at the dorsal posterior side, there was a low rate of successful manipulation (11.8%); however, when choosing the ventral cephalic side, the results showed a great rate of gene editing (93.3%). Trying to validate the previous results, the authors also targeted *pax6*. Based on the previously obtained results, they microinjected the CRISPR-Cas9 construct at the ventral cephalic side. The results showed an editing rate of 100% [67]. Targeting the same gene, *mrjp 1*, similar results have been obtained in another study [78]. Thus, functional analysis of *A. mellifera* genes can be effectively performed by using the CRISPR-Cas9 system.

Considering the same topic of gene function research, Nie et al. (2021) [79] used the CRISPR-Cas9 technology to determine if the *yellow-y* gene plays a crucial role in the process of cuticular melanin synthesis in *A. mellifera*. They targeted this gene due to its great potential for mutants screening, being a selectable marker. By disrupting it, the phenotype of worker cuticle has changed, mainly due to the black pigment decreasing, thus confirming the *yellow-y* gene critical role in melanin pigmentation. However, as future prospects, this could be a great genetic marker for upcoming genomic research [79].

Referring to *A. mellifera* sex determination, it is controlled by the heterozygosity at a particular locus that harbors the key *complementary sex determiner* (*csd*) gene. The bees that are heterozygous at this specific locus are females, while the males are homozygous or hemizygous [80]. In a recent study, Wang et al. (2021) [81] used the CRISPR-Cas9 tool in order to knock out the *csd* gene and, thus, eliminated the genetic difference between females and males. Subsequently, they aimed to observe the transcriptome difference between the two sexes in this particular genetic background. They also successfully induced target mutations in mutant haploid individuals. It was observed that the expression level of several male-biased genes was higher in the mutant males. On the other hand, the expression level of several specific female-biased genes was lower. Their data also confirmed that *csd* interacts with certain genes, such as *fruitless*, *troponin T*, and *transformer-2* just to mention a few [81].

3. BOMBYX MORI

For numerous reasons, *B. mori* is one of the most studied insects, especially because it presents a real interest for the scientific community. It has been completely domesticated and plays the pivotal role in sericulture, being reared principally for silk production on a large scale [82]. Considering the great role of silk in the textile industry and its use as a biomaterial in medicine, major efforts are being made to enhance silk's quality, but also to increase its quantity [12]. Moreover, there are numerous studies that describe the process of obtaining enhanced silk with appreciable properties by genetically manipulating the silkworms [83–86]. *B. mori*'s genome was manipulated in order to enhance silk's properties. For instance, by knocking in the major ampullate silk protein from spiders into *B. mori*'s genome, a research group obtained silk with superior mechanical characteristics [84].

B. mori is an oligophagous insect and its main food source are the mulberry leaves, a nutritional preference that influences its biological and economical parameters. There is a major drawback when it comes to this source of nutrition, namely the limited availability of mulberry leaves. In order to be able to rear silkworms not only in spring and summer, artificial diets are currently being used for their nourishment [77,87].

Regardless of human medicine advances, certain microorganisms developed survival strategies and numerous infectious diseases remain a crucial problem worldwide [88]. In this context, *B. mori* is a reliable experimental model and exhibits the great advantage of a short development cycle that consists of four different stages. The first period of growth is the egg phase, followed by the larvae, pupa, and moth phases. The larvae phase plays a crucial role in silkworm's development, being the longest phase and including five different stages. Another significant advantage of using *B. mori* is the fact that in the last instar larvae, the body size has nearly 5 cm; hence, it is facile to manipulate or to exploit it for various purposes. Moreover, this size also facilitates the body's dissection; thus, multiple target tissues or organs can easily be obtained [89,90].

B. mori as a Model Organism

The level of attention received as a model experimental organism increased, since several groups made available *B. mori*'s genome data [91]. Various mutant strains have been described and the genetic analysis confirmed the numerous genetic traits. Another important aspect is that its manipulation is not associated with ethical concerns [90].

Being susceptible to various infectious agents, *B. mori* is one of the most used experimental models for drug screening, evaluation of different virulence factors, and identifying the pathogen genes responsible for its virulence [90]. Hitherto, several studies used *B. mori* to evaluate the effectiveness of antibiotics against certain human pathogens [92–94].

In a recent study, silkworms were used to examine the efficacy of three different glycopep- tide antibiotics against *Staphylococcus aureus* infection. As a result, authors highlighted the great feasibility and efficacy of using *B. mori* to mimic bacterial infections in order to examine the therapeutic potential of antibiotics [95]. Moreover, recently, silkworms were used to evaluate the impact on several antibacterial compounds against *Cutibacterium acnes* [96]. As an experimental model organism, *B. mori* is currently being intensively used for various other purposes, as detailed in Table 2.

Table 2. Summary of various uses of B. mori as an experimental model organism.

Type of Model Organism	Brief Description	Purpose	References
Human disease model	Transgenic B. mori expressing hIR	Drug evaluation for diabetes treatment	[97,98]
	Bacterial (Listeria monocytogenes) infection model in B. mori	Evaluating the interaction between host and pathogen; investigating the activity of vitamin A against microbial infections	[99]
	Fungal (C. albicans) infection model in silkworms	Assessing the C. albicans biofilm development	[100]
	Inducing deletions in the BmSpr gene, leading to sepiapterin reductase deficiency	Treatment options evaluations against sepiapterin reductase deficiency	[100]
Model for pesticide toxicity	Exposing silkworms to phoxim	Identifying specific biomarkers for phoxim stress; evaluating the toxicity reaction and the pretreatment with nanoparticulate titanium dioxide	[102]
	Inducing genotoxicity by feeding the silkworms with different doses of avermectin	Exploration of certain genes that are required for the DNA repairing mechanism	[103]
	Treating *B. mori* larvae with Fenvalerate-20EC	Evaluation of Fenvalerate-20EC impact on several digestive enzymes	[104]

Model for drugs toxicity	Injecting the silkworms with three different pharmacologically active agents (4-methyl umbelliferone, 7-ethoxycoumarine)	Evaluation of the metabolic pathway of these compounds	[92]
	Exposing the silkworms to fungal infections	Exploring pharmacokinetic parameters of an antifungal agent, Voriconazole	[105]
	Injecting cytotoxic drugs into *B. mori* larvae	Evaluation of cytotoxic drugs impact	[106]
Model for nanomaterials toxicity	Spreading silver nanoparticles on mulberry leaves	Toxicity evaluation of silver nanoparticles	[107–110]
	Injecting subcutaneously zinc oxide nanoparticles	Evaluation of zinc oxide nanoparticles toxicity, accumulation, and distribution	[111]
	Injecting in the dorsal vein different nanoparticles with great interest in various life science branches	Investigation of different silicon and carbon nanomaterials toxicity level against hemocytes	[112]

4. APPLICATIONS OF CRISPR-CAS IN *B. MORI*

The first communication of successful manipulation of *B. mori*'s genome by using the CRISPR-Cas9 tool, was reported by Wang et al. (2013) [113]. The authors targeted an essential gene [113], *BmBlos2*, that is orthologous to the *Blos2* human gene [114]. Two sgR- NAs (23-bp) were designed to induce mutations leading to loss of target gene function. Each complex formed by one sgRNA and the Cas9 nuclease was injected in the preblastoderm embryonic stage. Ordinarily, the larval integument is opaque, but when the *BmBlos2* gene function is lost, the tegument becomes translucent. This effect could be viewed as a phenotypic marker for mutant's detection. Of all individuals, 94% and 95.6%, respectively were successfully edited by using the two sgRNAs. This study highlighted the feasibility of using CRISPR-Cas9 not only in *B. mori*, but also in other lepidopteran insects. These find- ings are of great interest by revealing CRISPR-Cas9 system's applicability in pest control approaches [113].

The multiplexable potential of CRISPR-Cas9 technology was highlighted by Liu et al. (2014) [115]. First, the *BmBlos2* gene was targeted for site specific mutagenesis, in order to confirm the feasibility of using the CRISPR-Cas9 system in *B. mori*. Following this, other six genes were targeted to confirm the system's multiplexable feature: *tyrosine hydroxy- lase*, *red egg*, *yellow-e*, *kynureninase*, *ebony*, and *flugellos*. Mutations were induced in each target gene, without evidence for the system's off-target activity [115]. The multiplexable feature of CRISPR-Cas9 is facilitating the process of genome engineering by simultane- ously and precisely inducing mutations in different sites. This property of CRISPR-Cas9 enables researchers to perform precise and elaborate target mutagenesis in a time-effective manner [116].

Other approaches aimed to knock out the *BmKu70* gene by using CRISPR-Cas9 in order to target its second exon [117]. The *BmKu70* gene is coding a highly conserved protein, Ku70, which plays a key role in numerous mechanisms: cell adhesion, apoptosis, the maintenance of telomeres length, etc. In addition, numerous studies reported that by inactivating the *BmKu70* gene, the frequency of homologous repair is increased. In order to test this hypothesis, the authors performed a transient analysis in genetically manipulated embryos. They knocked in the *Bm702* gene that is found on the Z chromosome. Their results confirmed that by knocking out *BmKu70* the homologous repair frequency is expanded. These promising results are of great interest for fundamental research in *B. mori*. [117].

It has been demonstrated by Fujinaga et al. (2017) [118] that in *B. mori*, the insulin- like growth factor-like peptide (IGFLP), is closely related to the genital disc, particularly involved in its growth. Due to the lack of studies performed in vivo on this topic, the same research group aimed to confirm the role of IGFLP by inactivating it [119]. For this purpose, they used the CRISPR-Cas9 genome editing tool. The absence of IGFLP leads to smaller ovaries and a lower number of laid eggs compared with the wild type. On the other hand, the size of laid eggs and its development were not affected. These findings indicated that this hormone has no impact on *B. mori*'s fertility. The ecdysteroids play a crucial role in IGFLPs production, by inducing gene expression. Furthermore, it has been shown previously that ecdysteroids have a key role in ovary development. Therefore, the authors initially appraised that a low ecdysteroids titre caused reduced ovary weights. However, by analyzing the transgenic females, their data showed that the ecdysteroids titre was the same. This study reveals insights on IGFLPs impact on ovarian development [119].

In a recent study [120], a research group explored the effect of activating the *BmFibH* gene in the *B. mori* embryonic cells. For this purpose, they constructed a complex that involved the inactive form of Cas9 nuclease (dCas9) and a VPR activating domain, driven by a certain promoter. This activating domain consists of several different activators: VP64, p65, and Rta, respectively. When it comes to sgRNAs, three specific constructs were designed to target the promoter of the gene of interest. In order to confirm the success of *BmFibH* activation, first they determined its expression in normal embryonic cells. Their results showed that the *BmFibH* gene is strongly downregulated in untransformed cells. On the other hand, the transfected cells exhibited a higher *BmFibH* expression level. Moreover, their data showed that the activation of the target gene impacted the cellular stress responses [120]. Cui et al. (2018) [121] targeted the *BmFibH* gene in order to explore its role in the development of the silk gland. After the CRISPR-Cas9 construct was designed, a total number of 630 eggs were microinjected, but only 12.5% hatched. After analyzing the unhatched eggs, they observed that all embryos were genetically edited. By knocking out the *BmFibH*, severe changes were observed, such as naked pupa or thin cocoons. All individuals that exhibited naked pupae died. Moreover, by inactivating this gene, several other genes involved in the processes of degradation, such as the autophagy, were upregulated. These findings offer a better understanding of FibH protein's role in silk gland development [121].

Keeping in mind the feasibility of expressing spider silk genes in *B. mori* [84], in order to obtain enhanced silk, Zhang et al. (2019) [122] used CRISPR-Cas9 to acquire high- performance fibers. By using this technique, the authors successfully knocked in spider silk genes in silkworms' genome. Accordingly, they designed two types of systems, FibL- CRISPR-Cas9 and FibH-CRISPR-Cas9. To avoid disrupting the protein production, spider silk genes were knocked in into one of the introns of *BmFibL* or *BmFibH*. They demonstrated the feasibility of employing CRISPR-Cas9 *in B. mori* to obtain silk with enhanced mechanical properties at industrial scale. The strategy described in this study can be further used to obtain numerous exogenous proteins that exhibit great interest for medical applications and beyond [122].

The microRNAs (miRNAs) are key regulators when it comes to gene expression. Notably, they recognize by complementarity specific mRNAs and inactivate them [123]. It has been revealed that *miR-2* is one of the most important miRNAs for wing morphogen- esis in *D. melanogaster* and exhibits great influence on the Notch pathway. The *BmAwd* and *BmFng* genes are known positive regulators of this signaling pathway and are also potential *miR-2* targets. On this topic, Ling et al. (2015) [124] used the CRISPR-Cas9 technology to investigate the function of *miR-2* in *B. mori*. In the first phase of the study, the authors used the Gal4/UAS system to overexpress the *miR-2*, resulting in deformed adults' wings. However, in the second phase of the research, the CRISPR-Cas9 system was used to knock out the two *miR-2* target genes. The loss of function of *BmFng* and *BmAwd* also led to deformed wings. Both phases of the study confirmed that in silkworms, the *miR-2* plays a crucial role in wing development [124].

In another study, Liu et al. (2020) [125] used the CRISPR-Cas9 system to explore the function of *miR-34*, another miRNA that exhibits a great impact on insect development. Firstly, they overexpressed *miR-34* in a transgenic line constructed by using a pBac plas- mid. The *miR-34* overexpression negatively impacted the body size and wing morphology. Secondly, the CRISPR-Cas9 system was used to inactivate the *miR-34* by using two dif- ferent gRNAs. The *miR-34* ablation led to larvae developmental delay. Using several bioinformatic tools, they predicted that the *miR-34* target genes could be *BmE74*, *BmCpg4*, *BmLcp*, *BmWcp11*, and *BmBrc-z2*. Various analyses revealed that miR-34 target genes are just *BmE74* and *BmCpg4*. While it is well known that *BmE74* plays a key role in growth and morphogenesis, functional analysis of the second gene had to be performed. Given this, the CRISPR-Cas9 system was used to knock out *BmCpg4* resulting in affected wings, thus highlighting the gene's role in wing development [125].

Regarding the silkworms' innate defense mechanisms, there are two major activation pathways involved in the expression of numerous evolutionarily conserved antimicrobial peptides (AMPs) genes, and the Toll and the Imd pathways [126,127]. When Gram-negative bacterial or fungal contamination occurs, the Toll pathway is activated. One of the key genes that is involved in the Toll pathway is *BmCactus*. Being a negative regulator, once the infection with the mentioned pathogens occurs, this gene is being phosphorylated and inactivated. Considering this, Park et al. (2018) [128] used the CRISPR-Cas9 technology to perform site-target mutagenesis targeting the *BmCactus* gene in a specific *B. mori* cell line. The authors designed six different gRNAs and transfected the CRISPR-Cas9 complex in the *B. mori* ovarian cell line by electroporation, but observed a very low survival rate of only 24%. Their data showed that all gRNAs determined site induced mutagenesis. By dis- rupting the *BmCactus* gene, the expression of several antimicrobial proteins (e.g., lysozyme and lebocin) was stimulated [128]. Keeping in mind the great importance of AMPs in clinical research [129], this study underlined the outstanding potential of *B. mori* usage in life science fields and the feasibility of using the advanced CRISPR-Cas9 genetic scissors to edit genomes for different purposes [128].

Ecdysteroids are steroid hormones that play a crucial role in the process of molting and metamorphosis in insects. The most important molting hormone is 20-hydroxyecdysone (20E) [130], but even if its biosynthesis processes have been intensively studied, the 20E metabolism has not been well-documented. However, there are several genes that are believed to be involved in the inactivation of 20E, but their biological functions have not been fully understood. Therefore, Li et al. (2015) [131] used *B. mori* to investigate the biological functions of one particular 20E inactivation enzyme, specifically the ecdysone oxidase (EO). Having a great impact on insects' key processes, it is crucial to regulate the ecdysteroids concentration. Accordingly, the EO participates in ecdysteroids' oxidation [132]. The CRISPR-Cas9 system was used to deplete the *BmEo* gene, and consequently, there was observed that the duration of the fifth instar larval phase was prolonged by 24h [131].

Another key element that is involved in insects' development is the juvenile hormone (JH) [133]. The central role in the JH degradation process is played by the juvenile hormone esterase (JHE). Zhang et al. (2017) [134] used CRISPR-Cas9 in *B.*

mori to deplete the encoding gene for JHE, specifically *BmJhe*, in order to investigate its function. Their data showed that by knocking out this gene, the fourth and fifth instar stages were prolonged due the fact that JH metabolism was delayed. These findings are not only important for functional analysis, but also for the sericulture industry. For silk production, it is of great impact to expand the larval stages, because this leads to larger larvae, and thus, larger cocoons are produced. These findings highlight the feasibility of using genome editing tools for economic purposes [134].

Moreover, CRISPR-Cas was used in *B. mori* to perform epigenetic modifications. In this context, Liu et al. (2019) [135] explored the impact of methylation on silkworms' development. This study is of great significance by providing a strategy for the investigation of DNA methylation importance in a target locus. Furthermore, their study represents a starting point for exploring the impact of DNA methylation status on different phenotypes in silkworms and beyond [135]. Furthermore, Xing et al. (2019) [136] used this technology for labeling endogenous regions in *B. mori* embryonic cell line and targeted the *BmFibH* gene. Using CRISPR-dCas9 as an imaging tool has extraordinary impact on performing fundamental research, but also on providing insights on insecticide resistance [136].

As for the detection of CRISPR-Cas induced mutations, there are a plethora of screen- ing methods. On this topic, in a recent study, Brady et al. (2020) [137] described a new approach and provided a protocol for screening, characterizing, and stabilizing the mutant silkworm lines. The provided protocol involves several molecular methods that allow the recognition of the induced mutations on both autosomes and sex chromosomes [137].

In Table 3, several studies are described that used the CRISPR-Cas technology in order to genetically edit *B. mori*.

Applicability of CRISPR-Cas in Anti-BmNPV Therapy

Viruses represent a major threat to numerous hosts, including humans or insects, being one of the most rapidly mutating biological entities. However, there is an urgent need to develop new methods to combat these pathogens. Being one of the most valued molecular tools, the CRISPR-Cas system is currently being used for developing antiviral strategies in various organisms [155–157]. Of all nucleases described in the specialized literature, it has been shown that different variants of Cas9, Cas12, and Cas13 exhibit the most promising results regarding the antiviral approaches [155].

In respect to insect virology, BmNPV was the first discovered pathogen (1998) [158]. In the sericulture industry, this baculovirus causes extensive economic losses; therefore, it has been intensively studied. Being part of *Baculoviridae* family, its infection leads to the most severe disease and only a few silkworm strains are resistant to this virus [159]. Currently, there are several traditional methods that help to enhance silkworms' resistance to BmNPV, but they exhibit serious limitations [158,160,161].

However, CRISPR-Cas has been successfully used as an antiviral therapy in *B. mori*, especially against BmNPV. Chen et al. (2017) [162] selected two genes that are involved in baculovirus' replication and propagation processes: *immediate early-1 (ie-1)* and *me53*, respectively. The authors designed two gRNA for each target gene. The engineered plasmid contained three expression cassettes. One cassette contained the Cas9 nuclease, the second one included the gRNAs and the last one harbored a selecting marker, specifically the enhanced green fluorescent protein (EGFP). Even if the silkworms' viability and fecundity was not impacted, the transgenic homozygotes experienced a delay of the larval stage devel- opment. After performing viral inoculation in both wild type group and transgenic group, it was observed that the second category posed great viral resistance against BmNPV [162]. Likewise, another group of researchers targeted two other genes, *ie-0* and *ie-2*, by using CRISPR-Cas9. After being inoculated with occlusion bodies of BmNPV, the survival rate reached 65% for the transgenic silkworms [163].

The multiplexable feature of the CRISPR-Cas9 technology was underlined in a study performed by Dong et al. (2019) [164]. The researchers targeted in BmNPV three different genes essential for viral replication: *ie-1*, the *major envelope glycoprotein* and the *late expression factor-11*. The success of this study revealed one promising strategy in inhibiting different *B. mori* by using the powerful CRISPR-Cas method.

Table 3. CRISPR-Cas applications in B. mori.

Target Gene	Mutation Type	Delivery Ap-proach	Objective	Gene Function	Refer-ences
BmTim	Deletions	Plasmid	Functional gene analysis	Exhibit an impact on the embryo hatching process	[138]
BmApp	Deletions, insertions	mRNA	Functional gene analysis	Regulates wing development and cell mitosis	[139]
BmFoxo	Deletions	Plasmid	Analysis of *BmFoxo* and JH interaction	Involved in JH degradation	[140]
BmKmo	Deletions	Protein	Phenotypic analysis	Involved in the pro-cess of egg formation and eye coloring	[141]

BmYki	Deletions	Plasmid	Functional gene analysis	Involved in organ development and regeneration	[142]
BmTorso	Deletions	Plasmid	Functional gene analysis	Maintain the steroid hormones balance	[143]
BmEsp	Deletions	Plasmid	Functional gene analysis	Involved in a female's reproducibility	[144]
BmTudor	Deletions, insertions	Plasmid	Investigating the frequency of homologous recombination	Included in the stress granule formation	[145]
BmIdgf	Deletions	mRNA	Analyzing the pigmentation mechanism	Plays a key role in the melanization mechanism	[146]
BmBngr-a2	Deletions	Plasmid	Exploring functional studies of certain ion-transport peptides	Involved in water homeostasis	[147]
BmTctp	Deletions	Plasmid	Functional analysis	Involved in different cell process, such as growth, development, and proliferatio	[148]
BmGr66	Deletions	Plasmid	A better understanding of the specific feeding preference	Involved in silkworms' specific feeding preferences	[149]
BmOvo	Deletions	Plasmid	Functional analysis	Involved in germline sex determination and wing metamorphosis	[150]
BmPhyhd1	Deletions	Protein	Functional analysis	Exhibits a great impact on certain features of the epithelial cells	[151
BmWnt1	Deletions	mRNA	Functional analysis	Involved in the embryogenesis	[152]
BmE75b	Deletions	mRNA	Functional analysis	Controls the developmental timing	[153]
BmOrco	Deletions	Plasmid	Exploration of adult mating behavior	Involved in silkworms' olfactory system, being an odorant receptor co-receptor	[154]

5. CONCISE SUMMARY

B. mori is a significant domesticated insect, valued for its substantial potential as a biotechnological platform for recombinant protein production and its performance as an experimental model organism. [92,110]. Due to its extraordinary prospects, a wide range of studies have been performed that accelerated fundamental research and beyond in silkworms. Being the most feasible technology in terms of genome editing, the CRISPR-Cas system is currently used in many laboratories specialized in medicine, agriculture, alimentary industry, and entomology research [34,165]. A hallmark of CRISPR- Cas is the relative ease of designing CRISPR-based experiments. As reviewed elsewhere, there are numerous bioinformatics tools that facilitate the guide RNA design, as well as the prediction and evaluation of editing results [166]. In our experience, a thorough guide RNA design can also be achieved by using standard sequence alignment tools and manual inspection of potential target regions. Considering the aforementioned, the utilization of the CRISPR-Cas system as a gene editing tool augmented the research in *B. mori*. Most of the studies have been focused on using the CRISPR-Cas system to perform functional gene analysis, to elucidate certain mechanisms [135,148], or to enhance silkworms' resistance to BmNPV [164]. By reviewing the most remarkable work in this field, we provide deep insights that offer support for future research not only in *B. mori*, but also in other insect experimental models.

The potential prospects for employing CRISPR-Cas technology in silkworms are intriguing. Currently, significant advancements in comprehending fibroin and sericin synthesis are being achieved by the genetic manipulation of the B. mori genome, alongside substantial progress in elucidating key mechanisms in silkworms. Significant applications of CRISPR-Cas in B. mori include the enhancement of silk fibers and the generation of recombinant proteins vital for numerous scientific disciplines. Despite comparisons with previous genome editing approaches, CRISPR-Cas9 demonstrates less off-target activity; however, it remains unverified that these adverse effects are entirely eradicated. Nonetheless, there is a significant necessity to eradicate or, at the very least, mitigate the off-target action.

REFERENCES

1. Moon, S.B.; Kim, D.Y.; Ko, J.H.; Kim, Y.S. Recent advances in the CRISPR genome editing tool set. *Exp. Mol. Med.* **2019**, *51*, 1–11. [CrossRef] [PubMed]
2. Guha, T.K.; Wai, A.; Hausner, G. Programmable Genome Editing Tools and their Regulation for Efficient Genome Engineering. *Comput. Struct. Biotechnol. J.* **2017**, *15*, 146–160. [CrossRef] [PubMed]
3. Li, H.; Yang, Y.; Hong, W.; Huang, M.; Wu, M.; Zhao, X. Applications of genome editing technology in the targeted therapy of human diseases: Mechanisms, advances and prospects. *Signal Transduct. Target. Ther.* **2020**, *5*, 1–23. [CrossRef] [PubMed]

4. Schuijff, M.; De Jong, M.D.T.; Dijkstra, A.M. AQ methodology study on divergent perspectives on CRISPR-Cas9 in the Netherlands. *BMC Med. Ethics* **2021**, *22*, 48. [CrossRef]
5. Zhang, D.; Hussain, A.; Manghwar, H.; Xie, K.; Xie, S.; Zhao, S.; Larkin, R.M.; Qing, P.; Jin, S.; Ding, F. Genome editing with the CRISPR-Cas system: An art, ethics and global regulatory perspective. *Plant Biotechnol. J.* **2020**, *18*, 1651–1669. [CrossRef]
6. Li, P.; Wang, L.; Yang, J.; Di, L.; Li, J. Applications of the CRISPR-Cas system for infectious disease diagnostics. *Expert Rev. Mol. Diagn.* **2021**, *21*, 723–732. [CrossRef]
7. Sezutsu, H.; Sumitani, M.; Kondo, M.; Kobayashi, I.; Takasu, Y.; Suzuki, T.; Yonemura, N.; Iizuka, T.; Uchino, K.; Tamura, T.; et al. Construction of a Platform for the Development of Pharmaceutical and Medical Applications Using Transgenic Silkworms. *Yakugaku Zasshi J. Pharm. Soc. Jpn.* **2018**, *138*, 863–874. [CrossRef]
8. Kurihara, H.; Sezutsu, H.; Tamura, T.; Yamada, K. Production of an active feline interferon in the cocoon of transgenic silkworms using the fibroin H-chain expression system. *Biochem. Biophys. Res. Commun.* **2007**, *355*, 976–980. [CrossRef]
9. Nakaya, H.; Tatematsu, K.I.; Sezutsu, H.; Kuwabara, N.; Koibuchi, N.; Takeda, S. Secretory expression of thyroid hormone receptor using transgenic silkworms and its DNA binding activity. *Protein Expr. Purif.* **2020**, *176*, 105723. [CrossRef]
10. Itoh, K.; Kobayashi, I.; Nishioka, S.; Hidaka, T.; Tsuji, D.; Sezutsu, H. A transgenic silkworm overexpressing human lysosomal enzyme as a novel resource for producing recombinant glycobiologics and its application to development of enzyme replacement therapy for lysosomal diseases. *Mol. Genet. Metab.* **2016**, *120*, S69. [CrossRef]
11. Xu, H.; O'Brochta, D.A. Advanced technologies for genetically manipulating the silkworm *Bombyx mori*, a model lepidopteran insect. *Proc. R. Soc. B Biol. Sci.* **2015**, *282*, 20150487. [CrossRef]
12. Montali, A.; Romanelli, D.; Cappellozza, S.; Grimaldi, A.; De Eguileor, M.; Tettamanti, G. Timing of autophagy and apoptosis during posterior silk gland degeneration in *Bombyx mori*. *Arthropod Struct. Dev.* **2017**, *46*, 518–528. [CrossRef] [PubMed]
13. Ma, S.Y.; Smagghe, G.; Xia, Q.Y. Genome editing in *Bombyx mori*: New opportunities for silkworm functional genomics and the sericulture industry. *Insect Sci.* **2018**, *26*, 964–972. [CrossRef] [PubMed]
14. Jinek, M.; Chylinski, K.; Fonfara, I.; Hauer, M.; Doudna, J.A.; Charpentier, E. A Programmable Dual-RNA—Guided DNA endonuclease in adaptive bacterial immunity. *Science* **2012**, *337*, 816–822. [CrossRef] [PubMed]

15. Nidhi, S.; Anand, U.; Oleksak, P.; Tripathi, P.; Lal, J.A.; Thomas, G.; Kuca, K.; Tripathi, V. Novel CRISPR—Cas Systems: An Updated Review of the Current Achievements, Applications, and Future Research Perspectives. *Int. J. Mol. Sci.* **2021**, *22*, 3327. [CrossRef] [PubMed]
16. Hahn, F.; Loures, L.S.; Sparks, C.A.; Kanyuka, K.; Nekrasov, V. Efficient CRISPR/Cas-Mediated Targeted Mutagenesis in Spring and Winter Wheat Varieties. *Plants* **2021**, *10*, 1481. [CrossRef] [PubMed]
17. Hesami, M.; Yoosefzadeh Najafabadi, M.; Adamek, K.; Torkamaneh, D.; Jones, A.M.P. Synergizing off-target predictions for in silico insights of CENH3 Knockout in Cannabis through CRISPR/Cas. *Molecules* **2021**, *26*, 2053. [CrossRef]
18. Ishino, Y.; Shinagawa, H.; Makino, K.; Amemura, M.; Nakata, A. Nucleotide Sequence of the iap Gene, Responsible for Alkaline Phosphatase Isozyme Conversion in Escherichia coli, and Identification of the Gene Product. *J. Bacteriol.* **1987**, *169*, 5429–5433. [CrossRef]
19. Gophna, U.; Brodt, A. CRISPR/Cas systems in archaea. *Mob. Genet. Elem.* **2012**, *2*, 63–64. [CrossRef]
20. Horvath, P.; Barrangou, R. CRISPR/Cas, the immune system of Bacteria and Archaea. *Science* **2010**, *327*, 167–170. [CrossRef]
21. Shabbir, M.A.B.; Shabbir, M.Z.; Wu, Q.; Mahmood, S.; Sajid, A.; Maan, M.K.; Ahmed, S.; Naveed, U.; Hao, H.; Yuan, Z. CRISPR-cas system: Biological function in microbes and its use to treat antimicrobial resistant pathogens. *Ann. Clin. Microbiol. Antimicrob.* **2019**, *18*, 21. [CrossRef] [PubMed]
22. Rath, D.; Amlinger, L.; Rath, A.; Lundgren, M. The CRISPR-Cas immune system: Biology, mechanisms and applications. *Biochimie* **2015**, *117*, 119–128. [CrossRef] [PubMed]
23. Faure, G.; Shmakov, S.A.; Yan, W.X.; Cheng, D.R.; Scott, D.A.; Peters, J.E.; Makarova, K.S.; Koonin, E.V. CRISPR–Cas in mobile genetic elements: Counter-defence and beyond. *Nat. Rev. Microbiol.* **2019**, *17*, 513–525. [CrossRef] [PubMed]
24. Mcdonald, N.D.; Regmi, A.; Morreale, D.P.; Borowski, J.D.; Boyd, E.F. CRISPR-Cas systems are present predominantly on mobile genetic elements in Vibrio species. *BMC Genom.* **2019**, *20*, 105. [CrossRef] [PubMed]
25. Haft, D.H.; Selengut, J.; Mongodin, E.F.; Nelson, K.E. A Guild of 45 CRISPR-Associated (Cas) Protein Families and Multiple CRISPR/Cas Subtypes Exist in Prokaryotic Genomes. *PLoS Comput. Biol.* **2005**, *1*, e60. [CrossRef] [PubMed]
26. Makarova, K.S.; Haft, D.H.; Barrangou, R.; Brouns, S.J.J.; Charpentier, E.; Horvath, P.; Moineau, S.; Mojica, F.J.M.; Wolf, Y.I.; Yakunin, A.F.; et al. Evolution and classification of the CRISPR-Cas systems. *Nat. Rev. Microbiol.* **2011**, *9*, 467–477. [CrossRef] [PubMed]

27. Makarova, K.S.; Koonin, E.V. Annotation and Classification of CRISPR-Cas Systems. In *CRISPR*; Springer: Berlin/Heidelberg, Ger- many, 2015; pp. 1–27. [CrossRef]
28. Koonin, E.V.; Makarova, K.S. Origins and evolution of CRISPR-Cas systems. *Philos. Trans. R. Soc. B* **2019**, *374*, 20180087. [CrossRef]
29. Alkhnbashi, O.S.; Shah, S.A.; Garrett, R.A.; Saunders, S.J.; Costa, F.; Backofen, R. Characterizing leader sequences of CRISPR loci. *Bioinformatics* **2016**, *32*, i576–i585. [CrossRef]
30. Alkhnbashi, O.S.; Costa, F.; Shah, S.A.; Garrett, R.A.; Saunders, S.J.; Backofen, R. CRISPRstrand: Predicting repeat orientations to determine the crRNA-encoding strand at CRISPR loci. *Bioinformatics* **2014**, *30*, i489–i496. [CrossRef]
31. McGinn, J.; Marraffini, L.A. Molecular mechanisms of CRISPR–Cas spacer acquisition. *Nat. Rev. Microbiol.* **2018**, *17*, 7–12. [CrossRef]
32. Sorek, R.; Lawrence, C.M.; Wiedenheft, B. CRISPR-Mediated Adaptive Immune Systems in Bacteria and Archaea. *Annu. Rev. Biochem.* **2013**, *82*, 237–266. [CrossRef] [PubMed]
33. Roberts, A.; Barrangou, R. Applications of CRISPR-Cas systems in lactic acid bacteria. *FEMS Microbio. Rev.* **2020**, *44*, 523–537. [CrossRef] [PubMed]
34. Hryhorowicz, M.; Lipin′ ski, D.; Zeyland, J.; Słomski, R. CRISPR/Cas9 Immune System as a Tool for Genome Engineering. *Arch. Immunol. Ther. Exp.* **2016**, *65*, 233–240. [CrossRef] [PubMed]
35. Terns, M.P.; Terns, R.M. CRISPR-based adaptive immune systems. *Curr. Opin. Microbiol.* **2011**, *14*, 321–327. [CrossRef] [PubMed]
36. Newsom, S.; Parameshwaran, H.P.; Martin, L.; Rajan, R. The CRISPR-Cas Mechanism for Adaptive Immunity and Alternate Bacterial Functions Fuels Diverse Biotechnologies. *Front. Cell. Infect. Microbiol.* **2021**, *10*, 1–10. [CrossRef]
37. Marraffini, L.A.; Sontheimer, E.J. CRISPR interference: RNA-directed adaptive immunity in bacteria and archaea. *Nat. Rev. Genet.* **2010**, *11*, 181–190. [CrossRef]
38. Ishino, Y.; Krupovic, M.; Forterre, P. History of CRISPR-Cas from Encounter with a Mysterious Repeated Sequence to Genome Editing Technology. *J. Bacteriol.* **2018**, *200*, e00580-17. [CrossRef]
39. Sontheimer, E.J.; Barrangou, R. The Bacterial Origins of the CRISPR Genome-Editing Revolution. *Hum. Gene Ther.* **2015**, *26*, 413–424. [CrossRef]
40. El-Mounadi, K.; Morales-Floriano, M.L.; Garcia-Ruiz, H. Principles, Applications, and Biosafety of Plant Genome Editing Using CRISPR-Cas9. *Front. Plant Sci.* **2020**, *11*, 56. [CrossRef]
41. Agustın-Pavon, C.; Isalan, M. Synthetic biology and therapeutic strategies for the degenerating brain. *Bioessays* **2014**, *36*, 979–990. [CrossRef]

42. Chen, L.; Tang, L.; Xiang, H.; Jin, L.; Li, Q.; Dong, Y.; Wang, W.; Zhang, G. Advances in genome editing technology and its promising application in evolutionary and ecological studies. *Gigascience* **2014**, *3*, 2047-217X. [CrossRef] [PubMed]
43. Tavakoli, K.; Pour-Aboughadareh, A.; Kianersi, F.; Poczai, P.; Etminan, A.; Shooshtari, L. Applications of CRISPR-Cas9 as an Advanced Genome Editing System in Life Sciences. *BioTech* **2021**, *10*, 14. [CrossRef]
44. Guha, T.K.; Edgell, D.R. Applications of Alternative Nucleases in the Age of CRISPR/Cas9. *Int. J. Mol. Sci.* **2017**, *18*, 2565. [CrossRef]
45. Khan, S.H. Genome-Editing Technologies: Concept, Pros, and Cons of Various Genome-Editing Techniques and Bioethical Concerns for Clinical Application. *Mol. Ther. Nucleic Acids* **2019**, *16*, 326–334. [CrossRef] [PubMed]
46. Chira, S.; Gulei, D.; Hajitou, A.; Zimta, A.A.; Cordelier, P.; Berindan-Neagoe, I. CRISPR/Cas9: Transcending the Reality of Genome Editing. *Mol. Ther. Nucleic Acids* **2017**, *7*, 211–222. [CrossRef]
47. Manghwar, H.; Lindsey, K.; Zhang, X.; Jin, S. CRISPR/Cas System: Recent Advances and Future Prospects for Genome Editing. *Trends Plant Sci.* **2019**, *24*, 1102–1125. [CrossRef] [PubMed]
48. Li, J.; Shi, Y.; Wu, J.; Li, H.; Smagghe, G.; Liu, T. CRISPR/Cas9 in lepidopteran insects: Progress, application and prospects. *J. Insect Physiol.* **2021**, *135*, 104325. [CrossRef] [PubMed]
49. Tyagi, S.; Kumar, R.; Das, A.; Won, S.Y.; Shukla, P. CRISPR-Cas9 system: A genome-editing tool with endless possibilities. *J. Biotechnol.* **2020**, *319*, 36–53. [CrossRef]
50. Zhang, Y.; Showalter, A.M. CRISPR/Cas9 Genome Editing Technology: A Valuable Tool for Understanding Plant Cell Wall Biosynthesis and Function. *Front. Plant Sci.* **2020**, *11*, 589517. [CrossRef] [PubMed]
51. Collias, D.; Beisel, C.L. CRISPR technologies and the search for the PAM-free nuclease. *Nat. Commun.* **2021**, *12*, 1–12. [CrossRef] [PubMed]
52. Yang, H.; Ren, S.; Yu, S.; Pan, H.; Li, T.; Ge, S.; Zhang, J.; Xia, N. Methods Favoring Homology-Directed Repair Choice in Response to CRISPR/Cas9 Induced-Double Strand Breaks. *Int. J. Mol. Sci.* **2020**, *21*, 6461. [CrossRef]
53. Bernheim, A.; Calvo-villamañán, A.; Basier, C.; Cui, L.; Rocha, E.; Touchon, M.; Bikard, D. Inhibition of NHEJ repair by type II-A CRISPR-Cas systems in bacteria. *Nat. Commun.* **2017**, *8*, 25–28. [CrossRef] [PubMed]
54. Di Stazio, M.; Foschi, N.; Athanasakis, E.; Gasparini, P.; d'Adamo, A.P. Systematic analysis of factors that improve homologous direct repair (HDR) efficiency in CRISPR/Cas9 technique. *PLoS ONE* **2021**, *16*, e0247603. [CrossRef]
55. Zhu, H.; Li, C.; Gao, C. Applications of CRISPR–Cas in agriculture and plant biotechnology. *Nat. Rev. Mol. Cell Biol.* **2020**, *21*, 661–677. [CrossRef] [PubMed]

56. Douris, V.; Denecke, S.; Van Leeuwen, T.; Bass, C.; Nauen, R.; Vontas, J. Using CRISPR/Cas9 genome modification to understand the genetic basis of insecticide resistance: Drosophila and beyond. *Pestic. Biochem. Physiol.* **2020**, *167*, 104595. [CrossRef] [PubMed]
57. Tyagi, S.; Kesiraju, K.; Saakre, M.; Rathinam, M.; Raman, V.; Pattanayak, D.; Sreevathsa, R. Genome Editing for Resistance to Insect Pests: An Emerging Tool for Crop Improvement. *ACS Omega* **2020**, *5*, 20674–20683. [CrossRef]
58. Yang, Y.; Xu, J.; Ge, S.; Lai, L. CRISPR/Cas: Advances, Limitations, and Applications for Precision Cancer Research. *Front. Med.* **2021**, *8*, 649896. [CrossRef]
59. Yang, Y.; Liu, X.; Li, S.; Chen, Y.; Zhao, Y.; Wei, Y.; Qiu, Y.; Liu, Y.; Zhou, Z.; Han, J.; et al. Genome-scale CRISPR screening for potential targets of ginsenoside compound K. *Cell Death Dis.* **2020**, *11*, 39. [CrossRef]
60. Xu, Y.; Li, Z. CRISPR-Cas systems: Overview, innovations and applications in human disease research and gene therapy. *Comput. Struct. Biotechnol. J.* **2020**, *18*, 2401–2415. [CrossRef]
61. Ma, X.; He, K.; Shi, Z.; Li, M.; Li, F.; Chen, X.-X. Large-Scale Annotation and Evolution Analysis of MiRNA in Insects. *Genome Biol. Evol.* **2021**, *13*, evab083. [CrossRef] [PubMed]
62. Brady, D.; Grapputo, A.; Romoli, O.; Sandrelli, F. Insect Cecropins, Antimicrobial Peptides with Potential Therapeutic Applications. *Int. J. Mol. Sci.* **2019**, *20*, 5862. [CrossRef] [PubMed]
63. Romoli, O.; Mukherjee, S.; Mohid, S.A.; Dutta, A.; Montali, A.; Franzolin, E.; Brady, D.; Zito, F.; Bergantino, E.; Rampazzo, C.; et al. Enhanced Silkworm Cecropin B Antimicrobial Activity against Pseudomonas aeruginosa from Single Amino Acid Variation. *ACS Infect. Dis.* **2019**, *5*, 1200–1213. [CrossRef]
64. Cui, Y.; Sun, J.; Yu, L. Application of the CRISPR gene-editing technique in insect functional genome studies—A review. *Entomol. Exp. Appl.* **2017**, *162*, 124–132. [CrossRef]
65. De Lazzari, F.; Sandrelli, F.; Whitworth, A.J.; Bisaglia, M. Antioxidant Therapy in Parkinson's Disease: Insights from Drosophila melanogaster. *Antioxidants* **2020**, *9*, 52. [CrossRef]
66. Taning, C.N.T.; Van Eynde, B.; Yu, N.; Ma, S.; Smagghe, G. CRISPR/Cas9 in insects: Applications, best practices and biosafety concerns. *J. Insect Physiol.* **2017**, *98*, 245–257. [CrossRef] [PubMed]
67. Hu, X.F.; Zhang, B.; Liao, C.H.; Zeng, Z.J. High-Efficiency CRISPR/Cas9-Mediated Gene Editing in Honeybee (*Apis mellifera*) Embryos. *G3 Genes Genomes Genet.* **2019**, *9*, 1759–1766. [CrossRef]
68. Gratz, S.J.; Cummings, A.M.; Nguyen, J.N.; Hamm, D.C.; Donohue, L.K.; Harrison, M.M.; Wildonger, J.; O'Connor-giles, K.M. Genome Engineering of Drosophila with the CRISPR RNA-guided Cas9 nuclease. *Genetics* **2013**, *194*, 1029–1035. [CrossRef]

69. Yu, Z.; Ren, M.; Wang, Z.; Zhang, B.; Rong, Y.S.; Jiao, R.; Gao, G. Highly Efficient Genome Modifications Mediated by CRISPR/Cas9 in Drosophila. *Genetics* **2013**, *195*, 289–291. [CrossRef] [PubMed]
70. Perry, T.; Batterham, P. Harnessing model organisms to study insecticide resistance. *Curr. Opin. Insect Sci.* **2018**, *27*, 61–67. [CrossRef]
71. Homem, R.A.; Davies, T.G.E. An overview of functional genomic tools in deciphering insecticide resistance. *Curr. Opin. Insect Sci.* **2018**, *27*, 103–110. [CrossRef]
72. Scott, J.G.; Buchon, N. Drosophila melanogaster as a powerful tool for studying insect toxicology. *Pestic. Biochem. Physiol.* **2019**, *161*, 95–103. [CrossRef] [PubMed]
73. Matsuoka, Y.; Nakamura, T.; Watanabe, T.; Barnett, A.A.; Noji, S.; Mito, T.; Extavour, C.G. Establishment of CRISPR/Cas9-based knock-in in a hemimetabolous insect: Targeted gene tagging in the cricket *Gryllus bimaculatus*. *bioRxiv* **2021**. [CrossRef]
74. Donoughe, S.; Extavour, C.G. Embryonic development of the cricket *Gryllus bimaculatus*. *Dev. Biol.* **2016**, *411*, 140–156. [CrossRef] [PubMed]
75. Evans, J.D.; Aronstein, K.; Chen, Y.P.; Hetru, C.; Imler, J.; Jiang, H.; Kanost, M.; Thompson, G.J.; Zou, Z.; Hultmark, D. Immune pathways and defence mechanisms in honey bees *Apis mellifera*. *Insect Mol. Biol.* **2006**, *15*, 645–656. [CrossRef] [PubMed]
76. Cucu, A.A.; Baci, G.M.; Moise, A.R.; Dezsi, S.; Marc, B.D.; Stângaciu, S.; Dezmiream, D.S. Towards a Better Understanding of Nutritional and Therapeutic Effects of Honey and Their Applications in Apitherapy. *Appl. Sci.* **2021**, *11*, 4190. [CrossRef]
77. Baci, G.; Cucu, A.; Moise, A.R.; Dezmirean, D.S. Applicability of Honey on Silkworms (*Bombyx mori*) and Quality Improvement of Its Biomaterials. *Appl. Sci.* **2021**, *11*, 4613. [CrossRef]
78. Kohno, H.; Suenami, S.; Takeuchi, H.; Sasaki, T.; Kubo, T. Production of Knockout Mutants by CRISPR/Cas9 in the European Honeybee, *Apis mellifera* L. *Zool. Sci.* **2016**, *33*, 505–512. [CrossRef] [PubMed]
79. Nie, H.; Liang, L.; Li, Q.; Li, Z.; Zhu, Y.; Guo, Y.; Zheng, Q.; Lin, Y.; Yang, D.; Li, Z.; et al. CRISPR/Cas9 mediated knockout of Amyellow-y gene results in melanization defect of the cuticle in adult *Apis mellifera*. *J. Insect Physiol.* **2021**, *132*, 104264. [CrossRef] [PubMed]
80. Gempe, T.; Hasselmann, M.; Schiøtt, M.; Hause, G.; Otte, M.; Beye, M. Sex Determination in Honeybees: Two Separate Mechanisms Induce and Maintain the Female Pathway. *PLoS Biol.* **2009**, *7*, e1000222. [CrossRef]
81. Wang, X.; Lin, Y.; Liang, L.; Geng, H.; Zhang, M.; Nie, H.; Su, S. Transcriptional Profiles of Diploid Mutant *Apis mellifera* Embryos after Knockout of csd by CRISPR/Cas9. *Insects* **2021**, *12*, 704. [CrossRef]

82. Horie, Y.; Watanabe, H. Recent advances in sericulture. *Annu. Rev. Entomol.* **1980**, *25*, 49–71. [CrossRef]
83. Saviane, A.; Romoli, O.; Bozzato, A.; Freddi, G.; Cappelletti, C.; Rosini, E.; Cappellozza, S.; Tettamanti, G.; Sandrelli, F. Intrinsic antimicrobial properties of silk spun by genetically modified silkworm strains. *Transgenic Res.* **2018**, *27*, 87–101. [CrossRef] [PubMed]
84. Xu, J.; Dong, Q.; Yu, Y.; Niu, B.; Ji, D.; Li, M.; Huang, Y.; Chen, X.; Tan, A. Mass spider silk production through targeted gene replacement in *Bombyx mori*. *Proc. Natl. Acad. Sci. USA* **2018**, *115*, 8757–8762. [CrossRef]
85. Tang, X.; Ye, X.; Wang, X.; Zhao, S.; Wu, M.; Ruan, J.; Zhong, B. High mechanical property silk produced by transgenic silkworms expressing the spidroins PySp1 and ASG1. *Sci. Rep.* **2021**, *11*, 1–9. [CrossRef]
86. Leem, J.W.; Fraser, M.J.; Kim, Y.L. Transgenic and Diet-Enhanced Silk Production for Reinforced Biomaterials: A Metamaterial Perspective. *Annu. Rev. Biomed. Eng.* **2020**, *22*, 79–102. [CrossRef]
87. Cappellozza, L.; Cappellozza, S.; Saviane, A.; Sbrenna, G. Artificial diet rearing system for the silkworm *Bombyx mori* (Lepidoptera: Bombycidae): Effect of vitamin C deprivation on larval growth and cocoon production. *Appl. Entomol. Zool.* **2005**, *40*, 405–412. [CrossRef]
88. Marrazzo, P.; Crupi, A.N.; Alviano, F.; Teodori, L.; Bonsi, L. Exploring the roles of MSCs in infections: Focus on bacterial diseases. *J. Mol. Med.* **2019**, *97*, 437–450. [CrossRef]
89. Abdelli, N.; Peng, L.; Keping, C. Silkworm, *Bombyx mori*, as an alternative model organism in toxicological research. *Environ. Sci. Pollut. Res.* **2018**, *25*, 35048–35054. [CrossRef]
90. Meng, X.; Zhu, F.; Chen, K. Silkworm: A promising model organism in life science. *J. Insect Sci.* **2017**, *17*. [CrossRef]
91. Yokoi, K.; Tsubota, T.; Jouraku, A.; Sezutsu, H.; Bono, H. Reference Transcriptome Data in Silkworm *Bombyx mori*. *Insects* **2021**, *12*, 519. [CrossRef]
92. Hamamoto, H.; Tonoike, A.; Narushima, K.; Horie, R.; Sekimizu, K. Silkworm as a model animal to evaluate drug candidate toxicity and metabolism. *Comp. Biochem. Physiol. Part C Toxicol. Pharmacol.* **2009**, *149*, 334–339. [CrossRef]
93. Hamamoto, H.; Kurokawa, K.; Kaito, C.; Kamura, K.; Razanajatovo, I.M.; Kusuhara, H.; Santa, T.; Sekimizu, K. Quantitative Eval- uation of the Therapeutic Effects of Antibiotics Using Silkworms Infected with Human Pathogenic Microorganisms. *Antimicrob. Agents Chemother.* **2004**, *48*, 774–779. [CrossRef]
94. Kaito, C.; Sekimizu, K. A silkworm model of pathogenic bacterial infection. *Drug Discov. Ther.* **2007**, *1*, 89–93.
95. Montali, A.; Berini, F.; Brivio, M.F.; Mastore, M.; Saviane, A.; Cappellozza, S.; Marinelli, F.; Tettamanti, G. A Silkworm Infection Model for In Vivo Study

of Glycopeptide Antibiotics. *Antibiotics* **2020**, *9*, 300. [CrossRef] [PubMed]

96. Matsumoto, Y.; Tateyama, Y.; Sugita, T. Evaluation of Antibacterial Drugs Using Silkworms Infected by Cutibacterium acnes. *Insects* **2021**, *12*, 619. [CrossRef] [PubMed]
97. Matsumoto, Y.; Ishii, M.; Ishii, K.; Miyaguchi, W.; Horie, R.; Inagaki, Y.; Hamamoto, H.; Tatematsu, K.-I.; Uchino, K.; Tamura, T.; et al. Transgenic silkworms expressing human insulin receptors for evaluation of therapeutically active insulin receptor agonists. *Biochem. Biophys. Res. Commun.* **2014**, *455*, 159–164. [CrossRef] [PubMed]
98. Matsumoto, Y.; Ishii, M.; Hayashi, Y.; Miyazaki, S.; Sugita, T.; Sumiya, E.; Sekimizu, K. Diabetic silkworms for evaluation of therapeutically effective drugs against type II diabetes. *Sci. Rep.* **2015**, *5*, 10722. [CrossRef] [PubMed]
99. Castillo, Y.; Suzuki, J.; Watanabe, K.; Shimizu, T.; Watarai, M. Effect of Vitamin A on Listeria monocytogenes Infection in a Silkworm Model. *PLoS ONE* **2016**, *11*, e0163747. [CrossRef]
100. Matsumoto, Y.; Kurakado, S.; Sugita, T. Evaluating Candida albicans biofilm formation in silkworms. *Med. Mycol.* **2021**, *59*, 201–205. [CrossRef]
101. Jiang, G.; Song, J.; Hu, H.; Tong, X.; Dai, F. Evaluation of the silkworm lemon mutant as an invertebrate animal model for human sepiapterin reductase deficiency. *R. Soc. Open Sci.* **2020**, *7*, 191888. [CrossRef]
102. Wang, L.; Su, M.; Zhao, X.; Hong, J.; Yu, X.; Xu, B.; Sheng, L.; Liu, N.; Shen, W.; Li, B.; et al. Nanoparticulate TiO 2 Protection of Midgut Damage in the Silkworm (*Bombyx mori*) Following Phoxim Exposure. *Arch. Environ. Contam. Toxicol.* **2015**, *68*, 534–542. [CrossRef] [PubMed]
103. Shen, W.; Zhao, X.; Wang, Q.; Niu, B.; Liu, Y.; He, L.; Weng, H.; Meng, Z.; Chen, Y. Genotoxicity evaluation of low doses of avermectin to hemocytes of silkworm (*Bombyx mori*) and response of gene expression to DNA damage. *Pestic. Biochem. Physiol.* **2011**, *101*, 159–164. [CrossRef]
104. Vyjayanthi, N.; Subramanyam, M.V.V. Effect of Fenvalerate-20EC on Sericigenous Insects: II. Digestive Enzymes in the Nutritive Physiology of Silkworm, *Bombyx mori* L. *Ecotoxicol. Environ. Saf.* **2002**, *53*, 212–220. [CrossRef]
105. Yasu, T.; Matsumoto, Y.; Sugita, T. Pharmacokinetics of voriconazole and its alteration by Candida albicans infection in silkworms. *J. Antibiot.* **2021**, *74*, 443–449. [CrossRef] [PubMed]
106. Inagaki, Y.; Matsumoto, Y.; Kataoka, K.; Matsuhashi, N.; Sekimizu, K. Evaluation of drug-induced tissue injury by measuring alanine aminotransferase (ALT) activity in silkworm hemolymph. *BMC Pharmacol. Toxicol.* **2012**, *13*, 13. [CrossRef]
107. Pandiarajan, J.; Jeyarani, V.; Balaji, S.; Krishnan, M. Silver Nanoparticles an Accumulative Hazard in Silkworm: *Bombyx mori*. *Austin J. Biotechnol. Bioeng.* **2016**, *3*, 1057.

108. Gad, A.A. Toxicity effect of Silver Nanoparticles to the Haemocytes and Antioxidant activity of Silkworm *Bombyx mori*. *Physiol. Entomol.* **2020**, *45*, 154–160. [CrossRef]
109. Meng, X.; Abdlli, N.; Wang, N.; Lü, P.; Nie, Z.; Dong, X.; Lu, S.; Chen, K. Effects of Ag Nanoparticles on Growth and Fat Body Proteins in Silkworms (*Bombyx mori*). *Biol. Trace Elem. Res.* **2017**, *180*, 327–337. [CrossRef]
110. Nouara, A.; Lü, P.; Chen, L.; Pan, Y.; Yang, Y.; Chen, K. Silver effects on silkworm, *Bombyx mori*. *J. Toxicol. Sci.* **2018**, *43*, 697–709. [CrossRef]
111. Xu, Y.; Wang, W.; Ma, L.; Cui, X.; Lynch, I.; Wu, G. Acute toxicity of Zinc Oxide nanoparticles to silkworm (*Bombyx mori* L.). *Chemosphere* **2020**, *259*, 127481. [CrossRef]
112. Li, K.L.; Zhang, Y.H.; Xing, R.; Zhou, Y.F.; Chen, X.D.; Wang, H.; Song, B.; Sima, Y.H.; Xu, S.Q. Different toxicity of cadmium telluride, silicon, and carbon nanomaterials against hemocytes in silkworm, *Bombyx mori*. *RSC Adv.* **2017**, *7*, 50317–50327. [CrossRef]
113. Wang, Y.; Li, Z.; Xu, J.; Zeng, B.; Ling, L.; You, L.; Chen, Y.; Huang, Y.; Tan, A. The CRISPR/Cas System mediates efficient genome engineering in *Bombyx mori*. *Cell Res.* **2013**, *23*, 1414–1416. [CrossRef] [PubMed]
114. Fujii, T.; Daimon, T.; Uchino, K.; Banno, Y.; Katsuma, S.; Sezutsu, H.; Tamura, T.; Shimada, T. Transgenic analysis of the BmBLOS2 gene that governs the translucency of the larval integument of the silkworm, *Bombyx mori*. *Insect Mol. Biol.* **2010**, *19*, 659–667. [CrossRef] [PubMed]
115. Liu, Y.; Ma, S.; Wang, X.; Chang, J.; Gao, J.; Shi, R.; Zhang, J.; Lu, W.; Liu, Y.; Zhao, P.; et al. Highly efficient multiplex targeted mutagenesis and genomic structure variation in *Bombyx mori* cells using CRISPR/Cas9. *Insect Biochem. Mol. Biol.* **2014**, *49*, 35–42. [CrossRef]
116. Sakuma, T.; Nishikawa, A.; Kume, S.; Chayama, K.; Yamamoto, T. Multiplex genome engineering in human cells using all-in-one CRISPR/Cas9 vector system. *Sci. Rep.* **2014**, *4*, 5400. [CrossRef]
117. Ma, S.; Chang, J.; Wang, X.; Liu, Y.; Zhang, J.; Lu, W.; Gao, J.; Shi, R.; Zhao, P.; Xia, Q. CRISPR/Cas9 mediated multiplex genome editing and heritable mutagenesis of BmKu70 in *Bombyx mori*. *Sci. Rep.* **2014**, *4*, 4489. [CrossRef]
118. Fujinaga, D.; Kohmura, Y.; Okamoto, N.; Kataoka, H.; Mizoguchi, A. Insulin-like growth factor (IGF)-like peptide and 20-hydroxyecdysone regulate the growth and development of the male genital disk through different mechanisms in the silkmoth, *Bombyx mori*. *Insect Biochem. Mol. Biol.* **2017**, *87*, 35–44. [CrossRef]
119. Fujinaga, D.; Shiomi, K.; Yagi, Y.; Kataoka, H.; Mizoguchi, A. An insulin-like growth factor-like peptide promotes ovarian development in the silkmoth *Bombyx mori*. *Sci. Rep.* **2019**, *9*, 1–12. [CrossRef]

120. Hu, W.; Wang, X.; Ma, S.; Peng, Z.; Cao, Y.; Xia, Q. CRISPR-Mediated Endogenous Activation of Fibroin Heavy Chain Gene Triggers Cellular Stress Responses in *Bombyx mori* Embryonic Cells. *Insects* **2021**, *12*, 552. [CrossRef]

121. Cui, Y.; Zhu, Y.; Lin, Y.; Chen, L.; Feng, Q.; Wang, W.; Xiang, H. New insight into the mechanism underlying the silk gland biological process by knocking out fibroin heavy chain in the silkworm. *BMC Genom.* **2018**, *19*, 215. [CrossRef]

122. Zhang, X.; Xia, L.; Day, B.A.; Harris, T.I.; Oliveira, P.; Knittel, C.; Licon, A.L.; Gong, C.; Dion, G.; Lewis, R.V.; et al. CRISPR/Cas9 Initiated Transgenic Silkworms as a Natural Spinner of Spider Silk. *Biomacromolecules* **2019**, *20*, 2252–2264. [CrossRef] [PubMed]

123. Michlewski, G.; Cáceres, J.F. Post-transcriptional control of miRNA biogenesis. *RNA* **2019**, *25*, 1–16. [CrossRef] [PubMed]

124. Ling, L.; Ge, X.; Li, Z.; Zeng, B.; Xu, J.; Chen, X.; Shang, P.; James, A.A.; Huang, Y.; Tan, A. MiR-2 family targets awd and fng to regulate wing morphogenesis in *Bombyx mori*. *RNA Biol.* **2015**, *12*, 742–748. [CrossRef] [PubMed]

125. Liu, Z.; Xu, J.; Ling, L.; Luo, X.; Yang, D.; Yang, X.; Zhang, X.; Huang, Y. miR-34 regulates larval growth and wing morphogenesis by directly modulating ecdysone signalling and cuticle protein in *Bombyx mori*. *RNA Biol.* **2020**, *17*, 1342–1351. [CrossRef]

126. Mahlapuu, M.; Håkansson, J.; Ringstad, L.; Björn, C. Antimicrobial Peptides: An Emerging Category of Therapeutic Agents. *Front. Cell. Infect. Microbiol.* **2016**, *6*, 194. [CrossRef]

127. Furukawa, S.; Tanaka, H.; Ishibashi, J.; Imanishi, S.; Yamakawa, M. Functional Characterization of a Cactus Homolog from the Silkworm *Bombyx mori*. *Biosci. Biotechnol. Biochem.* **2009**, *73*, 2665–2670. [CrossRef]

128. Park, J.W.; Yu, J.H.; Kim, S.W.; Kweon, H.Y.; Choi, K.H.; Kim, S.R. Enhancement of antimicrobial peptide genes expression in Cactus mutated *Bombyx mori* cells by CRISPR/Cas9. *Int. J. Ind. Entomol.* **2018**, *37*, 21–28.

129. Gan, B.H.; Gaynord, J.; Rowe, S.M.; Deingruber, T.; Spring, D.R. The multifaceted nature of antimicrobial peptides: Current synthetic chemistry approaches and future directions. *Chem. Soc. Rev.* **2021**, *50*, 7820–7880. [CrossRef]

130. Yang, H.; Wang, M.; Zhang, P.; Sabhat, A.; Malik, F.A.; Bhaskar, R.; Zhou, F.; Li, X.H.; Hu, J.-B.; Sun, C.G.; et al. Cloning and characterization of the *Bombyx mori* ecdysone oxidase. *Arch. Insect Biochem. Physiol.* **2011**, *78*, 17–29. [CrossRef]

131. Li, Z.; You, L.; Zeng, B.; Ling, L.; Xu, J.; Chen, X.; Zhang, Z.; Palli, S.R.; Huang, Y.; Tan, A. Ectopic expression of ecdysone oxidase impairs tissue degeneration in *Bombyx mori*. *Proc. R. Soc. B Biol. Sci.* **2015**, *282*, 20150513. [CrossRef]

132. Yamamoto, K.; Nagaoka, S. Identification and localization of a novel ecdysone oxidase in the silkworm, *Bombyx mori*. *J. Insect Biotechnol. Sericol.* **2017**, *53*, 49–53.

133. Tettamanti, G.; Casartelli, M. Cell death during complete metamorphosis. *Philos. Trans. R. Soc. B Biol. Sci.* **2019**, *374*, 20190065. [CrossRef] [PubMed]

134. Zhang, Z.; Liu, X.; Shiotsuki, T.; Wang, Z.; Xu, X.; Huang, Y.; Li, M.; Li, K.; Tan, A. Depletion of juvenile hormone esterase extends larval growth in *Bombyx mori*. *Insect Biochem. Mol. Biol.* **2017**, *81*, 72–79. [CrossRef] [PubMed]

135. Liu, Y.; Ma, S.; Chang, J.; Zhang, T.; Chen, X.; Liang, Y.; Xia, Q. Programmable targeted epigenetic editing using CRISPR system in *Bombyx mori*. *Insect Biochem. Mol. Biol.* **2019**, *110*, 105–111. [CrossRef]

136. Xing, W.; Ma, S.; Liu, Y.; Xia, Q. CRISPR/dCas9-mediated imaging of endogenous genomic loci in living *Bombyx mori* cells. *Insect Sci.* **2019**, *27*, 1360–1364. [CrossRef] [PubMed]

137. Brady, D.; Saviane, A.; Cappellozza, S.; Sandrelli, F. An Efficient Workflow for Screening and Stabilizing CRISPR/Cas9-Mediated Mutant Lines in *Bombyx mori*. *Methods Protoc.* **2020**, *4*, 4. [CrossRef]

138. Nartey, M.A.; Sun, X.; Qin, S.; Hou, C.X.; Li, M.W. CRISPR/Cas9-based knockout reveals that the clock gene timeless is indispensable for regulating circadian behavioral rhythms in *Bombyx mori*. *Insect Sci.* **2020**, *28*, 1414–1425. [CrossRef] [PubMed]

139. Yu, Y.; Liu, X.; Ma, X.; Zhang, Z.; Wang, T.; Sun, F.; Hou, C.; Li, M. A palmitoyltransferase Approximated gene Bm-app regulates wing development in *Bombyx mori*. *Insect Sci.* **2020**, *27*, 2–13. [CrossRef] [PubMed]

140. Zeng, B.; Huang, Y.; Xu, J.; Shiotsuki, T.; Bai, H.; Palli, S.R.; Huang, Y.; Tan, X.A. The FOXO transcription factor controls insect growth and development by regulating juvenile hormone degradation in the silkworm, *Bombyx mori*. *J. Biol. Chem.* **2017**, *292*, 11659–11669. [CrossRef]

141. Hong, J.W.; Jeong, C.Y.; Yu, J.H.; Kim, S.B.; Kang, S.K.; Kim, S.W.; Kim, N.S.; Kim, K.Y.; Park, J.W. *Bombyx mori* kynurenine 3-monooxygenase gene editing and insect molecular breeding using the clustered regularly interspaced short palindromic repeat/CRISPR associated protein 9 system. *Biotechnol. Prog.* **2020**, *36*, e3054. [CrossRef] [PubMed]

142. Xu, X.; Zhang, Z.; Yang, Y.; Huang, S.; Li, K.; He, L.; Zhou, X. Genome editing reveals the function of Yorkie during the embryonic and early larval development in silkworm, *Bombyx mori*. *Insect Mol. Biol.* **2018**, *27*, 675–685. [CrossRef]

143. Zhang, Z.J.; Liu, X.J.; Yu, Y.; Yang, F.Y.; Li, K. The receptor tyrosine kinase torso regulates ecdysone homeostasis to control developmental timing in *Bombyx mori*. *Insect Sci.* **2020**, *28*, 1582–1590. [CrossRef] [PubMed]

144. Xu, X.; Wang, Y.; Liu, Z.; Wang, Y.; He, L.; Li, K.; Huang, Y. Disruption of egg-specific protein causes female sterility in *Bombyx mori*. *Insect Sci.* **2021**, 1–11. [CrossRef]
145. Zhu, L.; Mon, H.; Xu, J.; Lee, J.M.; Kusakabe, T. CRISPR/Cas9-mediated knockout of factors in non-homologous end joining pathway enhances gene targeting in silkworm cells. *Sci. Rep.* **2015**, *5*, 18103. [CrossRef] [PubMed]
146. Gao, Y.; Liu, Y.C.; Jia, S.Z.; Liang, Y.T.; Tang, Y.; Xu, Y.S.; Kawasaki, H.; Wang, H.B. Imaginal disc growth factor maintains cuticle structure and controls melanization in the spot pattern formation of *Bombyx mori*. *PLoS Genet.* **2020**, *16*, e1008980. [CrossRef]
147. Sun, L.; Zhang, Z.; Zhang, R.; Yu, Y.; Yang, F.; Tan, A. Molecular Disruption of Ion Transport Peptide Receptor Results in Impaired Water Homeostasis and Developmental Defects in *Bombyx mori*. *Front. Physiol.* **2020**, *11*, 424. [CrossRef] [PubMed]
148. Liu, Z.L.; Xu, J.; Ling, L.; Zhang, R.; Shang, P.; Huang, Y.P. CRISPR disruption of TCTP gene impaired normal development in the silkworm *Bombyx mori*. *Insect Sci.* **2018**, *26*, 973–982. [CrossRef] [PubMed]
149. Zhang, Z.; Zhang, S.S.; Niu, B.L.; Ji, D.F.; Liu, X.J.; Li, M.W.; Bai, H.; Palli, S.R.; Wang, C.Z.; Tan, A.J. A determining factor for insect feeding preference in the silkworm, *Bombyx mori*. *PLoS Biol.* **2019**, *17*, e3000162. [CrossRef]
150. Bi, H.; Xu, X.; Li, X.; Zhang, Y.; Huang, Y.; Li, K.; Xu, J. CRISPR Disruption of BmOvo Resulted in the Failure of Emergence and Affected the Wing and Gonad Development in the Silkworm *Bombyx mori*. *Insects* **2019**, *10*, 254. [CrossRef]
151. Fujii, T.; Banno, Y. Enlargement of egg size by CRISPR/Cas9-mediated knockout of a sex-linked gene in the silkworm, *Bombyx mori*. *J. Insect Biotechnol. Sericol.* **2018**, *87*, 3_071–3_078. [CrossRef]
152. Zhang, Z.; Aslam, A.F.; Liu, X.; Li, M.; Huang, Y.; Tan, A. Functional analysis of Bombyx Wnt1 during embryogenesis using the CRISPR/Cas9 system. *J. Insect Physiol.* **2015**, *79*, 73–79. [CrossRef] [PubMed]
153. Li, K.; Tian, L.; Guo, Z.; Guo, S.; Zhang, J.; Gu, S.; Palli, S.R.; Cao, Y.; Li, S. 20-Hydroxyecdysone (20E) Primary Response Gene E75 Isoforms Mediate Steroidogenesis Autoregulation and Regulate Developmental Timing in *Bombyx*. *J. Biol. Chem.* **2016**, *291*, 18163–18175. [CrossRef] [PubMed]
154. Liu, Q.; Liu, W.; Zeng, B.; Wang, G.; Hao, D.; Huang, Y. Deletion of the *Bombyx mori* odorant receptor co-receptor (*BmOrco*) impairs olfactory sensitivity in silkworms. *Insect Biochem. Mol. Biol.* **2017**, *86*, 58–67. [CrossRef] [PubMed]
155. Baddeley, H.J.E.; Isalan, M. The Application of CRISPR/Cas Systems for Antiviral Therapy. *Front. Genome Ed.* **2021**, *3*, 745559. [CrossRef] [PubMed]
156. Murphy, B.G.; Wolf, T.; Vogel, H.; Castillo, D.; Woolard, K. An RNA-Directed Gene Editing Strategy for Attenuating the Infectious Potential of Feline Im-

munodeficiency Virus-Infected Cells: A Proof of Concept. *Viruses* **2020**, *12*, 511. [CrossRef] [PubMed]

157. Fareh, M.; Zhao, W.; Hu, W.; Casan, J.M.; Kumar, A.; Symons, J.; Zerbato, J.M.; Fong, D.; Voskoboinik, I.; Ekert, P.G.; et al. Reprogrammed CRISPR-Cas13b suppresses SARS-CoV-2 replication and circumvents its mutational escape through mismatch tolerance. *Nat. Commun.* **2021**, *12*, 1–16. [CrossRef]
158. Yao, Q.; Li, M.W.; Wang, Y.; Wang, W.B.; Lu, J.; Dong, Y.; Chen, K.P. Screening of molecular markers for NPV resistance in *Bombyx mori* L. (Lep., Bombycidae). *J. Appl. Entomol.* **2003**, *127*, 134–136. [CrossRef]
159. Guo, H.; Zhang, B.; Zheng, X.; Sun, J.; Guo, H.; Li, G.; Zhao, G.; Xu, A.; Qian, H. Pathogenicity Detection and Genome Analysis of Two Different Geographic Strains of BmNPV. *Insects* **2021**, *12*, 890. [CrossRef]
160. Isobe, R.; Kojima, K.; Matsuyama, T.; Quan, G.X.; Kanda, T.; Tamura, T.; Sahara, K.; Asano, S.I.; Bando, H. Use of RNAi technology to confer enhanced resistance to BmNPV on transgenic silkworms. *Arch. Virol.* **2004**, *149*, 1931–1940. [CrossRef]
161. Qiong, Y.; Dong Xu, X.; Qing Rong, L.; Yang, X.; Ming Qiang, Y. Standard method for detecting *Bombyx mori* nucleopolyhedrovirus disease-resistant silkworm varieties. *Rev. Bras. Entomol.* **2017**, *62*, 19–22. [CrossRef]
162. Chen, S.; Hou, C.; Bi, H.; Wang, Y.; Xu, J.; Li, M.; James, A.A.; Huang, Y.; Tan, A. Transgenic Clustered Regularly Interspaced Short Palindromic Repeat/Cas9-Mediated Viral Gene Targeting for Antiviral Therapy of *Bombyx mori* Nucleopolyhedrovirus. *J. Virol.* **2017**, *91*, e02465-16. [CrossRef] [PubMed]
163. Dong, Z.; Hu, Z.; Qin, Q.; Dong, F.; Huang, L.; Long, J.; Chen, P.; Lu, C.; Pan, M. CRISPR/Cas9 Mediated Disruption of the immediate early-0 and 2 as a Therapeutic Approach to *Bombyx mori* Nucleopolyhedrovirus in Transgenic Silkworm. *Insect Mol. Biol.* **2018**, *28*, 112–122. [CrossRef] [PubMed]
164. Dong, Z.; Qin, Q.; Hu, Z.; Chen, P.; Huang, L.; Zhang, X.; Tian, T.; Lu, C.; Pan, M. Construction of a One-Vector Multiplex CRISPR/Cas9 Editing System to Inhibit Nucleopolyhedrovirus Replication in Silkworms. *Virol. Sin.* **2019**, *34*, 444–453. [CrossRef] [PubMed]
165. Xu, X.; Bi, H.; Wang, Y.; Li, X.; Xu, J.; Liu, Z.; He, L.; Li, K.; Huang, Y. Disruption of the ovarian serine protease (Osp) gene causes female sterility in *Bombyx mori* and Spodoptera litura. *Pest Manag. Sci.* **2019**, *76*, 1245–1255. [CrossRef]
166. Sledzinski, P.; Nowaczyk, M.; Olejniczak, M. Computatonal Tools and Resources Supporting CRISPR-Cas Experiments. *Cells* **2020**, *9*, 1288. [CrossRef]

7 Chapter

A SYSTEMATIC REVIEW OF THE NUTRITIONAL COMPOSITION OF BOMBYX MORI PUPAE

1. INTRODUCTION: AN OVERVIEW

The sole completely domesticated insect cultivated by humans is the mulberry silk-worm (Bombyx mori L., 1758), originally bred in captivity approximately 7500 years ago. Its capacity to produce silk is what confers its importance, although it also serves as a source of food or feed, a pet, and a model organism in scientific research [2]. Moreover, the silkworm has been utilized as a functioning bioreactor to produce various compounds, mostly those having therapeutic significance. [3].

Pupae are the most commonly employed life stage for food or feed since they make up around 60% of the dry cocoon weight [4] and are the principal by-product of the silk industry. They can be used as food or feed, which helps to support circular production chains and prevent the release of potentially harmful waste into the environment. The consumption of silkworm pupae is already widespread in Asia, particularly in India, China, Japan, South Korea, and Thailand [5–9], where they are valued as street food, or as a nutraceutical in traditional medicine [10–12].

Before consumption, it is important to weigh the dangers associated with allergens, microorganisms, or chemicals, in addition to the nutritional advantages of silkworm pupae. Regarding Europe, an EFSA (the European Food Safety Agency) opinion [13] explored the possibility of silkworms as a novel food, reflecting the growing interest in this topic in Western nations.

This systematic review focused on assessing and summarizing the literature about the nutritional composition of B. mori pupae, henceforth referred to as SP for "silkworm pupae." While there are reviews concerning silkworm composition [14], a systematic review on silkworm pupa composition is absent. The primary difference between this systematic review and others is its comprehensive comparison of all currently available data regarding SP published in the literature during the specified time frame, facilitating an understanding of the potential and limitations of SP as a dietary source. Conversely, we may formulate hypotheses concerning the primary factors that could influence the nutritional value of SP, grounded in a comprehensive comparison of the findings from multiple authors.

Authorities may consider this comprehensive review beneficial for assessing SP and its derivative products as new foods. Comprehending the prospective applications of technology and the effects of substituting traditional nutrient sources in innovative formulations may be advantageous for food producers. This analysis may provide a foundation for future research exploring innovative methods to alter SP composition.

2. MATERIALS AND METHODS

This systematic review seeks to provide a food-based characterization of the nutritional composition of mulberry SP (*B. mori*). All articles published in peer-reviewed journals in the languages of English, French, and Spanish were taken into consideration for this scope at the onset of our study without regard to publication dates.

Using the keywords listed in Table 1, we conducted searches in PUBMED, Web of Science Core Collection, and EMBASE (Title/Abstract, Topic (TS), and Title, Abstract, Author keywords, respectively). The keywords also contained terms pertaining to the allergic and security aspects of SP. The records pertaining to these subjects were located and used for a second systematic review with a focus on these specific concerns. The search was performed on 14 May 2020, and it was updated on 5 May 2021, with a time limit beginning on 1 January 2020, to include all the documents published in the meantime. We utilized the software EPPI-4 Reviewer [15] to examine the retrieved records.

There were two main phases to the selection process. Six researchers (S.B., F.M., A.P., L.T., A.S., and S.C.) conducted the first phase using a double-screening approach, in which two researchers separately classified each record to determine whether it related to the study topic "composition" or "safety" and to determine whether it was a review.

This step's inclusion criteria were as follows: (1) the language used had to be either English, French, or Spanish; (2) the data had to be taken directly from research publications and not from other reviews. All the records with unclear titles or abstracts were included for evaluation in the following stage. Based on the titles and abstracts, the selected studies were then split into two groups: those dealing with nutritional composition and those dealing with biological/chemical risk related to SP use. Three reviewers (S.B., F.M., and P.A.) did this classification.

The second phase concentrated on original papers addressing SP composition. The task involved classifying and evaluating each paper's data typology in full text for each record. Then, pertinent information was retrieved and listed in tables. We specifically found publications that discussed the following topics: (1) the overall chemical composition (macronutrients and ash), (2) the AA profile, (3) the FA profile, and (4) the composition of the macro and microelements.

Table 1. Keywords employed to retrieve relevant records reporting data on silkworm food safety and composition. The first keyword refers to the species and any alternate names it may have. The second keyword refers to all the terms related to nutritional composition that have been investigated in this review, while the last keyword refers to food safety terms. The first keyword was searched in conjunction with the second or third, linking them with logical operators.

Keywords (Title/Abstract)				
Bombyx OR Silkworm	AND	nutrition OR composition OR centesimal OR nutrient OR nutrients OR protein OR proteins peptide OR peptides OR aminoacid OR aminoacids OR "amino acid" OR "amino acids" OR acid OR acids OR polypeptide OR polypeptides OR fat OR fats OR lipid OR lipids OR "fatty acid" OR "fatty acids" OR "fatty alcohols" OR sugar OR sugars OR carbohydrate OR carbohydrates OR disaccharide OR disaccharides OR monosaccharide OR monosaccharides OR polysaccharide OR polysaccharides OR ash OR mineral OR minerals OR macronutrient OR macronutrients OR micronutrient OR micronutrients OR oligoelement OR oligoelements OR microelement OR microelements OR vitamin OR vitamins OR oil OR oils OR "trace element" OR "trace elements"	OR	safety OR hazard OR hazards OR risk OR risks OR microorganism OR microorganisms OR pathogen OR pathogens OR contaminant OR contaminants OR contamination OR contaminations OR chemical OR chemicals OR toxic OR toxics OR toxicity OR metal OR metals OR toxin OR toxins OR allergy OR allergies OR allergen OR allergens OR allergic OR allergenic OR sensitization OR sensitisation OR cross-reactivity OR anaphylactic OR anaphylaxis OR poisoning OR poison OR compound OR compounds OR pesticide OR pesticides OR residual OR residue OR residues OR antibiotic OR antibiotics OR antiparasitic OR antiparasitics OR mycotoxin OR mycotoxins OR dioxin OR dioxins OR polluting OR pollutant OR pollutants

The following additional exclusion criteria were applied:

1. Research involving pupae as feed;
2. Research describing the composition of the larval stage;
3. Research published before the year 2000.

As a result, the entire time frame of the systematic review is from 1 January 2000, through 5 May 2021. Additionally, we excluded data from food labeling when SP had been purchased from suppliers as food and only kept original data from the selected trials that were gathered in the research through an analytical method. Furthermore, we did not include information on SP fed a particular diet supplement. The information obtained from the investigations was arranged into pre-defined tables. Three researchers conducted this second stage (L.T., A.S., and S.C.). Information on silkworm strain, feeding (mulberry leaves or artificial diet), and sample processing was provided when available.

Insect-killing techniques, sample storage conditions, and drying technologies used before the analytical stages are among the sample processing variables that may have an impact on compositional results. Another potential source of variation was the analytical approach itself, which varied throughout the articles, particularly for fiber and ash. The existence of experimental replicates and the standard deviation, which were mentioned in this review if they were accessible in the original manuscripts, are additional factors to take into account. The relative standard deviation was estimated in cases when the data were required to be summed up or standardized using the principles of error propagation.

Data about chemical composition were extracted and normalized, if necessary, to be comparable among themselves. This included information about total nitrogen or protein content, ash, lipid or fat content, crude fiber, nitrogen-free extracts (NFE), amino acid (AA) profile, fatty acid (FA) profile, and oligo-elements. All these extracted data were reported in Table 2.

Only a small percentage of the selected papers stated the composition on a wet basis, necessitating the calculation of the composition on a dry matter basis using the moisture values. Therefore, the nutritional values were normalized according to the dry matter content and expressed as a percentage or "g per 100 g of dry matter". Protein content was also expressed using the nitrogen-to-protein conversion factor suggested by Janssen et al. [16].

To normalize and compare data given in articles included for the final review, amino- acid absolute values for the AA profile were transformed into percentage values (Table 3). The FA profile is shown in Table 4 as a percentage of all FAME (Fatty Acid Methyl Esters). To make data comparison easier, the isomeric forms of the same FA were combined together. The last step was to express the macro and microelements as mg/g on a dry matter basis (Table 5). According to the principles of error propagation, the standard deviation of the data produced by elaboration (addition or normalization of the dry matter) of the original values was determined.

3. RESULTS

After removing duplicates and reading the abstracts of all 14,008 papers that were found using our search parameters and logical operators, 385 of them were retained for additional analysis. Among them, 185 of the chosen studies addressed safety issues, and 156 dealt with chemical composition. As previously said, the former were taken into consideration for a second evaluation of safety issues. Twenty-six papers out of 156 dealing with chemical composition met our inclusion criteria and were utilized to gather data on nutritional information (Figure 1).

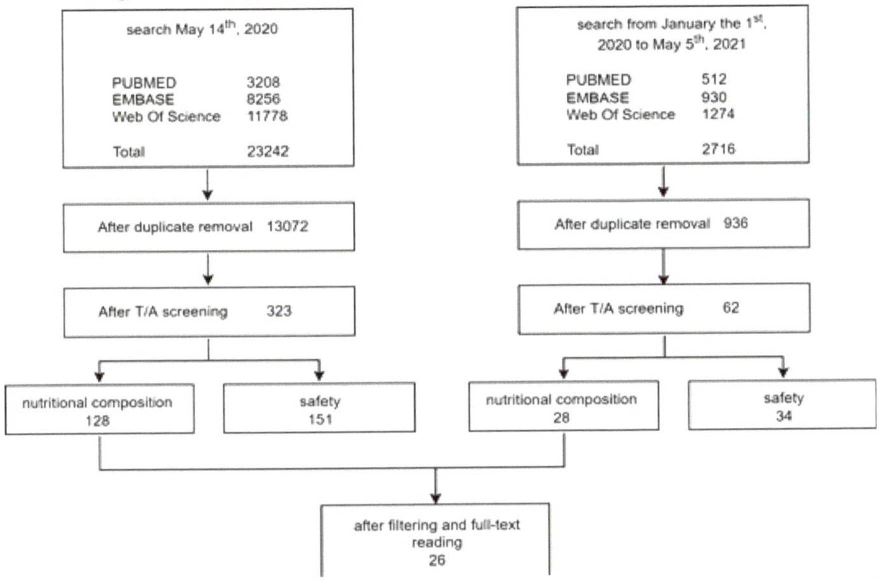

Figure 1. Representation of the procedure used to review and select the studies from which to extract the data for the current systematic review.

3.1 Macronutrients and Ash

A total of thirty-three data entries were retrieved from 16 studies reporting at least one nutrient value (Table 2). Two data entries referring to protein values on fresh weight [17] were not considered due to the absence of the moisture value that would have made the two values comparable. The two main types of reported data were "full-fat" SP and "defatted" SP. The values of protein (n = 33), lipids (n = 26), ash (n = 25), nitrogen-free extracts (n = 17), and crude fibers (n= 16) were reported in the 33 data entries.

An average nitrogen-to-protein conversion factor of 6.25 was used to calculate the protein values in each study. The corrected nitrogen-to-protein conversion factor of 4.76 proposed by Janssen et al. [16] was only adopted by Akande et al. [18] and Birman et al. [19]. The nitrogen- to-protein conversion factor has been "insect adjusted," and the protein values are as reported in the original studies shown in Table 2. The results discussed below, for simplicity's sake, apply to protein values obtained using the standard nitrogen conversion factor of 6.25. In three studies including defatted SP, protein levels ranged from 67.5 [20] to 82.9% [21], while residual lipids ranged from less than 0.2 [22] to 4.75% [20].

Table 2. Macro nutrient values and main processing variables (g/100 g) retrieved from selected articles. Data were normalized on the dry matter where necessary. * Protein values were originally calculated using Kp = 4.76 as a nitrogen conversion factor. ** the value were indicated in the original paper as "total fiber".

Reference	Treatment, Strain, Diet	Separa-tion	Protein (Kp = 6.25)	Protein (Kp = 4.76)	Fat	Crude Fiber	NFE (Often Indicates as Carbo-hydrates in the Original Paper)	Ash
Mishra et al., 2003 [5]	Cocoons were boiled at 100° C for 30 min. Pupae were taken out and dissected to remove the intestine and waste material. Composition was expressed on wet weight (100 g).	full-fat	34.38 ± 0.25	26.19 ± 0.19	57.64 ± 0.22	0.54 ± 00	5.16 ± 0.19	2.27 ± 0.17
Akande et al., 2020 [18]	Edible insects boiled without seasonings were prepared with 100 g of each of the edible insects with 150 mL of potable water and boiled for 15 min. Data are referred to as boiled silkworm pupae without seasoning (SWOS).	full-fat	47.17 ± 1.36	35.92 ± 1.04 *	32.16± 0.85	1.68 ± 0.02	28.15 ± 1.23	2.12 ± 0.03
David-Birman et al., 2019 [19	Finely milled silkworm pupae flour (SWF) was purchased from JR Unique Foods Ltd. (Udon Thani, Thailand).	full-fat	53.05 ± 0.09	40.4 ± 0.07 *	29.56 ± 0.08	n/a	n/a	n/a

Kim et al., 2016 [20]	Dried, untreated SP flour. The whole insects were cleaned with distilled water, sieved, ground using a hammer mill, and passed through a 20-mesh sieve.		full-fat	47.87 ± 1.72	36.46 ± 1.31	30.85 ± 3.81	6.38 ± 0.32 **	n/a	6.96 ± 0.61
		SP defatted using hexane. The insect flour was also acid-hydrolyzed and dried in the dry oven at 70° C	defatted	67.54 ± 0.97	51.44 ± 0.74	4.75 ± 1.79	9.54 ± 0.58 **	n/a	9.54 ± 1.22
		SP defatted using hexane. The insect flour was also acid-hydrolyzed and dried in the dry oven at 70° C	defatted	73.51 ± 0.74	55.99 ± 0.56	2.05 ± 0.5	3.24 ± 0.19 **	n/a	8.04 ± 0.18
			full-fat	56.64 ± 0.42	43.14 ± 0.32	34.07 ± 0.62	n/a	n/a	n/a
		The sample was further dried at 60 C for 18 h. Dried samples were ground and packed in an aluminum foil bag until use	full-fat	52.69 ± 0.46	40.13 ± 0.35	36.21 ± 0.23	n/a	n/a	n/a

Anoottha-to et al., 2019 [21]	Frozen silkworm pupae (20° C) were thawed at 4° C over-night and washed three times before blanching at 95° C for 10 min.	The sample was further dried at 60 C for 18 h, ground, and packed in an aluminum foil bag until use. Dried samples were mixed and defat-ted using 99.8% ethanol (1:3 w/v) at 50 ∘ C with stirring for 30 min. Then, each sample was filtered. Defatting was carried out twice before drying in the tray dryer at 50° C for 3 h. Defatted samples were ground to 150 μm before hydrolysis	defat-ted	82.88 ± 0.5	63.12 ± 0.38	1.47 ± 0.44	n/a	n/a	n/a
Felix et al 2020 [22]	The silkworm protein concentrate (SPC) used in this study was supplied ® SPC was defatted using n-hexane		full-fat	55.43± 1.10	42.22 ± 0.84	31.72 ± 0.22	n/a	5.60 ± 1.42	7.24± 0.88
			defat-ted	81.14± 2.06	61.80 ± 1.57	<0.2	n/a	8.23 ± 2.04	10.63 ± 1.27
Akande et al., 2020[23	Freshly harvested mulberry silkworm pupae (SWP) were stifled in an oven at 93° C for 1 h, cut out of their cocoons and dried in a hot air oven (Gen. Lab. oven)at 40°C for 8 h. The dried SWP was milled into flour using an electric blender (Binatone, Model No. 51-777) and kept airtight. The data are referred to as silkworm pupae powder(SWP)		full-fat	60.7± 0.2	46.23 ± 0.15	23.5 ± 0.21	1.1 ± 0.16	11.3 ± 0.42	0.9± 0.56
Kimetal., 2016[24]	Thedriededibleinsectswerepurchasedfromthree vendorslocatedin Chungnam,Korea.Nutrientswere originallyexpressedasg/kgdryweight.		full-fat	52.58± 0.36	40.04± 0.27	19.21 ± 0.34	n/a	23.41± 0.01	4.8± 0.71
Kuntadi et al., 2018 [25]	The SP were oven dried at 60–70 °C for 12–24 h and then ground.		full-fat	60.03	45.72	29.47	n/a	0.92	5.79

Lambertietal., 2019[26]	SP were collected 7 days after reaching the cocoon stage. Pupae were lyophilized and stored at -80 °C.	male silkworm reared on an artificial diet	full-fat	61.45	46.80	n/a	n/a	n/a	5.68
		male silkworm reared on leaf	full-fat	55.02	41.90	n/a	n/a	n/a	4.62
		female silkworm reared on an artificial diet	full-fat	72.36	55.11	n/a	n/a	n/a	5.46
		female silkworm reared on leaf	full-fat	56.29	42.87	n/a	n/a	n/a	5.02
Rodri-guez-Or-tega et al., 2016 [27]	Dried pupae. Larvae reared on mulberry tree leaves.		full-fat	64.31 ± 0.00 (true protein = 46.87 ± 0.00)	48.98 ± 0.00	20.63 ± 0.105	4.89 ± 0.107	3.92	6.25 ± 0.017
Tomotake et al., 2010 [28]	SP were purchased from Nishiki Food Ltd. (Nagano, Japan). SP were lyophilized followed by grinding into a fine powder.		full-fat	55.60	42.34.00	32.20	n/a	n/a	n/a
Pereira et al., 2003 [29]	Silkworm pupae fed with younger leaves of mulberry tree, and dried at 130 ◦C, until 4–5% of moisture.		full-fat	51.1± 1.8	38.92± 1.37	34.4± 0.8	n/a	n/a	3.64± 0.09
Ghosh et al., 2020 [30]	Pupae of B. mori were ground in a mortar and pestle and centrifuged adding some distilled water. Then the liquid centrifugate was spray-dried and ground into fine powdered material.		full-fat	55.87 ± 1.16	42.55 ± 0.88	23.45 ± 0.28	3.77 ± 1.16	1.76 ± 1.19	5.38 ± 0.07
Mishyna et al., 2020 [31]	Insects were stored at -20 °C prior to grinding or drying. After drying, the insects were ground in a food blender and stored in a double plastic bag at -20 °C prior to use, but no longer than 1 month.	freeze-dried	full-fat	60.29 ± 0.38	45.91 ± 0.29	n/a	n/a	n/a	n/a
		oven-dried	full-fat	59.33 ± 0.12	45.19 ± 0.09	n/a	n/a	n/a	n/a
		microwavedried	full-fat	61.57 ± 0.21	46.89 ± 0.16	n/a	n/a	n/a	n/a

Hirunyo-phat etal., 2021[32]	Frozen and stored at -18 °C.	Nangnoi (NN)strain	full-fat	54.83± 0.43	41.76± 0.33	23.61± 0.56	2.67± 0.06	14.99± 1.05	3.99± 0.11
		Siwtui (ST)strain	full-fat	55.39± 0.19	42.19± 0.14	21.26± 0.92	2.91± 0.92	17.45± 0.83	3.12± 0.06
		LuangSaraburi (LS)strain	full-fat	50.36± 0.21	38.35± 0.16	27.59± 0.43	3.29± 0.07	13.42± 0.43	5.41± 0.09
		Ubon Ratchathani 60–35(U60–35) strain	full-fat	48.31± 0.55	36.79± 0.42	29.93± 0.85	2.81± 0.03	14.87± 0.08	4.23± 0.47
	Frozen and stored at 18 ° C, thawed at 5 ° C for 12 h, and dried at 60 ° C for 16 h; ground to pass through 177-micron meshsieve;storedat -18 ° C;composition dryweight.	Nangnoi (NN)strain	full-fat	52.94± 0.11	40.32 ± 0.08	25.53 ± 0.01	4.78 ± 0.06	12.00 ± 0.09	4.69± 0.05
		Siwtui (ST)strain	full-fat	56.00± 0.4	42.65 ± 0.3	23.95 ± 0.15	5.58 ± 0.09	10.78 ± 0.7	3.94± 0.02
		Luang Saraburi (LS) strain	full-fat	49.19± 0.26	37.46 ± 0.2	30.47 ± 0.33	5.46 ± 0.14	9.93 ± 0.76	4.73± 0.06
		Ubon Ratchathani 60–35(U60–35) strain	full-fat	46.22± 0.15	35.2 ± 0.11	34.69 ± 0.27	4.92 ± 0.19	9.82 ± 0.03	4.25± 0.02

The principal method utilized to remove lipids from the samples was chemical extraction, utilizing ethanol or hexane as the organic solvent. Full-fat SP was the subject of 16 investigations, yielding a total of 31 data entries. Protein levels in full-fat SP ranged from 34.4 [5] to 72.4% [26]. The lowest value was reported by Mishra et al. [5] and related to SP that had been boiled, dissected, and had the inner content removed. Lamberti et al. [26] discovered female pupae raised on an artificial diet displayed the greatest protein value. Rodriguez-Ortega et al. [27] determined that the highest protein value was 64.3% when only pupae raised on mulberry leaves were used. Lamberti et al. [26] also investigated the protein differences between males and females, and the results highlighted the highest values for females. When microwave-dried SP was compared to oven-dried SP, freeze-dried SP, and other types of dried SP, Mishyna et al. [31] discovered that microwave-dried SP had a considerably higher protein value.

The total lipid content ranged from 19.2 [24] to 57.6% [5] and was available for 22 data entries pertaining to full-fat SP. The majority of the research that dealt with lipid extraction methods used standard laboratory extraction procedures using n-hexane, chloroform, methanol, and ethanol as the extraction solvents.

It should be noted that the study by Wei et al., was the only one to investigate supercritical CO_2 as the extraction solvent [33], while Hu et al. employed a microwave-assisted extraction using a mixture of ethanol and n-hexane as the extraction solvent [4]. The crude fiber content of full-fat pupae ranged from 0.54 [5] to 6.38% [24]. For this nutrient, only 14 data entries were available. Nitrogen-free extract content was provided for 17 data entries and ranged from 0.92% [25] to 28.2% [18]. In most of the studies, NFE was referred to as "carbohydrate" and calculated by difference. Ash content ranged from 0.9 [23] to 7.94% [22] for 22 data entries.

3.2 Amino Acid Profile

Table 3 presents a total of 11 data entries from six original investigations about the AA profile. The relative abundance of each AA was given as a percentage of the total amount, albeit the values were biased because only three studies quantified all the AA. A complete comparison was impossible since two studies failed to provide the tryptophan value and one failed to provide tryptophan, cysteine, and methionine values [21,30,31].

We further assumed that asparagine and glutamine values were included in aspartic and glutamic acid values, respectively, as a result of the widespread AA measurement methods [34]. However, research by Kwon et al. [35] emphasized how the value of certain AA varied over the course of pupae development; for instance, histidine values declined with pupal age while threonine and serine values increased. There are a few differences between the AA reported in this study for pupae that are 12–13 days old (D12–13) and for pupae that are 13 days old (D–13), but the comparison was not possible since the size and variability of the samples are unknown.

3.3 Fatty Acid Profile

Seventeen data points about the FA profiles were taken from nine papers and reported in Table 4. The single FA appears to be influenced by a number of variables, including SP strain, nutrition, pupal age, sex, drying and extraction procedure, and analytical method- ologies. According to research by Kwon et al. [35] on the influence of pupal age on the composition of fatty acids, linoleic acid was the most variable FA between days 10 and 13, when its relative abundance varied from 5.7 to 8.7% total FAME.

Table 3. Amino acid composition. When available in the original data, values given as a percentage of the total AA along with their

Reference		Khoeler et al., 2019 [17]		Akande et al., 2020 [18]	Anootthato et al., 2019 [21]	Tomotake et al., 2010 [28]	Kwon et al., 2012 [36]				Shi et al., 2018 [37]	
Treatment		SH (ready to-eat, deep fried in cooking oil. Samples were frozen and freeze-dried for 24 h, powdered andco served at -80ºC)	SM (non-fried snack, cooked using steam and hot air. Samples were frozen and freeze-dried for 24 h, powdered and conserved at -80ºC)	SP boiled without seasonings, blended	Frozen SP were thawed at 4ºC overnight and washed three times before blanching at 95ºC for 10 min. Then, the sample was dried in a tray dryer at 60 ºC for 18 h	SP were lyophilized, followed by grinding into a powder	Frozen SP were extracted and then the protein extract was hydrolyzed with hydrochloric acid at 110ºC for 24 h				SP protein hydrolysate, obtained through enzymatic hydrolysis	
							Pupal age: Day 12–13	Pupal age: Day D7	Pupal age: Day D10	Pupal age: Day D13	Spray dried	Freeze dried
Amino acid	Val	5.95	5.99	4.94 ± 0.04	6.18 ± 0.12	5.53	5.7	4.2	4	2.8	5.27	4.86
	Thr	4.82	4.72	4.62 ± 0.06	5.08 ± 0.07	4.59	3.3	4.6	8.5	9.7	3.89	5.19
	Ile	4.16	4.35	4.92 ± 0.09	4.84 ± 0.10	4	2.7	2.3	1.8	2.1	4.27	2.61
	Leu	7.44	7.39	8.16 ± 0.07	8.16 ± 0.07	7.29	3.5	2.4	2.1	2.5	7.07	6.04
	Lys	7.4	7.19	9.00 ± 0.05	9.09 ± 0.21	7.18	4.5	2.9	3.1	3.9	7.78	7.93
	Met	3.37	4.01	2.28 ± 0.22	-	4	5	6.5	6.4	5	2.42	2.3
	Phe	5.04	5.71	4.21 ± 0.07	5.25 ± 0.02	5.41	3	2.6	1.4	1.6	7.2	7.33

Amino acid	His	2.98	3.04	2.63 ± 0.06	3.25 ± 0.02	3.18	14.6	25.1	16	13.8	3.16	3.66
	Trp	1.49	1.55	1.28 ± 0.06	-	1.76	-	-	-	-	-	-
	Glu	10.99	11.31	17.02 ± 0.46	13.58 ± 0.21	11.18	18.3	18.9	21.5	18.1	12.44	12.68
	Gln	-	-	-	-	-	-	-	-	-	-	-
	Gly	4.6	4.18	5.16± 0.04	4.51± 0.05	4.24	7.3	5.2	7	6.1	8.73	10.09
	Ala	5.47	4.85	7.85± 0.04	5.80± 0.07	4.59	10.2	6.4	3.8	5.1	5.14	4.77
	Ser	4.95	4.61	4.63± 0.09	5.30± 0.10	4.35	5.2	3.4	9	10.6	3.78	3.1
	Pro	4.55	4.09	3.61± 0.05	4.87± 0.10	8.24	4.1	6.1	4.4	3.4	9.58	9.76
	Asp	11.69	11.48	8.57± 0.04	-	10.71	1.5	1.5	2.2	6	11.23	11.04
	Asn	-	-	-	12.12± 0.31	-	-	-	-	-	-	-
	Arg	5.47	5.77	6.62± 0.06	6.61± 0.10	5.53	3.6	1.3	1.5	2.5	3.25	3.68
	Tyr	8.23	8.46	2.62± 0.41	5.35± 0.14	6.59	6.2	6.2	6.8	6.7	3.79	4.09
	Cys	1.4	1.31	1.90± 0.05	-	1.65	1.5	0.2	0.6	0.4	1	0.88

Table 4. Fatty acid profile (% total FAME) of selected studies and data.

Reference	Treatment		C12:0	C14:0	C15:0	C15:1	C16:0	C16:1, C16:1n-7	C16:2n-6	C16:3n-3	C17:0	C17:1	C18:0	C18:1, C18:1n-7, C18:1n-9	C18:2, C18:2(n-6)	C18:3, C18:3(n-3)	C20:0	C20:1, C20:1n-9	C20:3 (and C20:3n-6, C20:3n-3)	C20:4n-6	C20:5, C20:5n-3	C22:0	C24:0	others
Hu et al., 2017 [4]	BMP were dried in a vacuum drier at 70°C for 24h to a moisture content of <5% and ground into a fine powder. The powder was sieved and stored at 4°C.	Microwave assisted e traction, using a mixed solvent consisting of ethanol and n-hexane (1:1, v/v)	-	0.18 ± 0	-	-	23.18 ± 0.52	1.07 ± 0.09	-	-	0.15 ± 0	0.1 ± 0	4.69 ± 0.17	28.32 ± 0.63	3.88 ± 0.13	38.25 ± 0.75	0.16	-	-	-	-	-	-	-
		Soxhlet extraction using n-hexane, at 80°C for 6 h. The oil was dried at 100°C ± 5 °C for 15 min. in a drying oven	-	0.19 ± 0	-	-	23.04 ± 0.58	1.05 ± 0.07	-	-	0.17 ± 0	nd ± nd	4.68 ± 0.19	28.15 ± 0.54	3.85 ± 0.54	38.06 ± 0.68	0.16	-	-	-	-	-	-	-
Tomotake et al., 2010 [28]	Silkworm pupae were lyophilized, followed by grinding into a powder. Lipid content was determined according to the methods of AOAC		-	0.1	-	-	24.2	1.7	-	-	-	-	4.5	26	7.3	36.3	-	-	-	-	-	-	-	-

Pereira et al., 2003 [29]	Chrysalis (worm) toast (originated from the silkworm pupae fed with younger mulberry tree leaves and dried at 130°C until 4–5% of moisture). Lipids were chemically extracted using chloroform-methanol.		-	0.164 ± 0.02	-	-	24.6 ± 2.1	0.656 ± 0.04	-	-	0.192 ± 0.06	-	7.56 ± 1.54	34.8 ± 3.3	7.03 ± 1.08	24.4 ± 6.7	-	-	0.275 ± 0.1	0.334 ± 0.08	-	-	-	-
Wei et al., 2009 [33]	Silkworm pupae were vacuum dried at 60°C to a moisture content of <5%, then finely ground to powder, and finally oil was extracted by supercritical CO_2 extraction.		-	-	-	-	21.77	-	-	-	-	-	7.02	7.02 33.26	7.12	27.99	-	-	-	-	-	-	-	2.84
Kwon et al., 2012 [35]	The frozen silkworm pupae were ground, blended, mixed with ethanol and stirred. The sample was filtered, and the particles were extracted with ethyl acetate. The extracts were vacuumconcentrated and dissolved in hexane. Activated carbon was added, and the mixture was heated for 1 h and then cooled and filtered before vacuum	Pupal period days 12–13	-	-	-	-	19.7	2.5	-	-	-	-	8.6	19.9	7.4	41.6	-	-	-	-	0.3	-	-	-
		Pupae at day 7	-	-	-	-	22.6	2	-	-	-	-	9.6	21.1	8.2	36.4	-	-	-	-	-	-	-	-
		Pupae at day 10	-	-	-	-	21	1.7	-	-	-	-	8.3	24.3	5.7	38.9	-	-	-	-	0.1	-	-	-
		Pupae at day 13	-	-	-	-	20	1.8	-	-	-	-	8.8	21.8	8.7	38.6	-	-	-	-	0.3	-	-	-
Chieco et al., 2019 [37]	Cocoons were dried for 3 days at 60 °C. The pupae were extracted and stored at -80 °C. Frozen pupae were milled to a fine powder. Lipids were chemically extracted using a chloroform/ methanol solution according to the Folch's method.	White polyhibrid on artificial diet	-	-	-	-	29.2 ± 0.4	1.5 ± 0	-	-	-	-	10.9 ± 0.1	35.1 ± 0.3	11.4 ± 0.5	11.7 ± 0.3	-	-	-	-	0.3	-	-	-
		Golden nistari on artificial diet	-	-	-	-	25.1 ± 0.3	1.5 ± 0	-	-	-	-	9.3 ± 0.2	39.8 ± 0.3	11.8 ± 0.1	12.3 ± 0.2	-	-	-	-	0.3	-	-	-
		White polyhibrid on mulberry leaf of the Florio cultivar	-	-	-	-	25.2 ± 0.7	0.8 ± 0.1	-	-	-	-	7.2 ± 0.2	35.2 ± 0.3	7.1 ± 0.4	24.3 ± 0.2	-	-	-	-	0.3	-	-	-
		Golden nistari on mulberry leaf of the Florio cultivar	-	-	-	-	21.6 ± 0.4	1.4 ± 0	-	-	-	-	4.8 ± 0.1	32.3 ± 0.1	10.4 ± 0.2	29.3 ± 0.2	-	-	-	-	0.3	-	-	-
Tong et al., 2011 [38]	Silkworms were directly frozen on the third day of the 5th instar and then lyophilized. Lipid content was measured using Soxhlet extraction.		-	-	0.15	-	16.03	0.39	-	-	-	-	9.45	28.12	12.24	31.91	0.7	0.15	-	-	-	0.7	0.15	-

Usub et al., 2008 [39]	Sun-dried pupae. Five grams were grounded and the lipid extracted with 50 mL of chloroform-methanol (2:1, v/v) containing 10 mg L-1 of butylated hydroxytoluene and 0.1 mg L-1 of tricosanoic acid.	0.1 ± 0	0.1 ± 0	0.1 ± 0	0.1 ± 0	17 ± 0.1	0.6 ± 0	-	-	0.2 ± 0.1	-	6.2 ± 0	20.4 ± 0.1	8.8 ± 0.1	45.6 ± 0.2	0.4 ± 0	0.1 ± 0	0.1 ± 0	0.1 ± 0	-	-	-	-
	Solar-tunnel dried pupae. Five grams were ground and the lipid extracted with 50 mL of chloroform-methanol (2:1, v/v) containing 10 mg L-1 of butylated hydroxytoluene and 0.1 mg L-1 of tricosanoic acid.	0.1 ± 0	0.1 ± 0	0.1 ± 0	0.1 ± 0	17 ± 0.1	0.6 ± 0	-	-	0.2 ± 0.1	-	6.2 ± 0	20.4 ± 0.1	8.8 ± 0.1	45.6 ± 0.2	0.4 ± 0	0.1 ± 0	0.1 ± 0	0.1 ± 0	-	-	-	-
Wang et al., 2020 [39]	Silkworm pupae were dried in a vacuum at 60°C for 24 h to a moisture content of <5% and then slightly ground and sieved through a 60-mesh sieve. Lipid extraction was conducted using the Folch and Sloane method.	-	-	-	-	22.04 ± 0.49	0.92 ± 0.01	-	-	-	-	6.84 ± 0.23	33.91 ± 0.08	5.48 ± 0.14	30.81 ± 0.21	-	-	-	-	-	-	-	-

Yu et al., 2018 [40]	Pupae were frozen in liquid nitrogen and stored at -80°C before analysis. Pupae were homogenized and oil was chemically extracted using hexane and boron trifluoridemethanol solution.	Three-day old silkworm pupae, strain Dazao, fed with an artificial diet (Silkmate 2S), Table 2 in the original paper	-	-	-	-	16.6 ± 2.3	0.51 ± 0.04	0.55 ± 0	0.38 ± 0	-	-	11.26 ± 0.7	28.53 ± 1.83	20.54 ± 1.68	21.13 ± 1.7	0.51 ± 0.12	-	-	-	-	-	-	-
		Three-day-old silkworm pupae, strain Dazao, fed with an artificial diet (Silkmate 2S), Table 3 in the original paper	-	-	-	-	15.61 ± 1.89	0.46 ± 0.09	0.46 ± 0.1	0.23 ± 0.17	-	-	12.87 ± 1.4	26.14 ± 1.41	22.24 ± 1.25	21.31 ± 1.21	0.67 ± 0.2	-	-	-	-	-	-	-
		Three-day-old silkworm pupae, strain Dazao, female	-	-	-	-	17.37 ± 0.53	0.53 ± 0.06	0.48 ± 0.06	0.33 ± 0.05	-	-	12.55 ± 1.32	26.63 ± 1.4	21.32 ± 0.53	20.26 ± 0.59	0.53 ± 0.19	-	-	-	-	-	-	-
		Three-day-old silkworm pupae, strain Dazao, male	-	-	-	-	14.34 ± 1.19	0.45 ± 0.08	0.55 ± 0.06	0.36 ± 0.05	-	-	12.54 ± 1.46	25.38 ± 1.72	23.11 ± 1.09	22.57 ± 0.92	0.71 ± 0.11	-	-	-	-	-	-	-
		Three-day-old silkworm pupae, strain 305, female	-	-	-	-	18.44 ± 1.59	1.19 ± 0.57	0.53 ± 0.03	0.29 ± 0.15	-	-	12.93 ± 2.14	28.23 ± 2.08	18.72 ± 1.05	19.14 ± 1.3	0.53 ± 0.18	-	-	-	-	-	-	-
		Three-day-old silkworm pupae, strain 305, male	-	-	-	-	15.23 ± 0.36	1.16 ± 0.58	0.57 ± 0.08	0.38 ± 0.06	-	-	13.12 ± 0.96	25.97 ± 1.57	21.2 ± 1.3	21.54 ± 0.64	0.83 ± 0.38	-	-	-	-	-	-	-

The two extraction techniques—microwave and Soxhlet—used in the study by Hu et al. [4] and the two drying techniques—sun drying and solar tunnel drying—tested by Usub et al. [39] appear to have minimal impact on the final FAprofile. Additionally, the FAprofile is significantly impacted by the use of an artificial diet during the larval stage. The use of an artificial diet, as well as the usage of various silkworm strains, were both examined in the study by Chieco et al. [37]. The former had a significant impact on the SP's FA composition by raising stearic acid and lowering alfa-linolenic acid. Regarding the silkworm strain, it was shown that the pupae of polyhybrid and Nistari (a tropical pure strain) show notable differences in FA profile, with the Nistari strain having the largest concentration of linoleic and linolenic acids.

Two distinct strains were raised on an artificial diet in the study by Yu et al. [40] and the FA composition was assessed in relation to strain and sex. The findings suggest that higher levels of unsaturated FA (UFA) characterize female pupae, and they probably account for the largest quantities of energy devoted to oviposition. According to this study, the investigated strains were similar in terms of FA profile.

When comparing the findings obtained by Yu et al. [40], to those from other research where silkworms were grown on mulberry leaves, it was found that the concentration of the palmitic (C16:0), linolenic (C18:3), and linoleic (C18:2) acids decreased while the content of linoleic (C18:2) increased. All the investigations came to the same conclusion about the five primary FAs that characterized the SP. In fact, the acids palmitic (C16:0), stearic (C18:0), oleic (C18:1), linoleic (C18:2), and alfa-linolenic (C18:3) make up between 97.1 percent [41] and 99.1 percent [39] of total FAME. Among the five FAs, linoleic and alfa-linolenic acids are essential FAs. Their amount fluctuates between 23.1% and 54.4% of total FAME, with pupae grown on an artificial diet having the lowest value [37].

Furthermore, palmitoleic acid (C16:1 n-7) contributes a small but consistent amount to total FA quantity (0.51%–2.00% total FAME). A few FAs were found in single studies. Among unsaturated FA, C16:2n-6 and C16:3n-3 were identified only by Yu et al. [40]. Considering saturated FA, C22:0, and C24:0 were identified only by Tong et al. [38], while C12:0 appears only in the study by Usub et al. [39]. The percentage of total saturated FA (SFA) in total FAME ranges from 24.1 to 40.1%. In particular, the highest values were reached by those silkworms reared on an artificial diet [37].

3.4. Mineral Content

From six studies, a total of seven data entries about mineral content were accumulated and presented in Table 5. Iron, zinc, manganese, magnesium, and calcium were among the minerals found and quantified in the majority of the publications. It is important to note that the magnitudes of the minerals identified in the study by Akande et al. [23] differed greatly from the other studies. This difference could be because pupae were boiled before analysis, and leaking mineral components might have affected the results. Iron concentrations ranged from 2.83 to 4.95 mg/100 g, calcium from 92.0 to 181 mg/100 g, manganese from 1.08 to 2.30 mg/100 g, zinc from 1.39 to 24.4 mg/100 g, and magnesium from 89 to 280 mg/100 g, excluding the values found in the study by Akande et al. [23]. Other minerals were quantified in a few studies and included sodium, which ranged from 29.6 [29] to 363 mg/100 g [17].

Three investigations showed potassium content, which ranged from 477 to 672 mg/100 g. Table 5 lists additional minerals that were mentioned in the selected papers.

Table 5. The mineral content of selected studies and data. Akande et al., 2020, originally expressed it as ppm. Kim et al., 2016, originally expressed it as mg/kg. Rodriguez-Ortega et al. 2016, originally expressed it as a percentage of dry matter. Kohler et al., 2019, expressed it as mg/100g of edible portion (EP).

		Kohler et al. 2019 [17]		Akande et al., 2020 [18]	David Birman et al., 2019 [19]	Kim et al., 2016 [24]	Kuntadi et al., 2018 [25]	Rodriguez-Ortega et al., 2016 [27]	Pereira et al., 2003 [29]
Ref.		mg/100 g, (Edible Portion)		M ± SD; mg/100 g	M ± SD; mg/100 g (dry weight)	M ± SD; mg/100 g (dry weight)	mg/100 g (dry weight)	Mean ± SD; mg/100 g (dry weight)	M ± SD; mg/100 Street g (dry weight)
		Street Hawker	Supermarket						
Minerals	Cr	-	-	-	N/D	-	-	-	0.253± 0.023
	Hg	<0.005	<0.005	-	N/D	-	-	-	-
	Pb	0.0138	0.0044	-	N/D	-	-	-	-
	Cd	<0.005	<0.005	-	N/D	-	-	-	-
	As	0.0432	0.0165	-	-	-	-	-	-
	Se	0.0534	0.2285	-	-	-	-	-	-
	Mo	0.0151	0.0516	-	-	0.02± 0.00	-	-	-
	B	-	-	-	-	1.46± 0.42	-	-	-
	Cu	0.711	0.943	-	-	0.94± 0.01	-	-	1.52± 0.07
	Na	128.4	362.6	-	41.86± 3.45	-	-	-	29.60± 2.10
	K	492.9	672	-	513.8± 1.13	-	-	-	477± 16
	Mg	157.7	174.7	-	280.8± 1.13	252.29± 10.26	-	-	89± 0
	Fe	2.83	3.18	0.142± 0.063	N/D	4.95± 0.05	3.54	-	3.00± 0.88
	Zn	9.8	15.4	0.203± 0.004	1.39± 0.07	14.70± 1.45	-	-	24.40± 0.96
	Mn	1.26	1.86	0.021± 0.021	1.08± 0.05	1.68± 0.06	-	-	2.30± 0.17
	Ca	92.2	107.6	0.249± 0.013	137.01± 1.34	98.72± 0.00	29.17	760± 0	181± 9
	P	492.9	672	5.57± 0.144	-	870.92± 15.77	-	720± 0	-

M ± SD = mean ± standard deviation; N/D = not detected.

4. DISCUSSION

Protein content was the nutritional value that was most frequently reported, which is indicative of the fact that silkworm pupae were primarily seen as a promising protein source. The removal of the inner material, an unusual method in insect processing that could have altered the nutrient content of SP, may have caused the high value of lipids and low value of protein discovered by Mishra et al. [5]. Most of the protein values for full-fat SP were between 45 and 60% (percentage of the dry matter), which was comparable to other raised insects such as *Tenebrio molitor and Grillus bimaculatus* (53.2 and 58.3%, respectively) [42] and chicken eggs (54.1%) [43].

For the majority (86%) of the samples, the full-fat SP's total lipid content ranges from 20 to 35%. These figures could be in line with those discovered for chicken (34.8%), swine (26.3%), tuna (24.9%), and T. molitor (34.5%) expressed on a dry matter basis [42,43]. It should be noted that when comparing nutritional data for insects to that of other traditional animal sources, the nutritional content of insects is expressed as referring to the full organism, as opposed to traditional foods of animal origin, where only a portion of the animal is consumed.

A nitrogen-to-protein conversion factor (Kp) is frequently used to determine the protein content from the total nitrogen determined using analytical techniques. Most of the studies used a Kp of 6.25, whereas only two reported an insect corrected Kp of 4.76. The 6.25 conversion factor is the most commonly used for protein estimation in meat. However, it is based on the assumption that all of the nitrogen found in a food matrix comes from protein and that nitrogen accounts for 16 percent of protein [44]. This could result in an overestimation of protein content, particularly in insects that are characterized by the presence of chitin. As a result, Janssen et al. [16] proposed using a Kp conversion factor of 4.76 based on the quantification of non-protein nitrogen (i.e., chitin and chitosan, nucleic acids, and urea) in three insect species: *Hermetia illucens, Tenebrio molitor*, and *Alphitobius diaperinus.* Moreover, because NFE is usually calculated by difference, an incorrect estimation of protein content could lead to an incorrect estimation of NFE content.

For these reasons, and in the absence of a specific Kp for SP, we agree to use the 4.76 Kp value proposed by Janssen et al. [16]. The presence of all essential AA, characterized SP's amino acid content, but the variability of the retrieved data precludes further considerations. Kwon et al. [35] found variation in AA profile as a function of pupal age, which should be taken into account during the setup of rearing and further processing of SP in relation to the desired AA profile in the final product.

The main issue with the AA dataset was that only two of the six studies performed replications of their analyses, making data comparison even more difficult. Despite the variability of FA identified in the selected studies, five main FAs (palmitic, stearic, oleic, linoleic, and alfa-linolenic acid) characterized SP oil, making it a good source of UFA, whose content ranges between 59.7 and 75.8%, with the lowest values (from 59.7 to 65.4%) in SP reared on an artificial diet [37], which also have the highest SFA.

The majority of the UFA retrieved values were between 65 and 75%; however, the proportion of monounsaturated FA (MUFA) and polyunsaturated FA (PUFA) that contributes to UFA content varies across papers. The SP reared on an artificial diet had the highest MUFA values as well as the lowest PUFA values [37], implying that the use of an artificial diet could change the proportion of FA classes, reducing PUFA in favor of MUFA and SFA. Nonetheless, the SP reared on an artificial diet by Yu et al. [40] had PUFA and MUFA values comparable to all other SP reared on mulberry leaves. The findings of Chieco et al. [37] and Yu et al. [40] indicate that the use of an artificial diet has a significant impact on the SP FA profile. Domesticated silkworms are strictly monophagous insects that feed primarily on fresh mulberry leaves (*Morus alba* L.) but can also be reared on artificial diets [44]. In artificial diets, mulberry leaf powder is typically mixed with plant-based meals (soybean, wheat, or corn meals), vitamins, antioxidants, and inorganic salts [44,45]. As previously demonstrated [46], the type and proportion of the different plant matrices included in the diet determine the FA profile of the diet and are likely to influence the SP FA profile.

When we looked at the proportion of n-3 and n-6 FA in SP fed mulberry leaves versus an artificial diet, we discovered that SP fed mulberry leaves have an n-6/n-3 FA ratio ranging from 2.80:1 to 9.88:1, whereas those fed an artificial diet have an n-6/n-3 FA ratio ranging from 0.96:1 to 1.05:1. This is primarily due to the comparable amounts of linoleic and alfa-linolenic acid and a much lower amount of n-3 FAs in the artificial diet fed to SP, indicating that the diet appears to be the main variable influencing the SP's FA profile.

Since they are not synthesizable by the human body, n-3 and n-6 FA play an important role in the human diet and are thus considered essential FA. Furthermore, their ratio influences body fat metabolism and systemic inflammation, with an increase in the n-6/n-3 ratio associated with an increased risk of obesity, atherosclerosis, and diabetes [47,48]. As a result, an n-6/n-3 FA ratio of 1:1 to 5:1 is thought to be optimal for human health [49]. The average Western diet has an n-6/n-3 ratio of 15.0:1 to 16.7:1, so it would be preferable to include food supplements or nutraceuticals with a lower n-6/n-3 ratio in theWestern diet. Furthermore, lowering the n-6/n-3 ratio in broiler diets has been shown to improve growth performance and immune response in these birds [50]. In the case of other farmed animals' meat, n-6/n-3 values of 1.3, 2.1, 7.2, and 13.7 were found in the muscle tissues of sheep, cattle, pigs, and chickens, respectively [51,52]. The n-6/n-3 ratio is always lower in marine species such as herring, tuna, pollock, salmon, and rainbow trout because their diet is based on marine phytoplankton [53]. The values discovered for mulberry leaves fed SP are thus comparable to those found in sheep, cattle, and pigs.

The linoleic and alfa-linolenic acid content of mulberry leaves fed with SP varied greatly, but it is unclear which variable could explain these results. Consequently, in order to achieve a better optimal n-6/n-3 ratio, we must identify the most relevant variables and their cause-effect relationships.

In terms of the mineral content of SP, the available literature indicates that mulberry is important in determining the mineral content of silkworms, which bioaccumulate minerals through nutrition. The high variability in SP mineral content observed could thus be linked to mineral availability in the soil where the mulberry tree grows. Mineral uptake in the mulberry tree may also vary in response to heavy metal contamination in the soil [54]. Mulberry cultivar and harvest time are also important factors in mineral accumulation [55]. The high sodium value discovered by Kohler et al. [17] was found in a supermarket sample, and it cannot be excluded that it was due to the addition of additives. The *B. mori* strain used is another variable that influences the SP nutrient composition, as demonstrated by Hirunyophat et al. [32]. This information is frequently missing, except in papers in which the strain was one of the primary factors investigated. However, when not reported, the analysis would have most likely been performed on hybrid silkworms, which are the most common type in experimental and production settings. Although the impact of several intrinsic factors on the nutritional value of SP has been investigated, it is not always clear how the analytical methods were implemented, and especially if and how the analyzed pupae were preprocessed.

When silkworms were not reared in a laboratory or under controlled conditions, the process they went through to reach the pupal stage was sometimes unclear. This is especially true for SP recovered from silk reeling industries (by-product pupae), for which precise data on industrial processing, storage time, and temperature were not available [56]. Silkworm pupae from reeling industries have often been dried and boiled, which may have altered the nutritional content. Furthermore, once recovered from the reeling plants, they were most likely dried and stored under highly oxidative conditions. These factors may have a significant impact on the composition and must be considered if the pupae's final destination is the feed or food market.

Finally, as reported by Kwon et al. [35], pupal age may also play a role in determining the composition. It is worth noting that because only 64% of the data has a standard deviation or other measures of variation associated with mean values, a large portion of them are statistically weak and difficult to compare.

5. CONCISE SUMMARY

The literature on SP as a food source has recently expanded, and greater investigation is anticipated due of the increasing interest in entomophagy. Our systematic review, concentrating on the nutritional composition of SP, underscores the extensive array of data acquired. While the variables related to sample preparation and analysis are often thoroughly documented in research, factors such as the mulberry cultivar utilized for larval feeding, silkworm strain, and rearing environmental conditions are never disclosed. The acquisition of SP from the silk reeling industry was accompanied by inadequately described or unknown pre-treatments and storage conditions. Moreover, research presented data in disparate units of measurement or formats, complicating rapid comparison.

Consequently, there is an urgent need for new research employing more precise and well-defined analytical procedures, utilizing normalized data on a dry matter basis, and addressing findings that considerably deviate from previously published literature. It is advisable to adopt or adapt standard methodologies from other product sectors to evaluate the nutritional properties of edible insects for both the food and feed markets, and to provide a reference benchmark for the most frequently employed analytical processes. SP is primarily characterized by its protein content, which is the most often cited variable in scholarly articles. The extensive variability in protein content and amino acid profiles documented in the literature necessitates additional investigation, as do the factors that affect these values. The research consistently identifies five principal fatty acids that constitute almost 97.1% of the total sphingolipid composition. The levels of these five primary fatty acids are affected by nutrition and pupal age; nonetheless, the total of these five fatty acids remains remarkably stable irrespective of upstream conditions.

REFERENCES

1. Yang, S.-Y.; Han, M.-J.; Kang, L.-F.; Li, Z.-W.; Shen, Y.-H.; Zhang, Z. Demographic History and Gene Flow during Silkworm Domestication. *BMC Evol. Biol.* **2014**, *14*, 185. [CrossRef]

2. Meng, X.; Zhu, Y.; Chen, K. Silkworm: A Promising Model Organism in Life Science. *J. Insect Sci.* **2017**, *17*, 97. [CrossRef]

3. Xu, S.; Wang, F.; Wang, Y.; Wang, R.; Hou, K.; Tian, C.; Ji, Y.; Yang, Q.; Zhao, P.; Xia, Q. A Silkworm Based Silk Gland Bioreactor for High-Efficiency Production of Recombinant Human Lactoferrin with Antibacterial and Anti-Inflammatory Activities. *J. Biol. Eng.* **2019**, *13*, 61. [CrossRef] [PubMed]

4. Hu, B.; Li, C.; Zhang, Z.; Zhao, Q.; Zhu, Y.; Su, Z.; Chen, Y. Microwave-Assisted Extraction of Silkworm Pupal Oil and Evaluation of Its Fatty Acid Composition, Physicochemical Properties and Antioxidant Activities. *Food Chem.* **2017**,

231, 348–355. [CrossRef]

5. Mishra, N.; Hazarika, N.C.; Narain, K.; Mahanta, J. Nutritive Value of Non-Mulberry and Mulberry Silkworm Pupae and Consumption Pattern in Assam, India. *Nutr. Res.* **2003**, *23*, 1303–1311. [CrossRef]
6. Zhang, F.; Zhang, Z. Study on Edible Insect Resources and Their Exploitation and Utilization. *Resour. Sci.* **2001**, *23*, 58–61.
7. Mitsuhashi, J. Insects as Traditional Foods in Japan. *Ecol. Food Nutr.* **1997**, *36*, 187–199. [CrossRef]
8. Han, R.; Shin, J.T.; Kim, J.; Choi, Y.S.; Kim, Y.W. An Overview of the South Korean Edible Insect Food Industry: Challenges and Future Pricing/Promotion Strategies: S. Korean Edible Insect Food Industry. *Entomol. Res.* **2017**, *47*, 141–151. [CrossRef]
9. Yhoung-Aree, J.; Puwastien, P.; Attig, G.A. Edible Insects in Thailand: An Unconventional Protein Source? *Ecol. Food Nutr.* **1997**, *36*, 133–149. [CrossRef]
10. Aznar-Cervantes, S.D.; Monteagudo Santesteban, B.; Cenis, J.L. Products of Sericulture and Their Hypoglycemic Action Evaluated by Using the Silkworm, *Bombyx mori* (Lepidoptera: Bombycidae), as a Model. *Insects* **2021**, *12*, 1059. [CrossRef]
11. Ratcliffe, N.A.; Mello, C.B.; Garcia, E.S.; Butt, T.M.; Azambuja, P. Insect Natural Products and Processes: New Treatments for Human Disease. *Insect Biochem. Mol. Biol.* **2011**, *41*, 747–769. [CrossRef] [PubMed]
12. Zimian, D.; Yonghua, Z.; Xiwu, G. Medicinal Insects in China. *Ecol. Food Nutr.* **1997**, *36*, 209–220. [CrossRef]
13. EFSA Scientific Committee. Risk Profile Related to Production and Consumption of Insects as Food and Feed. *EFSA J.* **2015**, *13*, 4257. [CrossRef]
14. Wu, X.; He, K.; Velickovic, T.C.; Liu, Z. Nutritional, Functional, and Allergenic Properties of Silkworm Pupae. *Food Sci. Nutr.* **2021**, *9*, 4655–4665. [CrossRef] [PubMed]
15. Thomas, J.; Brunton, J.; Graziosi, S. *EPPI-Reviewer 4.0: Software for Research Synthesis. EPPI-Centre Software*; Social Science Research Unit, Institute of Education, University of London: London, UK, 2010.
16. Janssen, R.H.; Vincken, J.-P.; van den Broek, L.A.M.; Fogliano, V.; Lakemond, C.M.M. Nitrogen-to-Protein Conversion Factors for Three Edible Insects: *Tenebrio Molitor*, *Alphitobius Diaperinus*, and *Hermetia Illucens*. *J. Agric. Food Chem.* **2017**, *65*, 2275–2278. [CrossRef]
17. Köhler, R.; Kariuki, L.; Lambert, C.; Biesalski, H.K. Protein, Amino Acid and Mineral Composition of Some Edible Insects from Thailand. *J. Asia-Pac.*

Entomol. **2019**, *22*, 372–378. [CrossRef]

18. Akande, O.A.; Falade, O.O.; Badejo, A.A.; Adekoya, I. Assessment of Mulberry Silkworm Pupae and African Palm Weevil Larvae as Alternative Protein Sources in Snack Fillings. *Heliyon* **2020**, *6*, e03754. [CrossRef]
19. David-Birman, T.; Moshe, H.; Lesmes, U. Impact of Thermal Processing on Physicochemical Properties of Silk Moth Pupae (*Bombyx mori*) Flour and in-Vitro Gastrointestinal Proteolysis in Adults and Seniors. *Food Res. Int.* **2019**, *123*, 11–19. [CrossRef]
20. Kim, H.-W.; Setyabrata, D.; Lee, Y.J.; Jones, O.G.; Kim, Y.H.B. Pre-Treated Mealworm Larvae and Silkworm Pupae as a Novel Protein Ingredient in Emulsion Sausages. *Innov. Food Sci. Emerg. Technol.* **2016**, *38*, 116–123. [CrossRef]
21. Anootthato, S.; Therdthai, N.; Ritthiruangdej, P. Characterization of Protein Hydrolysate from Silkworm Pupae (*Bombyx mori*). *J. Food Process. Preserv.* **2019**, *43*. [CrossRef]
22. Felix, M.; Bascon, C.; Cermeño, M.; FitzGerald, R.J.; de la Fuente, J.; Carrera-Sánchez, C. Interfacial/Foaming Properties and Antioxidant Activity of a Silkworm (*Bombyx mori*) Pupae Protein Concentrate. *Food Hydrocoll.* **2020**, *103*, 105645. [CrossRef]
23. Akande, A.O.; Jolayemi, O.S.; Adelugba, V.A.; Akande, S.T. Silkworm Pupae (*Bombyx mori*) and Locusts as Alternative Protein Sources for High-Energy Biscuits. *J. Asia-Pac. Entomol.* **2020**, *23*, 234–241. [CrossRef]
24. Kim, S.-K.; Weaver, C.M.; Choi, M.-K. Proximate Composition and Mineral Content of Five Edible Insects Consumed in Korea. *CyTA–J. Food* **2016**, *15*, 143–146. [CrossRef]
25. Kuntadi, K.; Adalina, Y.; Maharani, K.E. Nutritional Composition of Six Edible Insects in Java. *Indones. J. For. Res.* **2018**, *5*, 57–68. [CrossRef]
26. Lamberti, C.; Gai, F.; Cirrincione, S.; Giribaldi, M.; Purrotti, M.; Manfredi, M.; Marengo, E.; Sicuro, B.; Saviane, A.; Cappellozza, S.; et al. Investigation of the Protein Profile of Silkworm (*Bombyx mori*) Pupae Reared on a Well-Calibrated Artificial Diet Compared to Mulberry Leaf Diet. *PeerJ* **2019**, *7*, e6723. [CrossRef] [PubMed]
27. Rodríguez-Ortega, A.; Pino-Moreno, J.M.; Ángeles-Campos, S.C.; García-Pérez, Á.; Barrón-Yánez, R.M.; Callejas-Hernández, J. Valor Nutritivo de Larvas y Pupas de Gusano de Seda (*Bombyx mori*) (Lepidoptera: Bombycidae). *Rev. Colomb. Entomol.* **2016**, *42*, 69. [CrossRef]
28. Tomotake, H.; Katagiri, M.; Yamato, M. Silkworm Pupae (*Bombyx mori*) Are New Sources of High Quality Protein and Lipid. *J. Nutr. Sci. Vitaminol.*

2010, *56*, 446–448. [CrossRef]

29. Pereira, N.R.; Ferrarese-Filho, O.; Matsushita, M.; de Souza, N.E. Proximate Composition and Fatty Acid Profile of *Bombyx mori* L. Chrysalis Toast. *J. Food Compos. Anal.* **2003**, *16*, 451–457. [CrossRef]
30. Ghosh, A.; Ray, M.; Gangopadhyay, D. Evaluation of Proximate Composition and Antioxidant Properties in Silk-Industrial Byproduct. *LWT* **2020**, *132*, 109900. [CrossRef]
31. Mishyna, M.; Haber, M.; Benjamin, O.; Martinez, J.J.I.; Chen, J. Drying Methods Differentially Alter Volatile Profiles of Edible Locusts and Silkworms. *J. Insects Food Feed* **2020**, *6*, 405–415. [CrossRef]
32. Hirunyophat, P.; Chalermchaiwat, P.; On-nom, N.; Prinyawiwatkul, W. Selected Nutritional Quality and Physicochemical Properties of Silkworm Pupae (Frozen or Powdered) from Two Species. *Int. J. Food Sci. Technol.* **2021**, *56*, 3578–3587. [CrossRef]
33. Wei, Z.-J.; Liao, A.-M.; Zhang, H.-X.; Liu, J.; Jiang, S.-T. Optimization of Supercritical Carbon Dioxide Extraction of Silkworm Pupal Oil Applying the Response Surface Methodology. *Bioresour. Technol.* **2009**, *100*, 4214–4219. [CrossRef] [PubMed]
34. Walker, V.; Mills, G.A. Quantitative Methods for Amino Acid Analysis in Biological Fluids. *Ann. Clin. Biochem. Int. J. Lab. Med.* **1995**, *32*, 28–57. [CrossRef] [PubMed]
35. Kwon, M.-G.; Kim, D.-S.; Lee, J.-H.; Park, S.-W.; Choo, Y.-K.; Han, Y.-S.; Kim, J.-S.; Hwang, K.-A.; Ko, K.; Ko, K. Isolation and Analysis of Natural Compounds from Silkworm Pupae and Effect of Its Extracts on Alcohol Detoxification: Nutritional Value of Silkworm Component. *Entomol. Res.* **2012**, *42*, 55–62. [CrossRef]
36. Chieco, C.; Morrone, L.; Bertazza, G.; Cappellozza, S.; Saviane, A.; Gai, F.; Di Virgilio, N.; Rossi, F. The Effect of Strain and Rearing Medium on the Chemical Composition, Fatty Acid Profile and Carotenoid Content in Silkworm (*Bombyx mori*) Pupae. *Animals* **2019**, *9*, 103. [CrossRef]
37. Tong, L.; Yu, X.; Liu, H. Insect Food for Astronauts: Gas Exchange in Silkworms Fed on Mulberry and Lettuce and the Nutritional Value of These Insects for Human Consumption during Deep Space Flights. *Bull. Entomol. Res.* **2011**, *101*, 613–622. [CrossRef]
38. Usub, T.; Lertsatitthanakorn, C.; Poomsa-ad, N.; Wiset, L.; Yang, L.; Siriamornpun, S. Experimental Performance of a Solar Tunnel Dryer for Drying Silkworm Pupae. *Biosyst. Eng.* **2008**, *101*, 209–216. [CrossRef]
39. Wang, W.; Xu, L.; Zou, Y.; Pang, D.; Shi, W.; Mu, L.; Li, E.; Lan, D.; Wang, Y.; Liao, S. Comprehensive Identification of Principal Lipid Classes and Tocochromanols in Silkworm (*Antheraea pernyi* and *Bombyx mori*) Pupae

Oils. *Eur. J. Lipid Sci. Technol.* **2020**, *122*, 1900280. [CrossRef]

40. Yu, X.; Shen, Y.; Cui, Q.; Chen, Y.; Sun, W.; Huang, X.; Zhu, Y. Silkworm (*Bombyx mori*) Has the Capability to Accumulate $_{20}$ and $_{22}$ Polyunsaturated Fatty Acids. *Eur. J. Lipid Sci. Technol.* **2018**, *120*, 1700268. [CrossRef]

41. Ghosh, S.; Lee, S.-M.; Jung, C.; Meyer-Rochow, V.B. Nutritional Composition of Five Commercial Edible Insects in South Korea. *J. Asia-Pac. Entomol.* **2017**, *20*, 686–694. [CrossRef]

42. CREA Centro Alimenti e Nutrizione Tabelle Di Composizione Degli Alimenti, Aggiornamento. 2019. Available online: https://www.alimentinutrizione.it/tabelle-nutrizionali (accessed on 15 April 2022).

43. World Health Organization; Food and Agriculture Organization of the United Nations. *Nitrogen and Protein Content Measurement and Nitrogen to Protein Conversion Factors for Dairy and Soy Protein-Based Foods: A Systematic Review and Modelling Analysis*; World Health Organization: Geneva, Switzerland, 2019; ISBN 978-92-4-151698-3.

44. Cappellozza, L.; Cappellozza, S.; Saviane, A.; Sbrenna, G. Artificial Diet Rearing System for the Silkworm *Bombyx mori* (Lepi-doptera: Bombycidae): Effect of Vitamin C Deprivation on Larval Growth and Cocoon Production. *Appl. Entomol. Zool.* **2005**, *40*, 405–412. [CrossRef]

45. Li, J.; Chen, C.; Zha, X. Midgut and Head Transcriptomic Analysis of Silkworms Reveals the Physiological Effects of Artificial Diets. *Insects* **2022**, *13*, 291. [CrossRef] [PubMed]

46. Horie, Y.; Nakasone, S. Effects of the Levels of Fatty Acids and Carbohydrates in a Diet on the Biosynthesis of Fatty Acids in Larvae of the Silkworm, *Bombyx mori*. *J. Insect Physiol.* **1971**, *17*, 1441–1450. [CrossRef]

47. Patel, A.; Desai, S.S.; Mane, V.K.; Enman, J.; Rova, U.; Christakopoulos, P.; Matsakas, L. Futuristic Food Fortification with a Balanced Ratio of Dietary ω-3/ω-6 Omega Fatty Acids for the Prevention of Lifestyle Diseases. *Trends Food Sci. Technol.* **2022**, *120*, 140–153. [CrossRef]

48. Simopoulos, A. An Increase in the Omega-6/Omega-3 Fatty Acid Ratio Increases the Risk for Obesity. *Nutrients* **2016**, *8*, 128. [CrossRef]

49. Lupette, J.; Benning, C. Human Health Benefits of Very-Long-Chain Polyunsaturated Fatty Acids from Microalgae. *Biochimie* **2020**, *178*, 15–25. [CrossRef]

50. Ibrahim, D.; El-Sayed, R.; Khater, S.I.; Said, E.N.; El-Mandrawy, S.A.M. Changing Dietary N-6:N-3 Ratio Using Different Oil Sources Affects Performance, Behavior, Cytokines MRNA Expression and Meat Fatty Acid Profile of Broiler Chickens. *Anim. Nutr.* **2018**, *4*, 44–51. [CrossRef]

51. Wood, J.D.; Enser, M.; Fisher, A.V.; Nute, G.R.; Sheard, P.R.; Richardson, R.I.; Hughes, S.I.; Whittington, F.M. Fat Deposition, Fatty Acid Composition and

Meat Quality: A Review. *Meat Sci.* **2008**, *78*, 343–358. [CrossRef]

52. Trembecka, L.; Hascik, P.; Cubon, J.; Bobko, M.; Pavelkova, A. Fatty Acids Profile of Breast and Thigh Muscles of Broiler Chickens Fed Diets with Propolis and Probiotics. *J. Cent. Eur. Agric.* **2016**, *17*, 1179–1193. [CrossRef]
53. Strobel, C.; Jahreis, G.; Kuhnt, K. Survey of N- 3 and n-6 Polyunsaturated Fatty Acids in Fish and Fish Products. *Lipids Health Dis.* **2012**, *11*, 144. [CrossRef]
54. Zeng, P.; Guo, Z.; Xiao, X.; Peng, C.; Liu, L.; Yan, D.; He, Y. Physiological Stress Responses, Mineral Element Uptake and Phytoremediation Potential of *Morus alba* L. in Cadmium-Contaminated Soil. *Ecotoxicol. Environ. Saf.* **2020**, *189*, 109973. [CrossRef] [PubMed]
55. Levickiene˙, D.; Vaitkevicˇiene˙, N.; Jariene˙, E.; Mažeika, R. The Content of Macroelements in White Mulberry (*Morus alba* L.) Leaves. *Žeme˙s U¯ kio Moksl.* **2019**, *25*. [CrossRef]
56. Saviane, A.; Tassoni, L.; Naviglio, D.; Lupi, D.; Savoldelli, S.; Bianchi, G.; Cortellino, G.; Bondioli, P.; Folegatti, L.; Casartelli, M.; et al. Mechanical Processing of Hermetia Illucens Larvae and *Bombyx mori* Pupae Produces Oils with Antimicrobial Activity. *Animals* **2021**, *11*, 783. [CrossRef] [PubMed]

Chapter 8

A BOMBYX MORI INFECTION MODEL FOR THE EVALUATION OF ANTIBIOTICS TARGETING *STAPHYLOCOCCUS EPIDERMIDIS.*

1. INTRODUCTION: AN OVERVIEW

Antibiotics are regarded as one of the most significant discoveries of the 20th century and remain crucial in contemporary medicine, utilized for both the treatment and prevention of infections in high-risk patients [1]. The extensive usage and/or misuse of antibiotics has facilitated the emergence and proliferation of resistant microorganisms [2]. The imperative to discover and develop innovative antibacterial agents is critical due to the escalating issue of antibiotic resistance. In vivo screening of new antibiotics may expedite the identification of promising compounds, prioritizing those with an appropriate therapeutic index (i.e., high efficacy and low toxicity) [3]. The extensive utilization of mammalian animal models, chiefly mice and rats, during the preclinical phase is impractical due to the substantial costs associated with their care and the ethical constraints that restrict their use, as stipulated by the European directive on animal protection adhering to the 3Rs principles (i.e., Replacement, Reduction, and Refinement) [4]. Recent proposals for developing appropriate alternatives include several invertebrate models, such as amoebas and nematodes [5,6]. Holometabolous insects constitute a viable alternative owing to their low cost and ease of larval management [7]. Furthermore, the utilization of insects is not constrained by ethical limitations [8], and despite their absence of acquired immunity, their innate reaction is evolutionarily and functionally comparable to that of mammals. [9].

Various insect species, such as Drosophila melanogaster, Galleria mellonella, and Bombyx mori, have lately been employed to examine the effects of new antimicrobials and assess their efficiency [8,10]. An advantage of utilizing silkworms as an infection model is the availability of germplasm banks, which preserve genetic stock collections of B. mori; these facilities utilize artificial diets for rearing, thereby standardizing the quality of insect supplies. Furthermore, the genome sequencing of B. mori [11] has facilitated the creation of molecular tools and gene editing systems, including RNA interference-based gene silencing, transposon-mediated transgenesis, and CRISPR/Cas9 [12], which have been instrumental in establishing customized disease models for examining drug mechanisms of action. Silkworm infection models have been utilized to assess antimicrobial agents against Gram-negative and Gram-positive bacteria within the ESKAPE group, including Escherichia coli, Pseudomonas aeruginosa, Staphylococcus aureus, and Klebsiella pneumoniae, as well as other bacterial pathogens such as Francisella tularensis, Listeria monocytogenes, Bacillus cereus, and Mycobacterium abscessus, in addition to fungi like Candida spp. and Aspergillus fumigatus. To yet, no research has been conducted in this context utilizing one of the most prevalent causes of nosocomial infections, coagulase-negative Staphylococcus epidermidis. Staphylococcus epidermidis is a prevalent, non-pathogenic inhabitant of human skin and mucosal surfaces. Nonetheless, it transforms into an opportunistic infection capable of inducing virulence once entering the human body through medical and prosthetic devices, including peripheral or central intravenous catheters and orthopedic implants. Bacteremia predominantly occurs due to contamination of indwelling medical devices by S. epidermidis, which can also generate biofilms that are resistant to host defenses and antimicrobial therapies. [23,24].

This study examined the effects of three life-saving glycopeptide antibiotics (vancomycin, teicoplanin, and dalbavancin), utilized in clinical practice to treat severe infections caused by Gram-positive bacteria, on B. mori larvae infected with a clinical isolate of S. epidermidis. Through the assessment of insect survival rates and the evaluation of three pertinent immunological markers (namely, hemocyte metabolic rate, lysozyme activity, and phenoloxidase (proPO) system activity), we established the therapeutic efficacy of the three GPAs, thereby validating the reliability of this insect-based infection model for the evaluation of antimicrobial agents..

1. MATERIALS AND METHODS

1.1 Experimental Model

The larvae of *B. mori* (polyhybrid (126 × 57) (70 × 90)) used in this study were provided by CREA-AA Sericulture Laboratory (Padova, Italy). Insects were reared in glass Petri dishes at 25 ± 0.5 °C, with 70 ± 5% relative humidity, under a 12L:12D photoperiod. Larvae were fed on artificial diet [26] until the end of the 4th larval instar. After animals had ecdysed to the 5th larval instar, they were synchronized [27] and fed with a germ-free diet [28].

1.2 Bacterial Strain

S. epidermidis strain 4, a clinical isolate kindly provided by Laboratorio di Microbiologia Clinica-Ospedale di Circolo (Varese, Italy), was grown in 10 mL of Müller–Hinton Broth 2 (MHB2; VWR International Srl, Radnor, PA, USA), overnight at 37 °C under 200 rpm shaking. Cells harvested by centrifuging 1 mL of culture at

1900 *g* at 4°C for 10 min were then resuspended at the desired concentration with saline solution (0.6% *w/v* NaCl). To determine the volume of saline solution to be used, the optical density of the culture was measured, considering that one unit of $OD600_{nm}$ corresponded to 7.03×10^8 CFU (colony forming units)/mL.

1.3 Minimum Inhibitory Concentrations (MICs) and Minimum Bactericidal Concentrations (MBCs)

MICs of vancomycin (Sigma-Aldrich, St. Louis, MO, USA), teicoplanin (Sigma-Aldrich), and dalbavancin (kindly provided by Sanofi, Brindisi, Italy) towards *S. epidermidis* strain 4 were determined by the broth dilution method, following Clinical and Laboratory Standards Institute guidelines [29]. A total of 5×10^5 viable bacterial cells were inoculated into MHB2 medium (VWR International Srl) in 96-well plates and supplemented with increasing antibiotic concentrations (from 0 to 128 μg/mL) up to 100 μL final volume. Plates were then incubated for 16–20 h at 37°C and 100 rpm. MIC was defined as the minimal concentration of antibiotic at which no turbidity could be detected. For MBC determination, bacterial cultures used for the MIC test were plated onto Müller–Hinton Agar (MHA; VWR International Srl) and incubated at 37°C for 24 h. MBC was defined as the lowest concentration of antibiotic at which no visible growth was detected on plates. All experiments were performed at least in triplicate.

1.4 Injection of Larvae and Collection of Hemolymph

Insects were infected at the second day of the fifth larval instar by injecting bacteria in the second right proleg with autoclaved Hamilton 1702 LT 25 μL syringes (Hamilton Company, Reno, NV, USA), under a sterile hood. After the injection, larvae were reared at 37°C. Hemolymph collection was performed by puncturing the larvae or cutting the second left proleg 6 or 24 h after the infection depending on the analysis.

1.5 Lethal Dose 50 (LD_{50}) for S. epidermidis

To determine the LD_{50}, larvae were infected with *S. epidermidis* at different concen- trations (from 3×10 CFU to 3×10^5 CFU in a final volume of 10 μL). The mortality rate was monitored every 24 h for three days. Larvae were considered dead when no reaction to a stimulation with a plastic tip was observed. Uninjected larvae and larvae injected with 10 μl of saline solution (0.6% *w/v* NaCl) were used as controls. Forty larvae for each experimental condition were used. LD_{50} calculated by Probit analysis [30], was defined as the concentration of bacteria at which 50% of animals died within 3 days after the infection.

1.6 Administration of Antibiotics

The effects of GPA administration were evaluated by injecting 10 μL of *S. epidermidis* at LD_{50} into the larvae, and 10 μL of vancomycin, teicoplanin, or dalbavancin (at a concentration equal to 8.75 μ g/g body weight) after 2 h, as previously reported [15]. Control groups were represented by uninjected larvae, larvae injected once or twice with saline solution (10 μ L of 0.6% *w/v* NaCl eventually repeated after two hours), and healthy larvae injected with 10 μ L of antibiotic. Larval mortality was monitored for three days. Fifty larvae were used for each experimental condition.

1.7 Analysis of the Immunological Markers

Untreated larvae, larvae injected once or twice with 10 μ L of 0.6% *w/v* NaCl, larvae injected with 10 μ L of antibiotic (i.e., vancomycin, teicoplanin, or dalbavancin at 8.75 μ g/g body weight), larvae injected with 3 10^3 CFU of *S. epidermidis* in 10 μ L, and larvae injected with 3 × 10^3 CFU of *S. epidermidis* in 10 μ L and two hours later with 10 μ L of antibiotic (8.75 μ g/g body weight) were used for the evaluation of the immunological markers (i.e., cells viability, lysozyme, and prophenoloxidase activity).

1.7.1 Hemocyte Viability

Hemolymph was collected from larvae 24 h after the first injection and diluted 1:50 with Saline Solution for Lepidoptera (sucrose 210 mM, KCl 45 mM, Tris-HCl 10 mM, pH 7.0). CellTiter-Glo Luminescent Cell Viability Assay (Promega, Madison, WI, USA) was used to analyze the hemocyte viability. A total of 100 μ L of CellTiter-Glo were added to 100 μ l of diluted hemolymph into a 96-well plate, and then incubated for 5 min at room temperature on an orbital shaker. The measurement of luminescence was performed by using an Infinite F200 96-well plate-reader (Tecan, Männedorf, Switzerland). Ten surviving larvae for each experimental group were analyzed.

1.7.2 Prophenoloxidase (proPO) System Activation

Hemolymph was collected from pools of three larvae and centrifuged at 250× *g* for 5 min at 4°C. A total of 100 μL of hemolymph were loaded into 96-well plates and the absorbance was measured by reading OD_{450nm} every 10 min for 50 min using an Infinite F200 96-well plate-reader (Tecan) [31]. A total of 100 μL of hemolymph supplemented with 2.5 mM N-phenylthiourea (PTU, Sigma-Aldrich) dissolved in 100% *v/v* EtOH were used as negative controls. The activation of proPO system was evaluated by incubating hemolymph with β-glucans from *Saccharomyces cerevisiae* (Zymosan, Sigma-Aldrich) and 4 mM $CaCl_2$ [32], too. ΔOD_{450nm} was calculated subtracting the OD_{450nm} recorded at time zero from the values obtained at the different time points. To evaluate the melanization rate, linear regression was performed for each ΔOD_{450nm} measurement obtained versus time [31].

2.7.3. Lysozyme Activity

Hemolymph was extracted from a pool of three larvae 6 h after infection and added with a few crystals of PTU to avoid melanization. After two centrifugations at 250× *g* for 5 min and one at 1600× *g* for 10 min at 4°C, the supernatant was collected and diluted 1:10 with sterile phosphate-buffered saline (PBS: 1370 mM NaCl, 27 mM KCl, 100 mM $Na_2HPO_4 \times 12H_2O$, 19.8 mM KH_2PO_4). The activity of lysozyme was determined according to Bruno et al. [33]. Briefly, 100 μL of diluted hemolymph were added to 150 μL of 0.45 mg/mL lyophilized *Micrococcus lysodeikticus* (Sigma-Aldrich) in 30 mM phosphate buffer (38 mM KH_2PO_4, 61.4 mM K_2HPO_4, pH 7.2) (OD_{600nm} of 0.6–0.7). *M. lysodeikticus* and hemolymph both added with PBS were used as controls. Finally, absorbance at 450 nm was recorded every 30 s for 10 min by using an Infinite F200 96-well plate-reader (Tecan, Switzerland)

2.8. *Statistical Analysis*

Statistical analysis was performed using ANOVA, followed by Tukey's Honestly Significant Difference (HSD) test (significance $p < 0.05$).

Results

Larval Survival and Calculation of LD_{50}

To determine the LD_{50} for infection experiments, *B. mori* larvae were injected with in- creasing concentrations of *S. epidermidis*. Daily monitoring revealed a normal development of control groups (i.e., untreated larvae and larvae injected with 0.6% *w/v* NaCl), with a survival rate of 100% after 72 h (Figure 1). A correlation between the mortality rate of larvae and the concentration of injected bacteria was instead observed after *S. epidermidis* injection. In detail, 70% of the larvae survived 72 h after the injection of 3×10^2 CFU of *S. epidermidis* (Figure 1), while only 27% and 11% of insects were still alive after the infection with 3×10^3 and 3×10^4 CFU of bacteria, respectively. Finally, no larvae infected with 3×10^5 CFU survived after 72 h (Figure 1). The lethal bacterial dose that killed 50% of the infected larvae (LD_{50}) calculated with Probit analysis, proved to be 1.05×10^3 CFU.

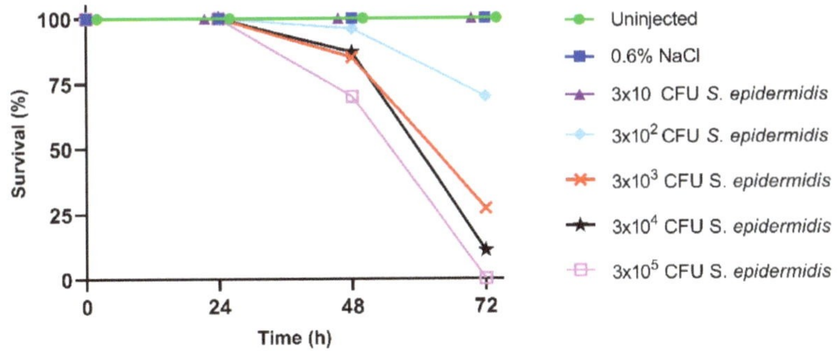

Figure 1. Survival rate of *B. mori* larvae injected with increasing CFU (colony forming unit) of *S. epidermidis*.

2.2 Effect of GPA Administration to the Larvae

The curative efficacy of the three selected GPAs (i.e., vancomycin, teicoplanin, and dalbavancin) was assessed by administering the antibiotics to the larvae infected with *S. epidermidis* at I_{50} Larvae injected with only GPAs were monitored to exclude any toxic effect. Moreover, insects were injected twice with saline solution to exclude any potential side effect due to the double injection (bacteria plus antibiotic). In accordance with previous observations [15], all the control groups showed 100% survival (Figure 2) confirming that GPAs were not toxic to larvae and that the experimental procedure of injection was reliable. All the GPAs tested improved the survival rate of infected silkworms. Indeed, while only 51% of the infected larvae survived 72 h after the infection, the administration of vancomycin, teicoplanin, and dalbavancin increased the survival rate to 90%, 93%, and 97%, respectively (Figure 2).

The minimum inhibitory concentration (MIC) and the minimum bactericidal con- centration (MBC) of the three GPAs towards *S. epidermidis* (Table 1) were in agreement with the results reported above. In fact, the selected strain was sensitive to all three GPAs and dalbavancin proved to be, among the three of them, the most effective antibiotic also in vitro.

Table 1. Minimum inhibitory concentration (MIC) and minimum bactericidal concentration (MBC) of vancomycin, teicoplanin, and dalbavancin towards S. epidermidis. Values represent the average of data from three independent experiments.

Antibiotic	MIC (µg/mL)	MBC (µg/mL)
Vancomycin	2	16
Teicoplanin	1	32
Dalbavancin	0.25	8

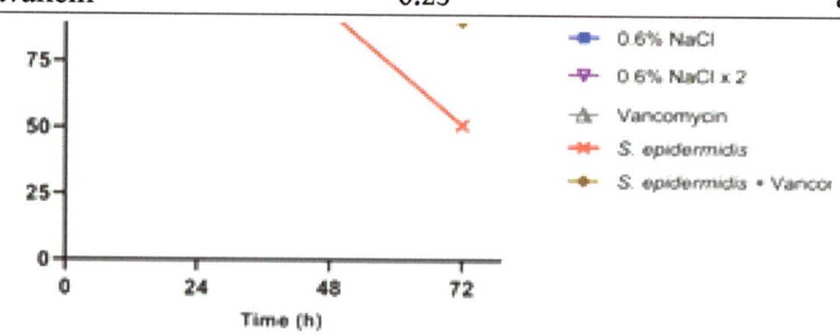

(a)

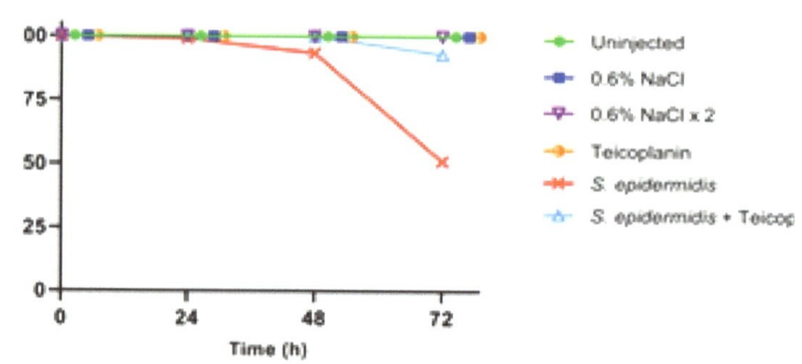

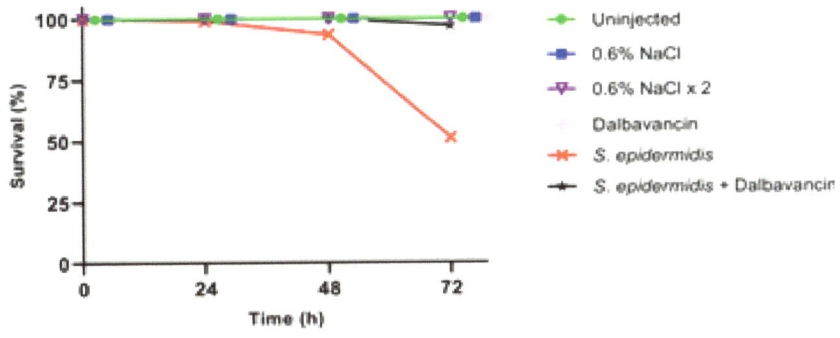

(c)

Figure 2. Survival rate. Curative efficacy of vancomycin (a), teicoplanin (b), and dalbavancin (c) (8.75 μg/g larval body weight) on larvae infected with *S. epidermidis* (1.05×10^3 CFU). Results are from the same experiment run in parallel, using the same control groups (uninjected larvae and healthy larvae injected once or twice with saline solution) and infected larvae.

2.3 Analysis of the Immunological Markers

The insect immune system consists of a fine and balanced crosstalk between hemocytes and humoral molecules, which aims at maintaining the hemolymph devoid of pathogens. Herein, we evaluated the cellular and humoral responses of B. mori larvae following *S. epidermidis* infection.

2.3.1 Hemocyte Viability

Hemocytes are the main mediators of the cellular immune response in insects [34].

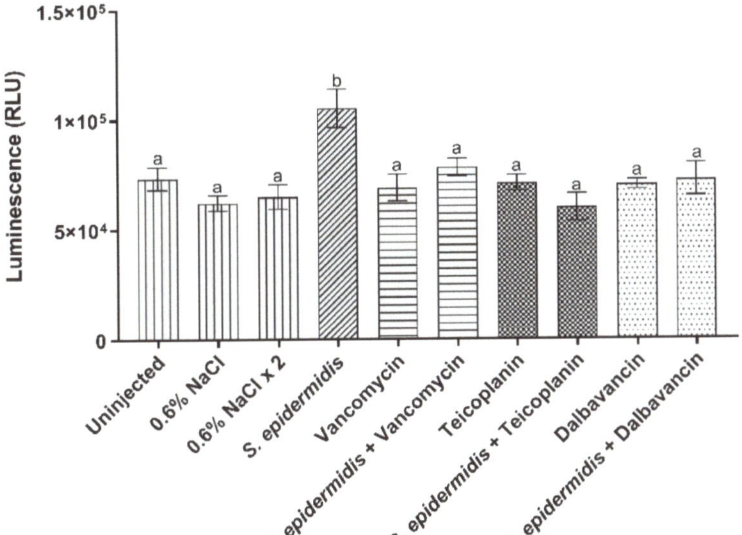

Figure 3. Hemocytes activity detected using a luminescence assay. Values represent mean ± s.e.m. Different letters indicate statistically significant differences among treatments ($p < 0.05$).

To evaluate the recruitment/activation of these immune cells following pathogen invasion, their viability was evaluated by quantifying the ATP content through a luminescence assay. In insects infected by *S. epidermidis,* a significant increase in luminescence was observed, whereas the treatment with the three GPAs restored the ATP amount to levels comparable to those of the control groups (Figure 3). These results indicated that the antibiotics blocked the infection and, consequently, the involvement of hemocytes in the immune response.

2.3.2 Activation of Prophenoloxidase System

In insects, the entrance of pathogens in the hemocoel can induce the activation of the proPO cascade, resulting in the production of melanin to isolate the foreign agents [35]. No significant variation in the activity of proPO system was instead observed after the infection of larvae with 3×10^5 CFU of *S. epidermidis*, compared to the control (Figure 4). Given these results, Zymosan, a specific activator of the proPO system, was used in combination with $CaCl_2$ to check if the enzyme activity responded to specific stimuli. A marked increase in the absorbance values confirmed a consistent activation of phenoloxidase under these last conditions (Figure 4), contrarily to what observed following bacterial infection.

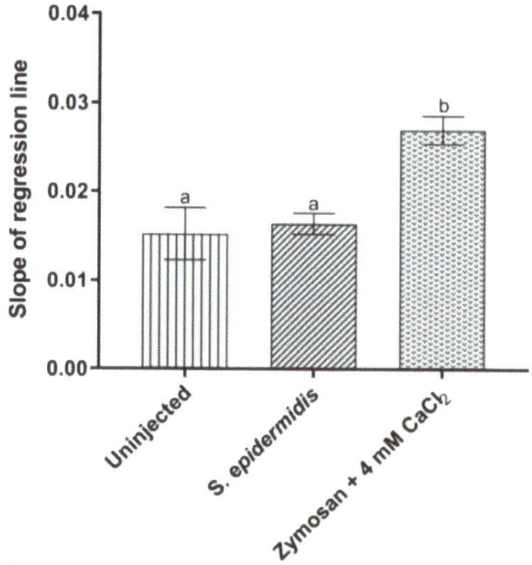

Figure 4. Analysis of proPO system activation. Values represent mean ± s.e.m. Different letters indicate statistically significant differences among treatments ($p < 0.05$).

2.3.3 Lysozyme Activity

Lysozyme cleaves the peptidoglycan component of the bacterial cell wall, leading to microbial cell death. The production of this enzyme generally increases after the pathogen attack [36]. In uninjected larvae, as well as in other control groups (i.e., healthy larvae injected once or twice with 0.6% *w/v* NaCl or with the antibiotics), the activity of lysozyme was comparable. The enzyme activity markedly increased after bacterial infection (Figure 5). Lysozyme activity was restored to the basal levels of the control groups following the administration of vancomycin, teicoplanin, or dalbavancin (Figure 5).

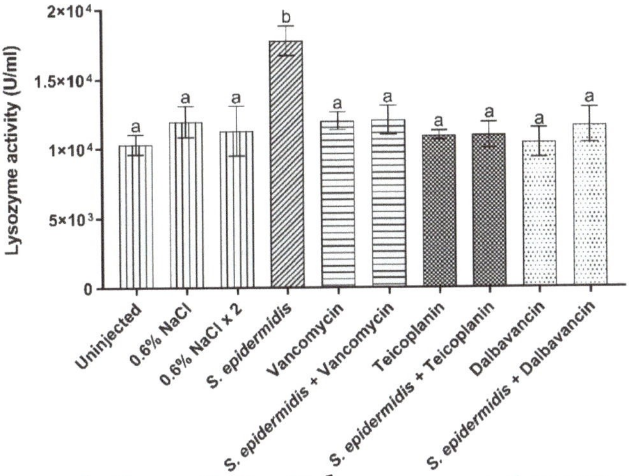

Figure 5. Lysozyme relative activity. Values represent mean ± s.e.m. Different letters indicate statistically significant differences among treatments ($p < 0.05$).

3. DISCUSSION

Various invertebrate models have been proposed over the years to overcome the drawbacks associated with the use of mammals in drug discovery and development, which include not only the ethical and regulatory issues, but also the high costs for animal han- dling, and the need of ad hoc facilities for their maintenance [37,38]. In this scenario, the silkworm *B. mori* has emerged as a suitable model for mimicking infections by various human pathogens, including bacteria [13,15,39–41] as well as filamentous fungi [22] and yeasts [39]. Moreover, this insect has been recently used to test the efficacy of antibac- terial agents, such as kanamycin, tetracycline, fluconazole, and vancomycin, in curing infections [7,13,15,39,42].

The optimal temperature for rearing silkworms is commonly 25°C [26], but this condition significantly differs from the human body temperature. The resistance of *B. mori* larvae to higher temperatures (37°C) for a limited period of time without any apparent side effect was recently proved [15]. This feature supported the development of a silkworm infection model, which can operate at the human body temperature, and better mimics in this way the trend of bacterial infections in humans [7,15,42]. Although a few papers reported on *B. mori* infected with the coagulase-positive *S. aureus* [13,15,39–41], to our knowledge, no publication has previously described infections determined by coagulase-negative staphylococci (CoNS), such as *S. epidermidis* and *S. haemolyticus*. Opportunistic CoNS, and *S. epidermidis* above all, are emerging as major sources of nosocomial infections, especially of foreign body-related infections (FBBIs) associated with indwelling or implanted devices and transmitted by medical and/or nursing procedures [43,44]. *S. epidermidis*-related infections include prosthetic valve endocarditis, keratitis associated with contact lens use, and intravascular catheter- and prosthesis-associated infections. Therapeutic options against CoNS are limited, due to the large number of methicillin-resistant strains or strains with reduced susceptibility to quinolones or to GPAs. In addition, these staphylococci form multilayered and difficult-to-eradicate biofilms, which are highly impenetrable to the majority of antibiotics [43,44].

In our work, we reported on the silkworm model infected at 37°C with a clinical isolate of *S. epidermidis* and, to validate it, we investigated the curative effect of three molecules belonging to the GPA class. GPAs are defined as 'drugs of last resort' for their role in treating life-threating infections caused by staphylococci, enterococci, and *Clostridium difficile* [25]. Their use in fighting clinical infections by *S. epidermidis* is consolidated, too [45]. The selected molecules for this study included the first-generation GPAs vancomycin and teicoplanin, introduced in 1958 and 1988, respectively, and still largely used in clinics, and the second-generation dalbavancin, approved in 2014 and designated as Qualified Infection Disease Product by the Food and Drug Administration (FDA) for its potency and extended dosing interval [25]. We determined the susceptibility profile of our *S. epidermidis* clinical isolate to the three selected GPAs by measuring their MICs and MBCs following in vitro standard protocols. Dalbavancin showed the highest efficacy in inhibiting the growth of the target strain, with MIC and MBC values lower than those recorded for vancomycin and teicoplanin, as also demonstrated in previous publications [46–49]. Consistently with previous data [46], teicoplanin was confirmed to have a weaker bactericidal effect on *S. epi- dermidis* cells than the other two GPAs. As regards the in vivo infection model, the dosage of antibiotics administered to the larvae (i.e., 8.75 μg/g body weight) was determined according to Montali et al. [15] and was comparable to the amount of vancomycin used to treat severe staphylococcal infections in humans. It is noteworthy that this dosage, as well as higher concentrations of all three GPAs (up to 35 μg/g body weight), did not exert any significant toxicity in the larvae [15]. Under the tested conditions, all the three antibiotics proved to be effective in counteracting the infection caused by our *S. epidermidis* clinical isolate.

Previous studies explored the use of the silkworm as infection model mainly monitoring the survival rate of the larvae and, only in some cases, evaluating additional markers, such as the quantification of bacteria and/or their localization in the insect body [13,39,42]. However, detailed studies on the immune system of *B. mori* [50] might provide new tools to finely dissect the physiological mechanisms triggered by invading bacteria. Although additional markers (e.g., analysis of the bacterial load and expression of antimicrobial peptides) were monitored in our previous study [15], the complexity of the procedure and the limited reproducibility of the results obtained by following their temporal trend, prompted us to look for new ones. Since our final goal was to use the silkworm model in the early antibiotic discovery and development phase as a means to select molecules with a promising therapeutic index, we looked for robust and trustable markers that could be easy and rapid to monitor. Our results demonstrate that the administration of GPAs in infected silkworm stopped the activation of the insect's immune response and that the protective effect of these antibiotics could be appreciated at different levels. First, hemocytes, which play a primary role against the invasion of microorganisms in insects [51], confirmed to be a reliable marker, showing a reduced activity in larvae treated with GPAs, caused by a block in their recruitment/activation [15]. In relation to the humoral response that coop- erates with the cellular processes to maintain the hemolymph devoid of pathogens [52], we examined the possible induction of two different enzyme activities, i.e., lysozyme and proPO. Lysozyme hydrolyzes peptidoglycan β-(1,4) glycosidic bonds in the cell wall of Gram-positive bacteria [52]. Our results confirm the activation of this enzyme following the injection of *S. epidermidis* in the larvae [53] and its return to control levels after GPA administration, demonstrating the antibiotic efficacy in reducing the effects of bacterial infection. To the best of our knowledge, this marker was used herein for the first time in an insect infection model. It is a rapid, simple, and reliable assay, which could be thus recommended for further studies on antibiotic response in insects. On the contrary, no difference in proPO activity between the control and infected larvae was observed, suggesting that this marker is not suitable for such type of investigations. In insects, proPO leads to melanin formation by generating quinones and other reactive intermediates, with the final aim of trophically isolating and killing pathogens and parasites [54]. The lack of activation of this enzymatic cascade might depend on the bacterial load in the hemolymph, since it is known that proPO system is induced only over a certain threshold concentration of bacteria [33] to avoid the unneeded production of melanin, which is toxic for the insect itself [55]. An alternative possible explanation might involve the ability of some specific pathogens to produce serpin-type inhibitors or non-proteinaceous factors (e.g., polyphenol derivatives), which specifically interfere with the proteolytic activation of proPO, increasing their survival in the hosts [54,56]. Subsequent inquiries in this area would be beneficial to elucidate the function of the proPO system following bacterial infections.

This work establishes the silkworm as a reliable infection model for evaluating antimicrobial agents with therapeutic efficacy against staphylococci. Alongside the assessment of larval survival rates, two immunological markers were verified (hemocyte viability and lysozyme activation), but a third marker (proPO system) was deemed ineffective for this purpose. The utilization of these markers may facilitate a more thorough analysis of the silkworm's positive or negative responses to various antimicrobial agents; however, a preliminary and precise assessment of their behavior at varying temperatures is advisable prior to their incorporation into the screening protocol. Furthermore, the diminished quantity of utilized insects and the decreased duration necessary for the analysis of these indicators may facilitate the identification of novel methods to enhance the procedures for large-scale screening of antimicrobial compounds. Our ultimate aspiration is that this insect infection model may facilitate the expedited identification and development of innovative chemicals urgently required to combat antibiotic resistance.

REFERENCES

1. Wencewicz, T.A. Crossroads of antibiotic resistance and biosynthesis. *J. Mol. Biol.* **2019**, *431*, 3370–3399. [CrossRef] [PubMed]
2. Berini, F.; Orlandi, V.; Gornati, R.; Bernardini, G.; Marinelli, F. Nanoantibiotics to fight multidrug resistant infections by·Gram- positive bacteria: Hope or reality? *Biotechnol. Adv.* **2022**, *57*, 107948. [CrossRef] [PubMed]
3. Paudel, A.; Panthee, S.; Urai, M.; Hamamoto, H.; Ohwada, T.; Sekimizu, K. Pharmacokinetic parameters explain the therapeutic activity of antimicrobial agents in a silkworm infection model. *Sci. Rep.* **2018**, *8*, 1578. [CrossRef] [PubMed]
4. Richmond, J. The 3Rs–Past, present and future. *Scand. J. Lab. Anim. Sci.* **2000**, *27*, 84–92.
5. Mylonakis, E.; Casadevall, A.; Ausubel, F.M. Exploiting amoeboid and non-vertebrate animal model systems to study the virulence of human pathogenic fungi. *PLoS Pathog.* **2007**, *3*, e101. [CrossRef]
6. Kong, C.; Eng, S.A.; Lim, M.P.; Nathan, S. Beyond traditional antimicrobials: A *Caenorhabditis elegans* model for discovery of novel anti-infectives. *Front. Microbiol.* **2016**, *7*, 1956. [CrossRef]
7. Matsumoto, Y.; Miyazaki, S.; Fukunaga, D.H.; Shimizu, K.; Kawamoto, S.; Sekimizu, K. Quantitative evaluation of cryptococcal pathogenesis and antifungal drugs using a silkworm infection model with *Cryptococcus neoformans*. *J. Appl. Microbiol.* **2012**, *112*, 138–146. [CrossRef]
8. Panthee, S.; Paudel, A.; Hamamoto, H.; Sekimizu, K. Advantages of the silkworm as an animal model for developing novel antimicrobial agents. *Front. Microbiol.* **2017**, *8*, 373. [CrossRef]
9. Buchmann, K. Evolution of innate immunity: Clues from invertebrates via

fish to mammals. *Front. Immunol.* **2014**, *5*, 459. [CrossRef]

10. Ménard, G.; Rouillon, A.; Ghukasyan, G.; Emily, M.; Felden, B.; Donnio, P.Y. *Galleria mellonella* larvae as an infection model to investigate srna-mediated pathogenesis in *Staphylococcus aureus*. *Front. Cell. Infect. Microbiol.* **2021**, *11*, 631710. [CrossRef]
11. The International Silkworm Genome Consortium. The genome of a lepidopteran model insect, the silkworm *Bombyx mori*. *Insect Biochem. Mol. Biol.* **2008**, *38*, 1036–1045. [CrossRef] [PubMed]
12. Xu, H.; O'Brochta, D.A. Advanced technologies for genetically manipulating the silkworm *Bombyx mori*, a model Lepidopteran insect. *Proc. Biol. Sci.* **2015**, *282*, 20150487.
13. Kaito, C.; Akimitsu, N.; Watanabe, H.; Sekimizu, K. Silkworm larvae as an animal model of bacterial infection pathogenic to humans. *Microb. Pathog.* **2002**, *32*, 183–190. [CrossRef] [PubMed]
14. Uchida, R.; Hanaki, H.; Matsui, H.; Hamamoto, H.; Sekimizu, K.; Iwatsuki, M.; Kim, Y.P.; Tomoda, H. In vitro and in vivo anti-MRSA activities of nosokomycins. *Drug Discov. Ther.* **2014**, *8*, 249–254. [CrossRef] [PubMed]
15. Montali, A.; Berini, F.; Brivio, M.F.; Mastore, M.; Saviane, A.; Cappellozza, S.; Marinelli, F.; Tettamanti, G. A silkworm infection model for in vivo study of glycopeptide antibiotics. *Antibiotics* **2020**, *9*, 300. [CrossRef]
16. Tuba, T.; Chowdhury, F.R.; Hossain, T.; Farzana, M.; Ahad, I.; Hossain, M.M.; Hossain, M.I.; Saleh, N.; Nawaar, N.; Uddin, M.A.; et al. *Klebsiella pneumoniae* pathogenicity in silk moth larvae infection model. *FEMS Microbial. Lett.* **2022**, *368*, 21–24. [CrossRef]
17. Usui, K.; Miyazaki, S.; Kaito, C.; Sekimizu, K. Purification of a soil bacteria exotoxin using silkworm toxicity to measure specificactivity. *Microb. Pathog.* **2009**, *46*, 59–62. [CrossRef]
18. Suzuki, J.; Uda, A.; Watanabe, K.; Shimizu, T.; Watarai, M. Sym-biosis with *Francisella tularensisprovides* resistance to patho-gens in the silkworm. *Sci. Rep.* **2016**, *6*, 31476. [CrossRef]
19. Castillo, Y.; Suzuki, J.; Watanabe, K.; Shimizu, T.; Watarai, M. Effect of vitamin A on *Listeria monocytogenes* infection in a silkwormmodel. *PLoS ONE* **2016**, *11*, e0163747. [CrossRef]
20. Hosoda, K.; Koyama, N.; Hamamoto, H.; Yagi, A.; Uchida, R.; Kanamoto, A.; Tomoda, H. Evaluation of anti-mycobacterial compounds in a silkworm infection model with *Mycobacteroides abscessus*. *Molecules* **2020**, *25*, 4971. [CrossRef]
21. Uchida, R.; Namiguchi, S.; Ishijima, H.; Tomoda, H. Therapeutic effects

of three trichothecenes in the silkworm infection assay with *Candida albicans*. *Drug Discov. Ther.* **2016**, *10*, 44–48. [CrossRef] [PubMed]

22. Nakamura, I.; Kanasaki, R.; Yoshikawa, K.; Furukawa, S.; Fujie, A.; Hamamoto, H.; Sekimizu, K. Discovery of a new antifungal agent ASP2397 using a silkworm model of *Aspergillus fumigatus* infection. *J. Antibiot.* **2017**, *70*, 41–44. [CrossRef] [PubMed]
23. Oliveira, W.F.; Silva, P.M.S.; Silva, R.C.S.; Silva, G.M.M.; Machado, G.; Coelho, L.C.B.B.; Correia, M.T.S. *Staphylococcus aureus* and *Staphylococcus epidermidis* infections on implants. *J. Hosp. Infect.* **2018**, *98*, 111–117. [CrossRef] [PubMed]
24. Otto, M. Molecular basis of *Staphylococcus epidermidis* infections. *Semin. Immunopathol.* **2012**, *34*, 201–214. [CrossRef] [PubMed]
25. Marcone, G.L.; Binda, E.; Berini, F.; Marinelli, F. Old and new glycopeptide antibiotics: From product to gene and back in the post-genomic era. *Biotechnol. Adv.* **2018**, *36*, 534–554. [CrossRef] [PubMed]
26. Cappellozza, L.; Cappellozza, S.; Saviane, A.; Sbrenna, G. Artificial diet rearing system for the silkworm *Bombyx mori* (Lepidoptera: Bombycidae): Effect of vitamin C deprivation on larval growth and cocoon production. *Appl. Entomol. Zool.* **2005**, *40*, 405–412. [CrossRef]
27. Franzetti, E.; Romanelli, D.; Caccia, S.; Cappellozza, S.; Congiu, T.; Rajagopalan, M.; Grimaldi, A.; de Eguileor, M.; Casartelli, M.; Tettamanti, G. The midgut of the silkmoth *Bombyx mori* is able to recycle molecules derived from degeneration of the larval midgut epithelium. *Cell Tissue Res.* **2015**, *361*, 509–528. [CrossRef]
28. Casati, B.; Terova, G.; Cattaneo, A.G.; Rimoldi, S.; Franzetti, E.; de Eguileor, M.; Tettamanti, G. Molecular cloning, characterization and expression analysis of *ATG1* in the silkworm, *Bombyx mori*. *Gene* **2012**, *511*, 326–337. [CrossRef]
29. CLSI. *Performance Standards for Antimicrobial Susceptibility Testing*, 28th ed.; Clinical and Laboratory Standards Institute: Wayne, PA, USA, 2018.
30. Finney, D.J. *Probit Analysis*, 3rd ed.; Cambridge University Press: Cambridge, UK, 1971.
31. Brady, D.; Saviane, A.; Romoli, O.; Tettamanti, G.; Sandrelli, F.; Cappellozza, S. Oral infection in a germ-free *Bombyx mori* model. In *Immunity in Insects*; Sandrelli, F., Tettamanti, G., Eds.; Humana: New York, NY, USA, 2020; pp. 217–231.
32. Ashida, M.; Ishizaki, Y.; Iwahana, H. Activation of pro-phenoloxidase by bacterial cell walls or β-1,3-glucans in plasma of the silkworm, *Bombyx mori*. *Biochem. Biophys. Res. Commun.* **1983**, *113*, 562–568. [CrossRef]
33. Bruno, D.; Montali, A.; Mastore, M.; Brivio, M.F.; Mohamed, A.; Tian, L.; Grimaldi, A.; Casartelli, M.; Tettamanti, G. Insights into the immune response

of the black soldier fly larvae to bacteria. *Front. Immunol.* **2021**, *12*, 745160. [CrossRef]

34. Eleftherianos, I.; Heryanto, C.; Bassal, T.; Zhang, W.; Tettamanti, G.; Mohamed, A. Haemocyte-mediated immunity in insects: Cells, processes, and associated components in the fight against pathogens and parasites. *Immunology* **2021**, *164*, 401–432. [CrossRef] [PubMed]
35. González-Santoyo, I.; Córdoba-Aguilar, A. Phenoloxidase: A key component of the insect immune system. *Entomol. Exp. Appl.* **2012**, *142*, 1–16. [CrossRef]
36. Hultmark, D. Insect lysozymes. *EXS* **1996**, *75*, 87–102. [PubMed]
37. Doke, S.K.; Dhawale, S.C. Alternatives to animal testing: A review. *Saudi Pharm. J.* **2015**, *23*, 223–229. [CrossRef]
38. Kaito, C.; Murakami, K.; Imai, L.; Furuta, K. Animal infection models using non-mammals. *Microbiol. Immunol.* **2020**, *64*, 585–592. [CrossRef]
39. Hamamoto, H.; Kurokawa, K.; Kaito, C.; Kamura, K.; Manitra Razanajatovo, I.; Kasuhara, H.; Santa, T.; Sekimizu, K. Quantitative evaluation of the therapeutic effects of antibiotics using silkworms infected with human pathogenic microorganisms. *Antimicrob. Agents Chemother.* **2004**, *48*, 774–779. [CrossRef]
40. Kaito, C.; Kurokawa, K.; Matsumoto, Y.; Terao, Y.; Kawabata, S.; Hamada, S.; Sekimizu, K. Silkworm pathogenic bacteria infection model for identification of novel virulence genes. *Mol. Microbiol.* **2005**, *56*, 934–944. [CrossRef]
41. Barman, T.K.; Arora, P.; Rao, M.; Bhadauriya, T.; Upadhyay, D.J. Utilization of *Bombyx mori* larvae as a surrogate animal model for evaluation of the anti-infective potential of oxazolidinones. *J. Infect. Chemother.* **2008**, *14*, 166–169. [CrossRef]
42. Tabuchi, F.; Matsumoto, Y.; Ishii, M.; Tatsuno, K.; Okazaki, M.; Sato, T.; Moriya, K.; Sekimizu, K. Synergistic effects of vancomycin and β-lactams against vancomycin highly resistant *Staphylococcus aureus*. *J. Antibiot.* **2017**, *70*, 771–774. [CrossRef]
43. Becker, K.; Heilmann, C.; Peters, G. Coagulase-negative staphylococci. *Clin. Microbiol. Rev.* **2014**, *27*, 870–926. [CrossRef]
44. Sabaté Brescó, M.; Harris, L.G.; Thompson, K.; Stanic, B.; Morgenstern, M.; O'Mahony, L.; Richards, R.G.; Moriarty, T.F. Pathogenic mechanisms and host interactions in *Staphylococcus epidermidis* device-related infection. *Front. Microbiol.* **2017**, *8*, 1401. [CrossRef] [PubMed]
45. Knafl, D.; Tobudic, S.; Cheng, S.C.; Bellamy, D.R.; Thalhammer, F. Dalbavancin reduces biofilms of methicillin-resistant *Staphylococcus aureus* (MRSA) and methicillin-resistant *Staphylococcus epidermidis* (MRSE). *Eur. J. Clin.*

Microbiol. Infect. Dis. **2017**, *36*, 677–680. [CrossRef] [PubMed]

46. Candiani, G.; Abbondi, M.; Borgonovi, M.; Romanò, G.; Parenti, F. In-vitro and in-vivo antibacterial activity of BI 397, a new semi-synthetic glycopeptide antibiotic. *J. Antimicrob. Chemother.* **1999**, *44*, 179–192. [CrossRef] [PubMed]
47. Chen, A.Y.; Zervos, M.J.; Vazquez, J.A. Dalbavancin: A novel antimicrobial. *Int. J. Clin. Pract.* **2007**, *61*, 853–863. [CrossRef] [PubMed]
48. Di Pilato, V.; Ceccherini, F.; Sennati, S.; D'Agostino, F.; Arena, F.; D'Atanasio, N.; Di Giorgio, F.P.; Tongiani, S.; Pallecchi, L.; Rossolini, G.M. In vitro time-kill kinetics of dalbavancin against *Staphylococcus* spp. biofilms over prolonged exposure times. *Diagn. Microbiol. Infect. Dis.* **2020**, *96*, 114901. [CrossRef] [PubMed]
49. Goldstein, B.P.; Draghi, D.C.; Sheehan, D.J.; Hogan, P.; Sahm, D.F. Bactericidal activity and resistance development profiling of dalbavancin. *Antimicrob. Agents. Chemother.* **2007**, *51*, 1150–1154. [CrossRef]
50. Chen, K.; Lu, Z. Immune responses to bacterial and fungal infections in the silkworm, *Bombyx mori*. *Dev. Comp. Immunol.* **2018**, *83*, 3–11. [CrossRef]
51. Li, T.; Yan, D.; Wang, X.; Zhang, L.; Chen, P. Hemocyte changes during immune melanization in *Bombyx mori* infected with *Escherichia coli*. *Insects* **2019**, *10*, 301. [CrossRef]
52. Tsakas, S.; Marmaras, V.J. Insect immunity and its signalling: An overview. *Invertebr. Surviv. J.* **2010**, *7*, 228–238.
53. Isao, M.; Tomoki, H.; Yoshiaki, Y. Lysozyme activity in immunized and non-immunized hemolymph during the development of the silkworm, *Bombyx mori*. *Comp. Biochem. Physiol.* **1994**, *108*, 311–314. [CrossRef]
54. Cerenius, L.; Lee, B.L.; Söderhäll, K. The proPO-system: Pros and cons for its role in invertebrate immunity. *Trends Immunol.* **2008**, *29*, 263–271. [CrossRef] [PubMed]
55. An, C.; Budd, A.; Kanost, M.R.; Michel, K. Characterization of a regulatory unit that controls melanization and affects longevity of mosquitoes. *Cell. Mol. Life Sci.* 2011, *68*, 1929–1939. [CrossRef] [PubMed]
56. Ma, Z.; Li, C.; Pan, G.; Li, Z.; Han, B.; Xu, J.; Lan, X.; Chen, J.; Yang, D.; Chen, Q.; et al. Genome-wide transcriptional response of silkworm (*Bombyx mori*) to infection by the microsporidian *Nosema bombycis*. *PLoS ONE* 2013, *8*, e84137. [CrossRef] [PubMed]

9
Chapter

THE IMPACT OF CENTRAL EUROPEAN LOCAL MULBERRY GENOTYPES ON SILKWORM DEVELOPMENT, HEALTH STATUS, AND RAW SILK QUALITY PARAMETERS.

1. INTRODUCTION: AN OVERVIEW

China and India currently dominate silk production, accounting for over 97% of worldwide sourcing of this natural fiber. However, projects are underway to revive silk production in European countries that were significant producers in previous centuries. During the Austro-Hungarian Empire, sericulture flourished in the regions that comprise present-day Slovenia and Hungary [1]. The intense competition in the worldwide silk market led to the cessation of advanced silkworm raising in these two countries around 60 years ago [2–4]. Currently, throughout these nations, one may still observe white mulberry (Morus alba L.) trees that are up to 600 years old—living testaments to sericultural history, since their leaves have been utilized for feeding the larvae of the domestic silk moth (Bombyx mori L.) for centuries. These mulberry genotypes are suited to the local climatic and environmental circumstances and were suitable for feeding silkworm larvae historically. Sericulture declined in earlier decades in Italy, but current production is on the rise. In Italy, high-performance hybrids of B. mori are presently utilized for silk production, typically nourished by chosen mulberry genotypes. A critical practical inquiry regarding the re-establishment of sericulture is whether locally adapted genotypes from other European nations are suitable for nourishing contemporary silkworm hybrids, or if reference varieties are essential to attain optimal performance, specifically in terms of silk cocoon quality that meets standards and market expectations, as well as quantities comparable to those of rearing farms in Italy [5].

A recent collaborative Slovenian-Hungarian fundamental research initiative was conducted to delineate the prerequisites for the reinstatement of ecologically and commercially viable sericulture. The primary research inquiry was whether the growth and health of silkworm larvae, silk cocoon production, and raw silk quality are influenced by the diversity and nutritional properties of locally adapted mulberry trees indigenous to Hungary and Slovenia, in comparison to reference varieties utilized for silkworm cultivation. Local mulberry genetic resources were documented, and their biochemical features were analyzed using a multidisciplinary research approach [6,7]. The objective of the research was to evaluate selected genotypes in a silkworm feeding experiment and to assess their impact on silkworm development, cocoon production, larval health, and raw silk parameters against reference mulberry cultivars.

2. MATERIALS AND METHODS

2.1Mulberry Material for Silkworm Feeding Experiment

Mulberry genotypes from the mulberry collection were chosen for the feeding experiment based on previous screening of metabolites in leaves [6,7], and were organized into three sections: the first part is represented by Italian sericultural *M. alba* varieties ('Giazzola', 'Florio', 'Morettiana') and the Japanese variety 'Kokusou-20' obtained from the gene bank at the Centro di Ricerca per l'Agricoltura e Ambiente (CREA-AA), Laboratorio di Gelsibachicoltura di Padova, Italy. The second part comprises vegetatively propagated trees derived from the local historical Slovenian and Hungarian mulberries, which were obtained during the classification of the mulberry gene pool. The third part of the collection is intended for growing recent varieties of *M. alba*, *M. nigra* and *M. australis* as well as hybrids of *M. alba* × *rubra* suitable for fruit production.

The choice of mulberry trees for reference plantation feeding experiments was conducted according to: (1) geographical distribution of the original old local mulberry geno- types in Hungary and Slovenia; (2) multivariate analyses of previous biochemical values of original trees that allowed us to define seven chemotypes in more details in terms of the composition of individual amino acids and phenolics [6,7], out of which genotypes rich in total proteins, total phenolics, chlorogenic acid, and particular flavonoids were chosen. For the experiment, three to five trees of the same genotype were harvested.

Mulberry trees are grouped into four main sections:

A. Slovenian mulberry genotypes: selected, locally adapted genotypes, obtained by cuttings from old (trunk diameter > 180 cm) Slovenian M. alba trees (n = 10);

B. Hungarian mulberry genotypes: selected, locally adapted genotypes, obtained by cuttings from old (trunk diameter > 180 cm) Hungarian M. alba trees (n = 16);

C. reference sericultural M. alba varieties, obtained from the mulberry gene bank of the CREA-AA ('Kokusou-20', 'Morettiana', 'Florio', 'Giazzola'); varieties of M. alba, M. alba × rubra, M. nigra and M. australis grown for fruit production.

D. The list of mulberry genotypes along with their location and geographic coordinates of historical mulberry trees are presented in Supplementary Table S1.

2.2 Determination of the Chemical Composition of Mulberry Leaves

For biochemical analyses, five to seven fully developed sun-exposed leaves (5th to 7th leaf below the apex) from a one-year branch were collected randomly per each sampled tree (3–5 trees per selected genotype) and used as one sample. The trees were sampled twice during the experiment on 16th and 23rd of June 2021. Samples were immediately stored on dry ice, later transferred to a freezer at - 80°C, and subsequently freeze-dried and grounded. The prepared samples were then stored in airtight vials at –20°C prior to biochemical analyses. The concentrations of the analyzed nutrients are calculated on a dry weight (DW) basis.

Total proteins were determined spectrophotometrically at 595 nm, following the Bradford procedure [8]. The total protein content was calculated based on a standard curve, which was prepared using bovine serum albumin (BSA, 0.05–0.25 mg mL^{-1}) and expressed as mg BSA equivalent per 100 g of dried mulberry leaves (mg BSA/100 g DW).

Total phenolics of the methanolic extracts (3% formic acid in 95% methanol) were determined using the Folin-Ciocalteu method following the procedure of Ainsworth and Gille-spie [9]. The absorbance was measured at 765 nm against a reagent blank (3% formic acid in 95% methanol). Gallic acid was used as the reference standard (GA, 0.025–0.25 mg mL^{-1}). The total PH was expressed as mg gallic acid equivalents per 100 g of dried leaf sample (mg GAE/100 g DW).

Furthermore, the methanolic extracts were subjected to the gradient HPLC analysis as described in detail in a previous study [6]. All considered phenolic compounds were previously identified by a mass spectrometer (Thermo Finigan, San Jose, CA, USA) with an electrospray interface (ESI) operating in negative ion mode as previously described in Šelih et al. [7]. The identification of compounds was confirmed by comparing their spectra, retention times, and fragmentation as well as by adding the standard solution to the sample. Quantification was achieved by comparing with corresponding external standards (chlorogenic acid, kaempferol, p-coumaric acid, rutin, and quercetin; all obtained from Sigma Aldrich) of known concentrations. For the compounds for which the standards were not available, related compounds were used as standards. Therefore, quercetin-3-O- glucoside (isoquercetin), quercetin dirhamnosylhexoside and quercetin malonylhexoside were quantified in quercetin equivalents, kaempferol acetylhexoside in kaempferol equiv- alents, caffeoylquinic acid derivatives in equivalent of chlorogenic acid and p-coumaric and p-coumaroylquinic acid derivatives in equivalent of p-coumaric acid. The contents of individual compounds were expressed in mg/g DW.

Macro- and microelement analyses (phosphorus (P), sulphur (S), chlorine (Cl), potassium (K), calcium (Ca), manganese (Mn), iron (Fe), nickel (Ni), zinc (Zn), rubidium (Rb) and strontium (Sr)) were performed with a tabletop X-ray fluorescence (XRF) spectrometer PEDUZO T02 (Jožef Stefan Institute, Slovenia) with rhodium-anode X-ray tube. X-ray fluorescence was detected by a silicon drift diode detector (Amptek Inc., Bedford, MA, USA). The energy resolution of the spectrometer at count rates below 1000 cps was 140 eV at 5.9 keV. Measurements were performed in air and the samples were irradiated for 1000–5000 s to ensure sufficiently low statistical error [10,11]. The spectra were analysed by software operating in LabVIEW [12]. Element analysis was validated using standard reference material NIST SRM 1573a (tomato leaves).

2.3 Silkworm Rearing

2.3.1 Silkworm Genetic Material

Certified, traceable silkworm genetic material (eggs of a polyhybrid strain) was obtained from CREA-AA, Laboratory of Sericulture in Padua, Italy. The polyhybrid (4-way-hybrid) strain was generated by crossing a parental Chinese strain (SC2 × SC3) × a parental Japanese strain (SG1 × SG3). Eggs proved to be negative to the microscopic pebrine analysis, which is routinely carried out to detect *Nosema bombycis* infection.

2.3.3 Silkworm Rearing Technology

Estimation of the optimal timing for the start of silkworm rearing was determined by observation of the vegetative status of mulberry trees according to the reference guidelines of Brion [13].

The eggs of silkworm hybrids were transported from CREA-AA to the Faculty of Agriculture and Life Sciences, University of Maribor, Slovenia in controlled temperature conditions (25°C). The eggs were weighed to have similar quantities for each lot, i.e., about 5000 larvae per thesis. In all five instars of silkworm development silkworms were reared in faculties facility near reference plantation in cardboard boxes, changed after each moult. Photoperiod was 12 h day-night. Rearing temperature and relative humidity all the time varied between 23–26°C and 45–60%, respectively. Rearing place was constantly aerated.

Until the beginning of the 4th developmental instar larvae were reared together in a common box and fed with the mixture of leaves of Slovenian (Slo), Hungarian (Hu) mulberry genotypes and reference sericultural varieties (ref) of *Morus alba* trees from the mulberry collection (groups 1–4). After the 3rd moult larvae were randomly distributed to 38 cardboard boxes (25 × 55 × 5 cm) in groups of 30 larvae to test individual mulberry genotypes. For studies on gut microbiome analysis and haemocyte parameters, groups of larvae were fed with mixtures of leaves according to the four main groups (i.e., Slo, Hu, ref, and fruit varieties).

Larvae were fed ad libitum with leaves of selected genotypes/species of mulberry trees. When in the rearing box about 10% of leaves from the previous feeding were left, a new portion of food was added every 3–4 h during the day and every 6 h, during the night. Leaves were offered until the last larva started to spin or for 12 days after the last moult to those larvae which did not start to spin. After the start of each moult, food was not offered for two days, in order to obtain complete moulting from all the individuals of each lot and to start the following instar homogeneously.

The cumulative body weight of the larvae within one group was measured every 24 h. Larvae were observed for activities (e.g., eating, moulting, spinning) and health status (signs of any disease).

Cocoons were collected eight days after beginning of spinning. The number and cumulative weight of the cocoons were measured in each experimental group. Cocoons were dried to 42–45% of the original fresh weight at 60°C.

2.3.3 Monitoring the Health and Microbiological Status of Silkworm Larvae

The activity of larvae was checked at least six times per day, during feeding and bodyweight measuring. Larvae showing signs of illness (e.g., inactivity, cessation of eating, sluggishness, or flaccidity, colour changes, swollen body, fragile intersegmental membranes, diarrhoea) or dead larvae were immediately removed from the group and were stored at −20°C until processing.

Molecular Detection of Bombyx Mori Nucleopolyhedrovirus

Bombyx mori nucleopolyhedrovirus (BmNPV, *Baculoviridae*) (the causative agent of Grasserie disease) has previously been detected in larvae at CREA-AA [14]. Vertical (transovar- ial) transmission occurs with BmNPV, and the virus can persist in some larvae as sub- lethal infection [15,16]. Therefore, diseased larvae were tested for infection by in-house developed, qualitative real-time PCR (rt-PCR) assay using the TaqMan technology. Larval specimens were homogenised in sterile ceramic mortars, and 10 volume sterile phosphate-buffered saline (PBS) was added. Homogenates were centrifuged at 1500 × g for five minutes and total DNA was extracted from 200 μL supernatants, using the Qiagen Viral DNA extraction kit (Qiagen, Hilden, Germany). The rt-PCR assay was targeting the polymerase gene region (genomic primer: 5′-GCCACCGTAATCACRCGTCTTT-3′, complementary primer: 5′-CGATAACCCGGGCAAAAA-3′ and TaqMan probe: 5′-FAM- ACCTTCATTATTATCGTCAGCCGATTTGCG-TAMRA-3′. Samples were tested in duplicates. The specificity of the assay was confirmed by direct sequencing of the amplification products and identification through BLAST search in the gene bank databases. Relative quantification of BmNPV DNA loads in the samples was based on cT values.

Identification of Gut Microbiome and Potential Bacterial Pathogens by Metagenomic Studies

Three randomly selected, healthy 5th instar larvae were collected from groups fed with Slovenian and Hungarian mulberry genotypes, reference varieties, *M. australis*, and *M. nigra*. The guts of larvae from each pool were removed, homogenized, and total DNA was extracted. The DNA samples were submitted to a metagenomic investigation targeting the 16S rRNA gene of bacteria to reveal the bacterial gut microbiomes of silkworms, fed with different genotypes and species of mulberry leaves.

After merging the paired-end reads by PEAR quality-based filtering, trimming was performed by Trimmomatic, using 20 as the quality threshold, and reads longer than 50 bp were retained only [17,18]. The remaining reads after deduplication by VSEARCH [19] were taxonomically classified using Kraken2 (k = 35) with the NCBI non-redundant nucleotide database for the shotgun sequenced samples [20,21]. In the case of 16S rRNA sequenced samples for the classification the Greengenes database was used following chimera filtering by VSEARCH [19,22]. The taxon classification data was managed in R (R Core Team) using functions of package phyloseq [23,24].

Comparison of Selected Qualitative and Quantitative Hemogram Parameters in the Main Feeding Groups of Larvae

For the description of the immune cells, 10 fifth instar larvae were collected from different experimental groups. Haemolymph was collected by cutting a proleg into Schneider's insect medium supplemented with 5% foetal bovine serum and phenylthiourea. Tenfold dilution of the haemolymph in the same medium was used to determine total cell count in a Bürker counting chamber and 30 μL aliquots of the diluted cell suspensions were placed on the spots of Hendley Essex—Diagnostic Microscope Slides for phenotype analysis. The haemocytes were allowed to settle and adhere to the slide for 60 min in a humidity chamber. After the one-hour incubation, the haemocytes were fixed with acetone for six minutes and were classified according to morphological criteria [25,26]. The proportions of three main haemocyte classes, the granular cells, the plasmacytes, and the prohemocytes were determined.

2.4 Methods for Evaluation of Quality of Silk Cocoons

The number and cumulative weight of the cocoons were measured in each experimental group. Cocoons were dried until they reached approximately 45% of their original weight at 60°C and sent to CREA-AA, Laboratory of Sericulture, for the analysis.

The whole cocoon weight, the silk shell weight, and the silk percentage were calculated, after cutting cocoons and extracting pupae on around 20 cocoons per sample. The silk percentage was calculated according to Lee [27].

Ten cocoons for each experimental sample were used for experimental reeling on a testing reeling machine apt to reel individual cocoons, by recording the silk thread length, the title in deniers, the number of breakages, and the wastes.

2.5 Statistical Data Analysis

The results of biochemical analyses of mulberry leaves are shown as mean (average) values (± standard deviation, SD) of the analyses on two sampling dates (16th and 23rd of June 2021) during the feeding experiments, in which three to five trees were sampled and used as one sample. Measurements were performed at least four times for each sample and in duplicate. Assumptions of normality for all chemical traits were checked with Kolmogorov–Smirnov test.

The chemical traits of mulberry leaves, silkworm weight (5–7th day/5th instar), cocoon weight (fresh), silk thread parameters, and reeling wastes were presented by means and SD and were statistically evaluated by one-way analysis of variance (ANOVA), followed by post-hoc comparison according to Duncan. Letters describe significant differences among genotypes and origin-dependent genotype groups.

The Pearson correlation coefficient (Sig. 2-tailed) was calculated between evaluated chemical parameters, silkworm weight, cocoon weight, and silk thread parameters, to analyze the correlative relationship between the measured parameters and to find out the most effective differentiating traits.

Principal component analysis (PCA) enabled us to perform a comprehensive assessment of chemical traits of individual trees, silkworm weight, cocoon weight, and silk thread parameters by discriminating geographical distribution with respect to reference varieties. IBM SPSS Statistics 25 (Armonk, New York, NY, USA; 2017), StatSoft, Inc. Statistica 8.0 (Victoria, Australia; 2007) and Past 3.17 (Zürich, Switzerland; 2020) software were used for statistical analysis [28].

3. RESULTS

3.1 Chemical Composition of the Leaves of Selected Slovenian and Hungarian Mulberry Genotypes, Reference Sericultural and Fruit Varieties

The results of biochemical analyses of the total proteins, total phenolics and individual phenolic components in leaves are presented in Figure 1A,B and Supplementary Table S2. The highest total protein content was determined in genotype SM 101.1 (239.42 mg/g DW), the lowest in the Bulgarian fruit genotype. Phenolics were the highest in *M. alba rubra* (21.51 mg/g DW) and the lowest in Slovenian genotype SM 6 (11.26 mg/g DW) (Figure 1A, Supplementary Table S2).

By evaluating the individual phenolics in mulberry leaves, eight different hydroxycinnamic acids and eleven flavonols were identified (Figure 1B, Supplementary Table S2). The main phenolic acids, from the hydroxycinnamic group were caffeoylquinic derivatives (with chlorogenic acid predominating) and p-coumaroylquinic acid derivatives. The predominant flavonoids were quercetin and kaempferol glycosides. The main quercetin glycosides were rutin, quercetin malonyl-hexoside, and quercetin-3-glucoside (isoquercetin), whereas the predominant kaempferol glycoside was kaempferol acetyl-hexoside (Figure 1B, Supplementary Table S2).

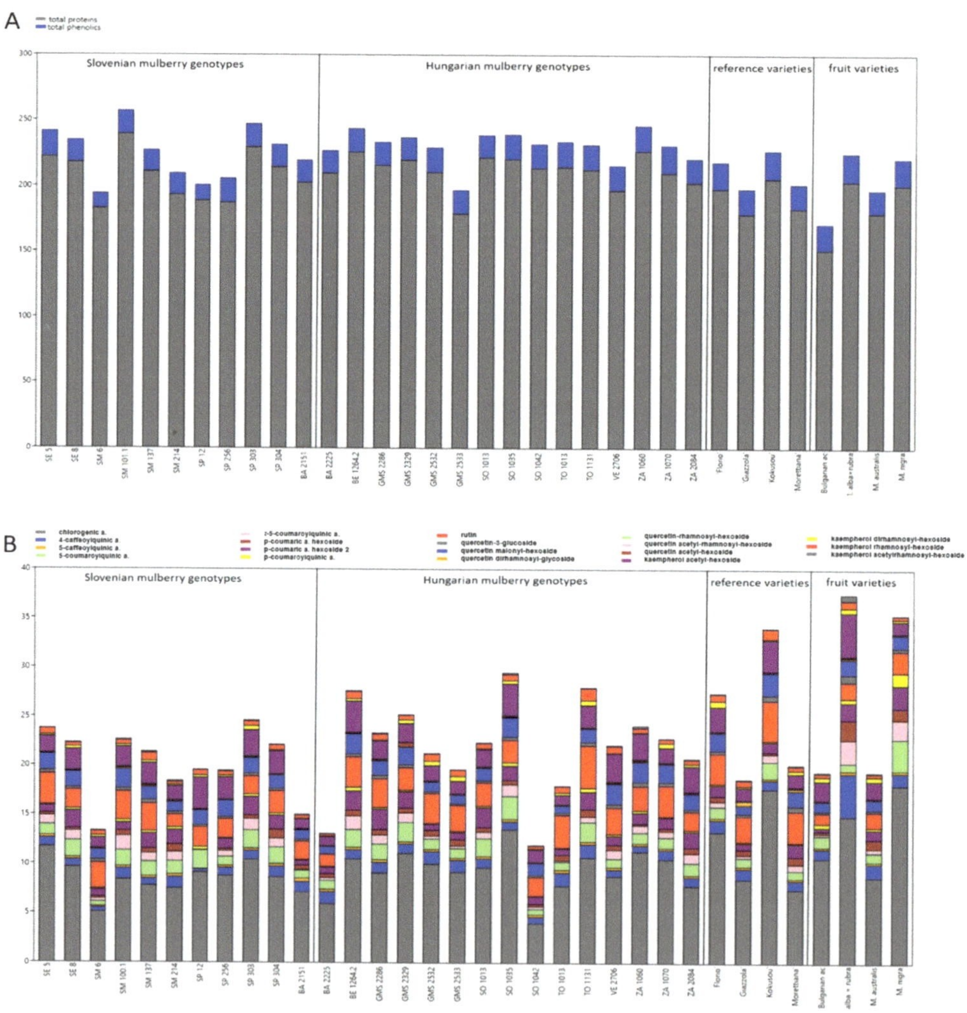

Figure 1. The mean concentrations of the (A) total proteins and total phenolics (mg/g DW), and the concentrations of individual phenolics (mg/g DW) in leaves of Slovenian, Hungarian old mulberry genotypes, reference sericultural, and fruit varieties. For detailed data and statistics see Table 1 and Supplementary Table S2.

The maximum concentration of chlorogenic acid was determined in *M. nigra* (18.05 mg/g DW), the minimum concentration in Hungarian genotype SO 1042 (4.00 mg/g DW). Within the Slovenian mulberry genotypes, the highest concentration was determined in SE 5 (11.73 mg/g DW), whereas within the Hungarian genotypes in SO 1035 (13.52 mg/g DW). Among the reference varieties 'Kokusou-20' yielded the highest chlorogenic acid content (17.67 mg/g DW), whereas the concentration of 4-caffeoylquinic acid (4.36 mg/g DW) was the highest in *M. alba* × *rubra*. The above-mentioned genotypes superior in chlorogenic acid were also characterized by the highest total caffeoylquinic acid derivatives (Figure 1B, Supplementary Table S2).

Table 1. Mean concentrations (±SD) of the total proteins, total phenolics, and predominant phenolics (mg/g DW) in Slovenian, Hungarian mulberry varieties, reference sericultural varieties, and fruit varieties. Different letters (a–c) indicate significant differences ($p < 0.05$) in the concentrations of specific compounds between the analysed groups, as determined by the post hoc Duncan test. *n*, number of repetitions.

	n	Total Proteins	Total PH	Chloro-genic a.	4-CQA	c-CQA	Rutin	Q-3-glu	QMH	KAH
Slovenian genotypes	19	211.48± 18.51a	17.03± 2.07b	9.07± 1.66c	0.92± 0.30b	1.50± 0.32a	2.32± 0.63b	0.26± 0.06b	1.51± 0.30ab	2.15± 0.48a
Hungarian genotypes	25	211.89± 13.77a	18.19± 2.02b	9.97± 1.61bc	0.91± 0.25b	1.52± 0.48a	2.81± 0.64a	0.31± 0.08ab	1.63± 0.43a	2.29± 0.65a
reference varieties	4	191.38± 12.72ab	19.95± 1.21a	11.72± 4.36ab	1.03± 0.14b	1.12± 0.37a	3.27± 0.57a	0.40± 0.15a	1.63± 0.53a	2.09± 0.87a
fruit varieties	4	183.95± 24.20b	20.07± 2.03a	13.43± 3.61a	2.47± 1.64a	1.53± 1.07a	1.63± 0.40c	0.41± 0.26a	1.26± 0.24b	2.35± 1.31a

The predominant coumaroylquinic acid derivative was 5-coumaroylquinic acid, whose concentration was the highest in *M. nigra*, yielding a total coumaroylquinic acid amount of 10 mg/g DW, followed by *M. alba* × *rubra* with 7.39 mg/g DW.

Among quercetin glycosides, rutin concentration was the highest in Hungarian geno-type TO 1131. Maximum quercetin malonylhexoside was determined in 'Kokusou-20' (2.25 mg/g DW). Quercetin-3-glucoside and kaempherol acetyl-hexoside were the highest *M. alba* × *rubra*. This hybrid was also characterized by the highest total kaempherol glycoside derivatives (6.27 mg/g DW) (Figure 1B, Supplementary Table S2).

The results of the element analysis of mulberry leaves are shown in Figure 2. P concentrations ranged between 1.07 and 2.04 g/kg DW and were the highest in genotype SE 5. Some mulberries contained more than 1 g/kg DW of S with the highest concen-tration in genotype GMS 2532 (1.26 g/kg DW). K reached maximum concentration in genotype BA 2225 (22.1 g/kg DW) and Ca in genotype GMS 2532 (20 g/kg DW) (Figure 2A, Supplementary Table S3).

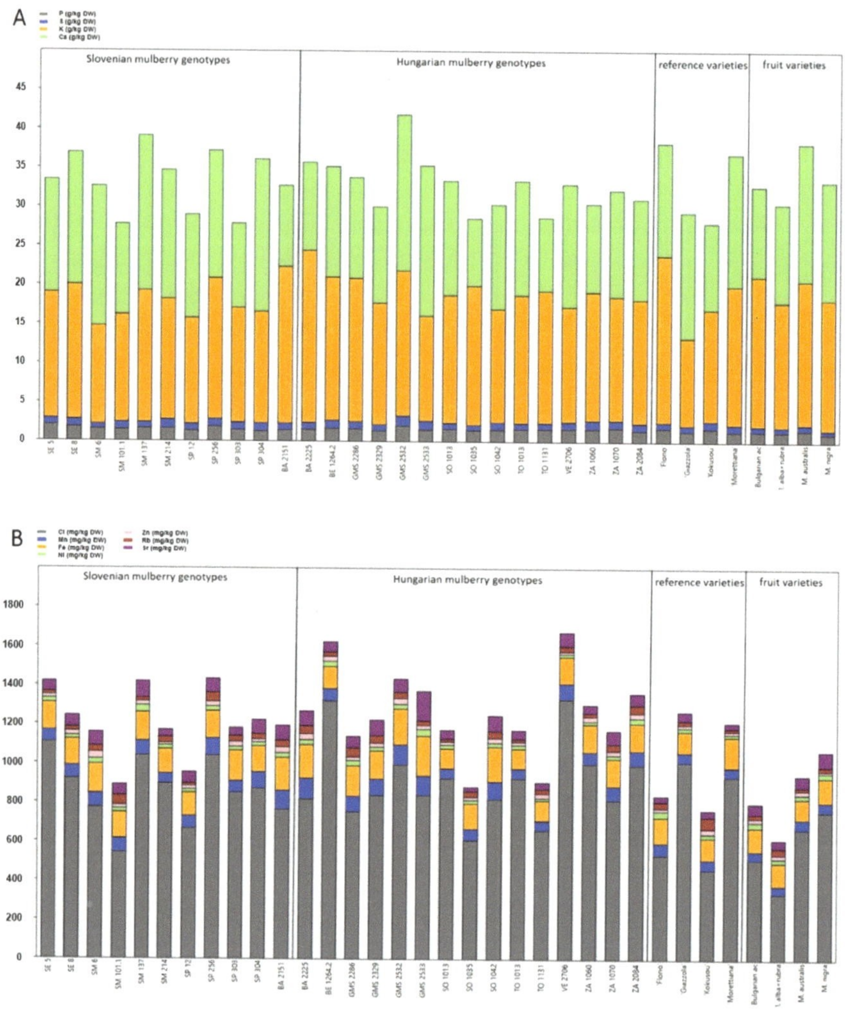

Figure 2. The mean concentrations of the (A) macroelements (mg/kg DW) and (B) microelements (mg/kg DW) in leaves of Slovenian, Hungarian old mulberry genotypes, reference sericultural and fruit varieties. For detailed data and statistics see Table 2 and Supplementary Table S3.

Table 2. Mean concentrations (± SD) of the macroelements (g/kg DW) and the predominant microelements (mg/kg DW) in mulberry leaves in Slovenian, Hungarian mulberry varieties, reference sericultural varieties, and fruit varieties. Different letters (a, b) indicate significant differences ($p < 0.05$), which were determined using the post hoc Duncan test. n, number of repetitions.

	n	P	S	K	Ca	Cl
Slovenian genotypes	19	1.75± 0.75	0.88± 0.23a	15.92± 3.59	17.14± 4.90	788.16± 36.26
Hungarian genotypes	25	1.42± 0.32	0.83± 0.19ab	16.47± 3.16	14.03± 3.47	953.20± 34.69
reference varieties	4	1.52± 0.23	0.84± 0.80ab	16.15± 4.39	14.63± 2.57	735.25± 27.79

fruit varieties	4	1.30± 0.18	0.62± 0.12b	17.58± 1.48	14.20± 2.72	569.25± 18.45

		Mn	Fe	Ni	Zn	Rb	Sr
Slovenian genotypes	19	66.04± 14.58a	147.42± 26.27ab	28.35± 11.88	23.91 ± 11.34	36.23 ± 28.72	54.02± 25.16ab
Hungarian genotypes	25	67.32± 21.29a	154.36± 29.89a	24.19± 9.40	19.98 ± 5.66	26.75 ± 9.73	72.01± 31.40a
reference varieties	4	52.03± 6.16ab	128.25± 21.42ab	21.68± 8.06	18.20 ± 5.32	31.20 ± 20.27	35.10± 7.16b
fruit varieties	4	45.38± 4.22b	119.50± 7.05b	23.65± 4.20	18.40± 4.59	22.40± 7.73	57.53± 15.07ab

Among microelements, maximum Cl concentrations were found in genotype BE 1264.2 (1320 mg/kg DW). Fe and Ni concentrations were the highest in genotype GMS 2533. The highest Zn concentration was found in Slovenian genotype SM 6. 'Kokusou-20' was found to be rich in Rb (59.4 mg/kg DW), whereas Sr concentrations ranged from 22.8 to 152 mg/kg DW with the maximum value in GMS 2533 (Figure 2B, Supplementary Table S3).

3.2 Origin Dependent Differences in Chemical Composition of Mulberry Leaves

The mulberry genotypes were categorized according to their origin to Hungarian and Slovenian old mulberry genotypes and compared to the reference and fruit varieties. Based on the statistical evaluation of the chemical compounds we determined the significant highest amounts of proteins in both Slovenian and Hungarian genotypes of *M. alba*, when compared to fruit varieties, which had the lowest protein contents. The sericultural reference varieties were intermediate in proteins (Table 1, Supplementary Table S2).

The significant highest amount of total phenolics was observed in classical contemporary sericultural genotypes and fruit varieties. In Table 1 predominant individual phenolics are summarised as mean values of Slovenian, Hungarian local genotypes, reference sericultural, and fruit varieties, whereas the ranges of all analysed phenolics of individually selected mulberry genotypes are presented in Supplementary Table S2. The least mean content of chlorogenic acid was found in Slovenian genotypes, the most in fruit varieties. Similarly, the highest amount of 4-caffeoylquinic acid was found in fruit varieties. The highest amount of rutin was analysed in reference and Hungarian genotypes of *M. alba*, the lowest in fruit varieties. Reference sericultural genotypes and fruit varieties have the highest amount of quercetin-3-glucoside, the least concentration was determined in Slovenian genotypes of mulberry trees. The highest amount of quercetin malonyl-hexoside was found in the leaves of Hungarian and reference mulberry trees, the lowest in fruit varieties (Table 1, Supplementary Table S2).

Significant differences between the mulberry groups in the contents of the main elements were determined for S, Mn, Fe, and Sr. Slovenian mulberry genotypes were characterized by the highest S content, whereas the lowest was determined in fruit varieties. Both, Slovenian and Hungarian genotypes were rich in Mn content. Fe was significantly highest in Hungarian genotypes and lowest in fruit varieties. Sr was significantly enhanced in Hungarian genotypes, whereas the lowest concentrations were found in reference varieties (Table 2, Supplementary Table S3).

3.3 Description of the Influence of Feeding Silkworms with Leaves of Various Genotypes of Mulberry Trees on Larval Development

The highest silkworm weight was obtained when silkworms were fed with the Hungarian genotype BA 2151 and the lowest with ZA 1060 (Supplementary Table S4). Out of Slovenian varieties, the maximum weight was obtained when larvae were fed with leaves of SP12 mulberry genotype, the lowest weight was determined for larvae that were fed with Submediterrean varieties (SM6, SM101.1, SM137). Among the reference varieties, significantly higher weight was obtained by feeding them with 'Florio'. Among the fruit varieties, larvae fed with *M. nigra* gave the highest weight (Figure 3, Supplementary Table S4).

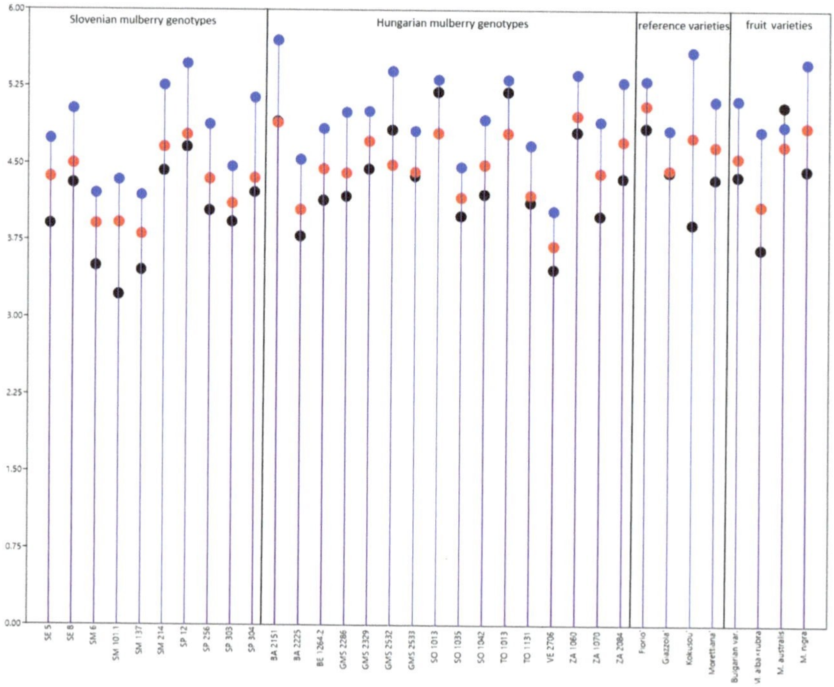

Figure 3. Mean weight of a larvae (±SD) on day 5 (black dots), 6 (red dots), 7 (blue dots) of the fifth instar when fed with Slovenian, Hungarian mulberry varieties, reference sericultural varieties, and fruit varieties. For detailed data and statistics see Table 3 and Supplementary Table S4.

Table 3. Mean weight (±SD) of a larvae and cocoon fresh weight (FW, g) when fed with Slovenian, Hungarian mulberry varieties, reference sericultural varieties, and fruit varieties. Different letters (a, b) indicate significant differences ($p < 0.05$), which were determined using the post hoc Duncan test. FW, fresh weight. n, number of repetitions.

	n	5th Instar/5th D	5th Instar/6th D	5th Instar/7th D	Cocoon FW
Slovenian mulberry genotypes	19	3.97± 0.46b	4.28± 0.33b	4.77± 0.46b	2.14± 0.12a
Hungarian mulberry genotypes	25	4.32± 0.46ab	4.46± 0.34b	4.96± 0.43b	2.26± 0.09a
reference sericultural varieties	4	4.54± 0.38a	4.85± 0.23a	5.36± 0.36a	2.25± 0.14a
fruit varieties	4	4.39± 0.57ab	4.55± 0.33ab	5.07± 0.30ab	2.20± 0.14a

3.4 Origin Dependent Differences in the Growth of the Silkworm Larvae and Cocoon Weight

Based on statistical evaluation of the four main groups (Slovenian, Hungarian, reference sericultural and fruit varieties) reference sericultural varieties were characterized by the highest larval growth, whereas no statistical differences were found between mean cocoon fresh weight (Table 3).

3.5 Description of the Influence of Feeding Silkworms with Leaves of Various Mulberry Genotypes on the Quality of Raw Silk

Among Slovenian varieties, the maximum length was obtained when larvae were fed with Slovenian variety SP 256. Superior length was also obtained with GMS 2286, whereas the lowest was obtained with the reference varieties 'Giazzola', Bulgarian fruit genotype, and M. alba × rubra, which also gave the lowest raw silk weight. Significant highest weight was obtained with Hungarian genotype SO 1013. The Hungarian genotype BE 1264.2 was characterized by the highest thickness and the Slovenian genotype SM 137 with the lowest. When analysing the reeling wastes, the highest struse was found in most of the Hungarian varieties and the significant highest telette was determined in Hungarian variety GMS 2533 (Supplementary Table S4).

3.6 Origin Dependent Differences in the Quality of Raw Silk

There was a trend for longer silk thread production when larvae were fed with Slovenian and Hungarian varieties in comparison to reference sericultural varieties, although the trend was insignificant. Fruit varieties were characterized by a shorter length and weight of raw silk. Struse was significantly largest when larvae were fed with Hungarian varieties, thickness and telette did not significantly differ among the groups. When analysing reeling wastes struse was highest in Hungarian genotypes and lowest in Slovenian genotypes and fruit varieties. There were no significant differences in telette (Table 4).

Table 4. Mean (±SD) raw silk and reeling waste parameters (g) of Slovenian, Hungarian mulberry varieties, reference sericultural varieties, and fruit varieties. Different letters (a, b) indicate significant differences ($p < 0.05$), which were determined using the post hoc Duncan test. n, number of repetitions.

	n	Silk Thread Parameters			Reeling Wastes	
		Length	Weight	Thickness	Struse	Telette
Slovenian genotypes	95	1419.09± 64.14	0.43± 0.03	2.72± 0.13	0.036± 0.013	0.017± 0.009
Hungarian genotypes	211	1429.18 ± 94.82 a	0.44 ± 0.03 a	2.82 ± 0.15 a	0.047 ± 0.022 a	0.021 ± 0.019 a
Reference varieties	70	1363.20± 135.94a	0.43± 0.02a	2.84± 0.19a	0.043± 0.016ab	0.017± 0.016a
Fruit varieties	30	1244.80± 44.03b	0.37± 0.03b	2.73± 0.13a	0.040± 0.020b	0.022± 0.016a

3.7 Correlation between Mulberry Metabolites, Silkworm, and Raw Silk Parameters

A high correlation was found between silkworm weight and raw silk thickness, as well as raw silk length, and weight; whereas a negative between silk thread parameters (length, weight) and reeling waste (telette, struse). The results of the chemical analysis of mulberry leaves indicate a strong correlation between total proteins and raw silk length and weight. Among phenolic acids, t-5-coumaroylquinic acid correlated strongly with the weight of silkworms. There was a medium negative correlation between p-coumaric acid hexoside and raw silk length. Among flavonols, rutin correlated positively with struse. There was a negative correlation between kaempherol-dirhamnosyl-hexoside and raw silk length and weight, whereas a positive correlation between kaempherol-rhamnosyl-hexoside and struse (Table 5).

Table 5. Pearson's correlation coefficient between silkworm, cocoon, and raw silk parameters in relation to the main phenolics (n = 32) in mulberry leaves listed according to their retention times on HPLC.

	Silkworm Weight (Larvae/g)				Cocoon Weight		Silk Thread			Reeling Waste	
Correlations	5th C/5th D	6th D	7th D	Mean 3d Weight	Fresh	Dry	Length	Weight	Thickness	Struse	Telette
5th C/5th D	1	00.856**	0.764 **	0.934**	0.777 **	0.766 **	0.11	0.191	0.438*	0.234	0.121
5th C/6th D	0.856 **	1	0.907 **	0.967**	0.713 **	0.715 **	0.088	0.21	0.452**	0.18	0.066
5th C/7th D	0.764 **	0.907**	1	0.935**	0.607 **	0.618 **	0.032	0.166	0.298	0.226	0.133
mean 3d weight	0.934 **	0.967**	0.935 **	1	0.744 **	0.745 **	0.082	0.199	0.418*	0.229	0.116
cocoon FW	0.777 **	0.713**	0.607 **	0.744**	1	0.992 **	0.024	0.376*	0.562**	0.193	0.081
cocoon DW	0.766**	0.715**	0.618**	0.745**	0.992**	1	0.016	0.378*	0.583**	0.189	0.076
length	-0.11	-0.088	-0.032	-0.082	0.024	0.016	1	0.785**	-0.075	-00.368*	-0.183
weight	0.191	0.21	0.166	0.199	0.376*	0.378*	0.785**	1	0.555**	-00.455**	-0.07
thickness	0.438*	0.452**	0.298	0.418*	0.562**	0.583**	-0.075	0.555**	1	-0.269	0.099
struse	0.234	0.18	0.226	0.229	0.193	0.189	-0.368*	-0.455**	-0.269	1	0.176
telette	0.121	0.066	0.133	0.116	0.081	0.076	-0.183	-0.07	0.099	0.176	1
total proteins	-0.198	-0.139	-0.116	-0.164	0.083	0.073	0.448**	0.495**	0.175	-0.024	-0.262
total phenolics	-0.001	0.131	0.137	0.087	0.096	0.108	-0.197	-0.12	0.069	0.191	0.13
chlorogenic a.	0.02	0.202	0.228	0.148	0.205	0.205	-0.198	-0.068	0.165	0.211	0.15

4-caffeoyl-QA	-0.092	-0.12	0.006	-0.071	-0.125	-0.136	-0.329	-0.312	-0.098	0.211	0.246
5-caffeoyl-QA	-0.031	-0.068	-0.002	-0.034	0.004	0.019	0.109	0.079	-0.023	0.046	-0.15
total caffeoyl-QA	-0.003	0.154	0.206	0.117	0.157	0.155	-0.248	-0.127	0.126	0.236	0.186
c-5-CQA	0.023	0.034	0.037	0.033	0.231	0.225	0.21	0.272	0.176	0.034	-0.167
t-5-CQA	-0.451**	-0.328	-0.281	-0.382	-0.265	-0.28	-0.01	0.002	0.003	-0.1	-0.092
p-CAH	-0.048	-0.15	-0.059	-0.084	-0.123	-0.13	-0.378*	-0.322	-0.045	0.158	0.196
p-CAH2	-0.044	-0.057	-0.1	-0.07	0.028	0.008	0.15	0.168	0.066	0.05	-0.281
p-CQA	-0.147	-0.181	-0.154	-0.168	-0.094	-0.09	-0.283	-0.313	-0.097	0.121	-0.034
total coumaroyl-QA	-0.183	-0.171	-0.146	-0.177	-0.04	-0.055	0.007	0.05	0.065	0.044	-0.145
rutin	-0.179	-0.054	-0.05	-0.108	0.081	0.101	-0.02	-0.041	-0.033	0.364*	-0.018
Q-3-glu	-0.326	-0.105	-0.056	-0.185	-0.163	-0.153	-0.247	-0.342	-0.225	0.298	0.033
QMH	-0.268	0.02	-0.047	-0.122	0.098	0.11	0.085	0.177	0.193	-0.029	-0.161
Q-diR-gly	-0.082	-0.098	-0.132	-0.109	-0.155	-0.128	-0.209	-0.228	-0.088	0.086	0.011
QRH	-0.299	-0.301	-0.235	-0.295	-0.25	-0.27	-0.16	-0.304	-0.303	0.127	0.155
Q-acetyl-RH	-0.207	-0.185	-0.084	-0.168	-0.044	-0.065	-0.082	-0.125	-0.1	0.046	-0.047
QAH	0.009	0.135	0.006	0.045	0.284	0.275	-0.132	0.054	0.264	0.189	-0.12
total quercetin-gly	-0.27	-0.044	-0.071	-0.15	0.077	0.097	-0.037	-0.018	0.033	0.298	-0.077
KAH	-0.197	-0.071	-0.007	-0.105	-0.043	-0.044	-0.077	-0.028	0.066	0.086	0.054
K-diRH	0.126	0.054	-0.044	0.052	0.05	0.01	-0.525	-0.407*	0.042	0.284	0.231
KRH	-0.199	-0.175	-0.075	-0.16	0.003	0.012	-0.152	-0.204	-0.125	0.482**	0.106
K-acetyl-RH	-0.192	-0.175	-0.091	-0.163	-0.216	-0.23	-0.247	-0.324	-0.22	0.163	0.167
total K-gly. deriv. *	-0.194	-0.101	-0.038	-0.123	-0.046	-0.053	-0.202	-0.162	0.011	0.231	0.118

** Correlation is significant at the 0.01 level (2-tailed); * Correlation is significant at the 0.05 level (2-tailed). FW, fresh weight; DW, dry weight. * 4-caffeoyl-QA, 4-caffeoylquinic acid; 5-caffeoyl-QA, 5-caffeoylquinic acid; 5-total CQA, total caffeoylquinic acid derivatives; c-5-CQA, c-5-coumaroylquinic acid; t-5-CQA, t-5-coumaroylquinic acid; p-CAH, p-coumaric acid hexoside; p-CAH2, p-coumaric acid hexoside 2; p-CQA, p-coumaroylquinic acid; total CQA, total coumaroylquinic acid derivatives; Q-3-glu, quercetin-3-glucoside; QMH, quercetin malonyl-hexoside; Q-diR-gly, quercetin dirhamnosyl-glycoside; QRH, quercetin rhamnosyl-hexoside; Q-acetyl-RH, quercetin acetyl- rhamnosyl hexoside; QAH.

There was no significant correlation between the macroelements and silkworm weight, silk thread parameters, and reeling waste. Among microelements, Cl correlated positively with silk weight, while Fe and Sr correlated with telette (Table 6).

Table 6. Pearson's correlation coefficient between silkworm, cocoon, and raw silk parameters in relation to the macro- and microelements (n = 32) in mulberry leaves.

	Silkworm Weight (Larvae/g)				Cocoon Weight		Silk Thread			Reeling Waste	
Correlations	5th C/5th D	6th D	7th D	Mean 3d Weight	Fresh	Dry	Length	Weight	Thickness	Struse	Telette
5th C/5th D	1	00.856**	0.764**	0.934**	0.777**	0.766**	0.11	0.191	0.438*	0.234	0.121
5th C/6th D	0.856**	1	0.907**	0.967**	0.713**	0.715**	0.088	0.21	0.452**	0.18	0.066
5th C/7th D	0.764**	0.907**	1	0.935**	0.607**	0.618**	0.032	0.166	0.298	0.226	0.133
mean 3d weight	0.934**	0.967**	0.935**	1	0.744**	0.745**	0.082	0.199	0.418*	0.229	0.116
cocoon FW	0.777**	0.713**	0.607**	0.744**	1	0.992**	0.024	0.376*	0.562**	0.193	0.081
cocoon DW	0.766**	0.715**	0.618**	0.745**	0.992**	1	0.016	0.378*	0.583**	0.189	0.076
length	-0.11	-0.088	-0.032	-0.082	0.024	0.016	1	0.785**	-0.075	-00.368*	-0.183
weight	0.191	0.21	0.166	0.199	0.376*	0.378*	0.785**	1	0.555**	-00.455**	-0.07
thickness	0.438*	0.452**	0.298	0.418*	0.562**	0.583**	-0.075	0.555**	1	-0.269	0.099
struse	0.234	0.18	0.226	0.229	0.193	0.189	-0.368*	-0.455**	-0.269	1	0.176
telette	0.121	0.066	0.133	0.116	0.081	0.076	-0.183	-0.07	0.099	0.176	1
P	-0.016	0.001	-0.001	-0.006	0.144	0.129	0.269	0.261	0.052	-0.068	-0.097
S	0.207	0.212	0.347	0.271	0.075	0.08	0.228	0.287	0.135	-0.017	0.331
K	0.288	0.236	0.208	0.262	0.282	0.296	0.178	0.234	0.176	-0.14	-0.155
Ca	0.074	-0.108	-0.086	-0.032	-0.103	-0.141	-0.16	-0.1	0.021	-0.108	0.166
Cl	-0.003	-0.136	-0.194	-0.11	0.042	0.063	0.237	0.352*	0.249	-0.211	-0.022
Mn	-0.058	-0.152	-0.05	-0.086	-0.035	-0.041	0.334	0.324	0.058	-0.106	0.226
Fe	-0.074	-0.115	-0.062	-0.086	-0.184	-0.202	0.167	0.113	-0.062	-0.019	0.434
Ni	-0.035	-0.027	-0.049	-0.04	-0.161	-0.211	0.027	0.012	-0.02	-0.151	0.185
Zn	-0.052	-0.048	0.025	-0.027	-0.139	-0.126	0.081	0.037	-0.032	-0.105	-0.016
Rb	-0.367	-0.094	0.056	-0.159	-0.276	-0.267	0.26	0.097	-0.185	-0.119	-0.16
Sr	-0.083	-0.205	-0.168	-0.154	-0.083	-0.123	-0.003	0.005	-0.024	-0.042	0.533**

** Correlation is significant at the 0.01 level (2-tailed); * Correlation is significant at the 0.05 level (2-tailed); FW, fresh weight; DW, dry weight.

3.8 Principal Component Analysis of Mulberry Metabolites, Silkworm, and Raw Silk Parameters

To enable a comprehensive assessment of the chemical composition of mulberry leaves, with respect to their genotype, larval growth and silk parameters, a principal component analysis (PCA) has been conducted. The discriminant function 1, which accounts for 78.14% of the variance, as explained by the model, was weighted most strongly by total phenolics, followed by rutin, quercetin glucoside, larval weight, and chlorogenic acid. It is further negatively associated with silk length, total proteins, Mn and Sr.

Thus, function 1 clearly (over 89%) separates reference varieties, which are character- ized by higher larval weight, which coincides with high values of chlorogenic acid, rutin, and other quercetin glycosides, from Hungarian and Slovenian varieties, being higher in proteins, length, and weight of silk thread. Furthermore, it must be pointed out that some old Hungarian and Slovenian varieties coincide with high concentrations of Cl, Mn, Zn, and Sr.

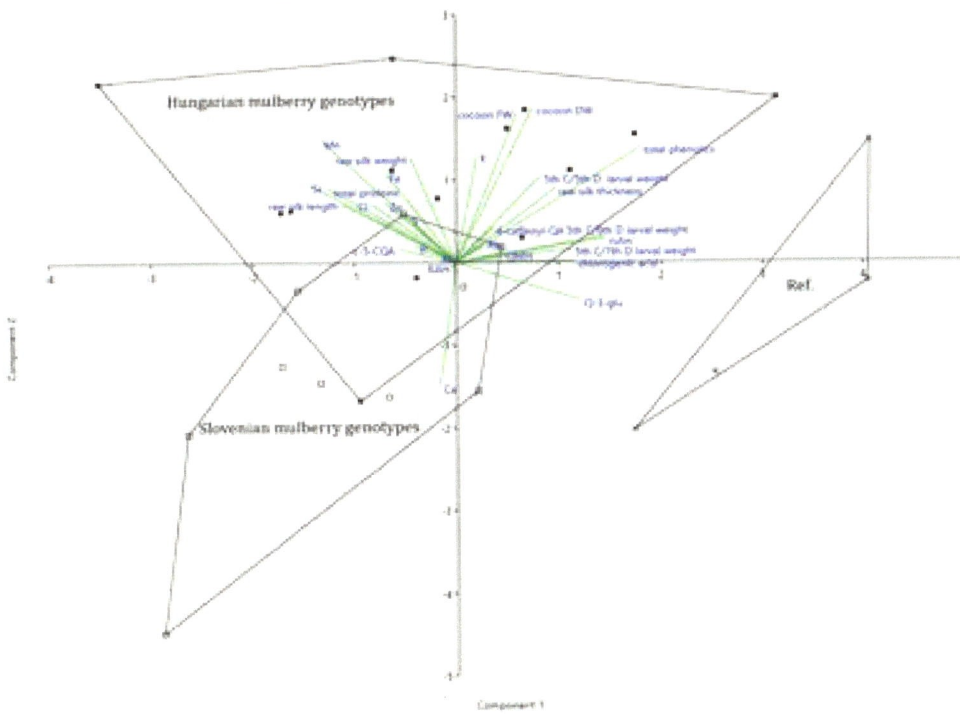

Figure 4. Principal component analysis of silkworm, cocoon, and raw silk parameters along with the chemical composition of leaves of different origins. The convex hulls delimit the space that includes the samples of mulberry trees from different Slovenian regions (squares), Hungarian regions (dots) compared to traditional high yielded sericultural varieties (Ref, crosses). * 4-caffeoyl-QA, 4-caffeoylquinnic acid; c-5-CQA, c-5-coumaroylquinnic a.; Q-3-glu, quercetin-3-glucoside; QMH, quercetin malonyl hex; KAH, kaemph acetyl-hexoside.

The second discriminant function accounts for another 21.86% of the variance. It is positively associated with cocoon fresh and dry weight, length, total phenolics, Mn, and K and negatively with Ca and quercetin glucoside. Function 2 clearly separates Hungarian mulberry (by over 87%), which is characterized by higher cocoon fresh and dry weight and the above-mentioned elements from Slovenian genotypes being high in Ca (Figure 4, Supplementary Table S5).

Genotypes with the highest average cocoon fresh and dry weight are GMS 2532 and GMS 2329. Genotype TO1131 contained high total phenolic concentration, in particular quercetin-3-rutinoside which coincided with high 5th instar/5th day weight. BA 2225 is highest in Mn and K concentrations. Slovenian genotype A6 was high in Ca and certain quercetin glycosides as well as NG 214 in Ca content and caffeoylquinic acid glycosides (Figure 4, Table 1).

3.9 Health and Microbiological Status of Silkworm Larvae

All larvae were moulting synchronously, no prolonged instar duration was observed. Obvious unequal growth and development were observed by three larvae of various groups. During rearing healthy, but cachectic larvae were found only in groups fed with Hungarian (Hu) mulberry genotypes of trees GMS 2532 (fourth instar), SO 1042 (fourth instar), and TO 1131 (fifth instar).

Six ill or dead larvae (0.47% of all larvae) were found in the following developmental stages: 1st day of third instar (one dead larva); 2nd day of third instar (one ill cachectic larva); 1st day of fourth instar (one dead larva); 1st day of fifth instar (one dead larva in group Hu-BE 1264/2); end of fifth instar (one dead larva from group Hu-ZA 2084; one dead larva from group Hu-mix).

In the groups of larvae fed with Slovenian mulberry genotypes (n = 10) all larvae successfully spun in three groups. In five groups 5%, in one group 15% and in one group 20% of the larvae failed to spin. All larvae successfully spinned in the group fed with the mixture of Slovenian mulberry genotypes.

In the groups of larvae fed with Hungarian *M. alba* genotypes (n = 16) all larvae successfully spinned in seven groups. In another seven groups 5% and in two groups 15% of the larvae failed to spin. In the group fed with the mixture of Hungarian mulberry genotypes 20% of the larvae failed to spin.

The groups fed with reference sericultural *M. alba* varieties 'Florio' and 'Morettiana' had 10% losses, while the larvae fed with leaves of variety 'Kokusou' successfully spinned. All larvae in groups fed with *M. alba* Bulgarian fruit genotype and *M. australis* leaves have successfully spinned. 5% losses were observed in the group fed with the

M. alba × *rubra*. None of the larvae fed with *M. nigra* leaves spinned cocoons, however, food was offered until 12 days after the last moult.

3.9.1 Molecular Detection of BmNPV by Real-Time PCR

The rt-PCR assays resulted in amplification signals (cT values at 17–21) in DNA extracts of diseased larvae. The nucleotide sequences of the amplification products have shown >99% similarity to an 81 bp long region of the polymerase gene region (between nt positions 47,965 and 48,045) of the BmNPV reference sequence (GenBank accession number: NC_001962). No amplification was detected in the negative control samples (healthy larvae).

Clinically healthy larvae were collected randomly from each experimental group before each moulting and spinning and were tested for BmNPV infection by real-time PCR: all real time-PCR assays in regularly sampled healthy larvae from each group were negative.

BmNPV DNA was detected in the dead or cachectic body of third and fourth instar larvae from the common box:

- 1st day, third instar: 1 dead (liquid rotting) larva; cT mean 40.63, standard deviation ·(SD) 0.84;
- 2nd day, third instar: 1 cachectic larva; cT mean 40.70, ·SD 1.49;
- 1st day, fourth instar: 1 larva died during moulting, showed signs of Grasserie; cT mean 18.32, SD 0.17.
- In the fifth instar, BmNPV DNA was detected in three out of the 1340 larvae (0.22%). Positive larvae were ill or were found dead:
- 1st day: Hu-BE 1264/2 (signs of Grasserie); cT mean 37.38, cT SD 0.69; end of 5th instar: Hu-ZA 2084.2 (moribund larva with black stripes), cT mean 20-09, ·SD 0.14;
- end of 5th instar: Hu-mix (dead larva, signs of Grasserie), CT mean 37.43, SD 0.92. BmNPV DNA was not detected in any of the 39 larvae extracted from cocoons showing signs of failed spinning.

3.9.2 Identification of Gut Microbiome and Potential Bacterial Pathogens by Metagenomic Studies

In the group of Slovenian genotypes mixture, approximately 102,489 different bacterial species were detected. In the group of Hungarian genotypes mixture, the number of detected species exceeded 106,000, while in the group reference genotypes mixture it exceeded 123,958. In group *M. australis* approximately 143,366, and in group *M. nigra* 144,551 different bacterial species were detected. Core bacterial microbiomes are shown in Figure 5.

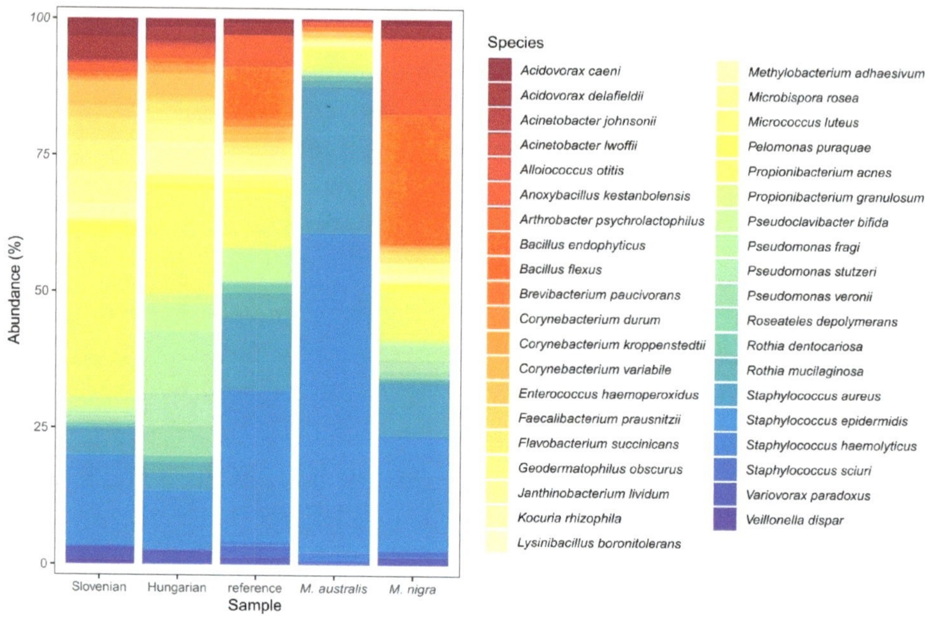

Figure 5. Species compositions of core gut bacterial microbiomes in the different experimental groups.

While on the 5th day of fifth instar all larvae were still eating, the gut microbiomes differed considerably in the samples of larvae fed with mixtures of Slovenian, Hungarian, and reference genotypes of *Morus alba*, as well as the fruit varieties *M. australis* and *M. nigra*.

Major differences in silkworm gut bacterial microbiome in different groups on the level of bacterial classes and Chloroplast are the followings:

1. 57% of representatives of Chloroplast (99% Streptophyta) were found by Slovenian genotypes and 47% by Hungarian genotypes, almost half less (22%) by reference genotypes, while *M. australis* and *M. nigra* had only 16% and 12%.
2. The range was quite opposite by class Betaproteobacteria (mainly representatives from order Burkholderiales): 44% by *M. nigra*, 39% by reference genotypes, 24% by Slovenian genotypes, 22% by Hungarian genotypes, and 18% by *M. australis*.
3. The representatives from Alphaproteobacteria (99% of representatives coming from order Rickettsiales) were in the following ranges: *M. nigra* and Hungarian genotypes 8%, Slovenian genotypes 6%, reference genotypes 3%, and *M. australis* 2% (representatives mainly from orders: Rhodobacterales 40%, Rhizobiales 24%, Sphingomonadales 12%, Rhodospirillales 12%, 9% Rickettsiales).

4. The representatives from class Gammaproteobacteria were in following ranges: Hungarian mulberry genotypes 4% (Pseudomonadales 3%, Enterobacteriaceae 1%, Xanthomonadales 0.1%), *M. nigra* 3% (Pseudomonadales 2%, Enterobacteriaceae 0.07%, Xanthomonadales 0.1%, Oceanospirillales 0.1%, Alteromonadales 0.1%), Slovenian genotypes 1% (Pseudomonadales 1%, Enterobacteriaceae 0.06%), reference genotypes 1% (Pseudomonadales 0.7%, Enterobacteriaceae 0.4%), *M. australis* 1% (Pseudomonadales 0.8%, Enterobacteriaceae 0.3%).
5. The representatives from class Actinobacteria were in the following ranges: reference genotypes 14%, Hungarian genotypes 9%, *M. australis* 9%, *M. nigra* 5%, and Slovenian genotypes 4%.
6. The representatives from class Bacilli were in the following ranges: Slovenian geno- types 4%, Hungarian genotypes 9%, reference genotypes 14%, *M. australis* 9%, and *M. nigra* 5%.
7. The representatives from class Clostridia were in the following ranges: *M. nigra* 3%, *M. australis* 2%, Slovenian genotypes 1%, Hungarian genotypes 1%, reference genotypes 0.8%.
8. The representatives from class Flavobacteriales were in the following ranges: *M. nigra* 3%, *M. australis* 2%, Slovenian genotypes 1%, Hungarian genotypes 1%, reference genotypes 0.8%.
9. The representatives from other classes were in the following ranges: Slovenian genotypes 4%, Hungarian genotypes 3%, reference genotypes 2.2%, *M. nigra* 2%, *M. australis* 1.2 %.

In comparison with other groups, Slovenian genotypes had the highest percentage of Chloroplast and lowest percentage of Gammaproteobacteria, Actinobacteria, Bacilli, Clostridia, and Flavobacteriale.

Hungarian genotypes had the highest percentages of Chloroplast, Alphaproteobacteria, and Gammaproteobacteria and the lowest of Bacilli, Clostridia, and Flavobacteriales.

In reference genotypes, the highest percentage of Actinobacteria and low percentage of Gammaproteobacteria, as well as Clostridia and Flavobacteriales, were described.

The group of *M. australis* had the lowest percentage of Chloroplast, Betaproteobacteria, Alphaproteobacteria, Gammaproteobacteria, and Flavobacteriales while the highest percentage of Bacilli of all groups. The group of *M. nigra* had the highest percentage of Alphaproteobacteria and Clostridia, second highest of Betaproteobacteria, and second lowest percentage of Chloroplast. Rikettsiales were the most representative order of class Alphaproteobacteria by all groups. The highest percentage of order Enterobacteriaceae was found by the Hungarian genotypes group, the lowest by *M. nigra*. The highest percentage of Lactobacillales was described by the group fed with *M. australis*. The lowest percentage of family Microbacteriaceae was found by Slovenian genotypes and *M. nigra*, and the highest by reference genotypes and Hungarian genotypes. The highest percentage of family Micrococcaceae was found by reference genotypes and *M. australis*; the highest percentage of family Bifidobacteriaceae was found by *M. australis* and the lowest by Hungarian genotypes and reference genotypes groups. The lowest percentage of family Staphylococcaceae was found by Slovenian genotypes and Hungarian genotypes, highest by *M. australis*.

3.9.3 Comparison of Selected Qualitative and Quantitative Hemogram Parameters in the Main Feeding Groups of Larvae

The highest total number of haemocytes was found in larvae fed with Hungarian mulberry genotypes (74.3×10^4/mL; SE 5.8×10^4/mL), followed by Slovenian genotypes (62.8×10^4/mL; SE 7×10^4/mL), reference genotypes (61.2×10^4/mL; SE 2.6×10^4/mL), *M. nigra* (58.3×10^4/mL; SE 11.2×10^4/mL) and *M. australis* (45.9×10^4/mL; SE 6.6×10^4/mL).

The qualitative analysis of the different haemocyte subsets revealed the dominance of granular cells, followed by plasmatocytes and small spheric cells (Figure 6).

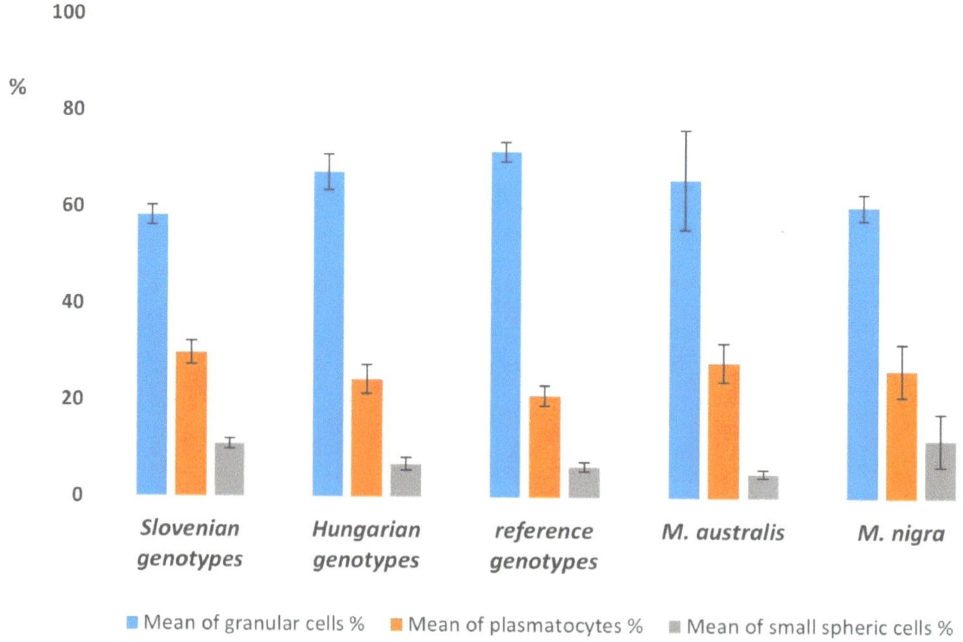

Figure 6. Differences in average haemocyte counts of larvae, fed with Slovenian, Hungarian, and reference sericultural genotypes and fruit varieties of mulberry trees.

4. DISCUSSION

The main factors that contribute to the success of cocoon harvest are mulberry feed, environmental conditions, silkworm cultivation techniques, and silkworm strains [29]. Mulberry leaves are known as a rich source of proteins with unique amino acid composition and are highly palatable and digestible for herbivorous animals [6,30–32]. It has been previously confirmed that high protein content in leaves has a direct impact on the growth of larvae and cocoon production [28,33–36].

Mulberry leaves are further known by their unique composition of phenolic com- pounds, which were shown to have biological properties [6,37–44]. The identification of phenolics in mulberry leaves with UPLC-MS has been the subject of only a few recent studies so far [44,45].

The main phenolic compound in mulberry leaves was identified as chlorogenic acid followed by other hydroxycinnamic acids, including caffeoylquinic acid, p-coumaric acid derivatives, and p-coumaroylquinic acid derivatives [6,7,44–47]. The flavonol's fraction mostly contains quercetin-3-O-rutinoside (rutin) as well as other quercetin and kaempferol glycosides [6,7,46,48–52].

It has been previously confirmed that chlorogenic acid enhances the rate of silkworm development. Furthermore, quercetin and kaempferol glycosides have been found to be transferred from the larval diet into the hemolymph and cocoons, where they act as UV shields and antimicrobial barriers and might therefore increase the survival rate of devel- oping insects inside the cocoons [53–58]. However, when treated artificially, flavonoids may act as antinutrients by binding to amino acids and proteins as well as digestive gut enzymes they may reduce the nutrient value of mulberry leaves [50].

Based on mulberry inventory in Slovenia and Hungary, we established a collection of historical mulberry trees and screened their biochemical patterns regarding important primary (proteins and amino acids) metabolites and phenolics; we aimed for a definition of high-yielding and nutritive richer mulberry genotypes from the local Slovenian and Hungarian gene pool, which were included in the presented feeding experiment and compared with reference sericultural and fruit varieties. The analysis of study result data indicates possible influences of mulberry chemical parameters on silkworm development and production.

4.1 The Influence of Proteins and Phenolics on Larval, Cocoon and Raw Silk Parameters

The silkworm's growth, cocoon, and raw silk quality can be affected by their feed sources to a great extent. The results of the screening of the metabolites in the leaves of Slovenian and Hungarian genotypes from their local origin were previously published by Urbanek Krajnc et al., and Šelih et al. [6,7]. These data served as the basis for the current research into the impact of feeding silkworms with selected mulberry genotypes out of the local gene pool. Based on biochemical analyses, we selected those genotypes of old local mulberries that proved to be the most favorable for silkworm rearing in terms of nutritional value and leaf yield.

The protein content of mulberry leaves ranges between 13 to 31% [29,51,59,60]. In the current experiment, the highest total protein content was determined in the Slovenian genotype SM 101.1 (239.42 mg/g DW), the lowest in the fruit variety Bulgarian accession (151.50 mg/g DW). The results further coincide with our previous study comprising local trees of the Gorizian region alone and Slovenian old mulberry genotypes sampled at the place of their origin with respect to pruning management and eco-geographical origin [6,7]. We confirmed a strong positive correlation between total proteins and raw silk length and weight. It has been previously confirmed that high protein content in leaves has a direct impact on the growth of larvae and cocoon production [29,61].

Mulberry leaves are also known to have high contents of phenolics, that in our varieties ranged between 11.26 and 21.51 mg/g DW. They were highest in the *M. alba* × *rubra* variety (21.51 mg/g DW) and lowest in SM 6 (11.26 mg/g DW). The analyzed concentrations were in accordance with those of Sánchez-Salcedo et al., who found phenolics in mulberry leaves in the range between 12.8 and 15.5 g GAE/g DW [62].

As previously reported, the predominant caffeoylquinic acid derivative was chlorogenic acid followed by 4-caffeoylquinic acid. In our feeding experiment, the average concentration of chlorogenic acid ranged between 4.00 and 18.05 mg/g DW, with the maximum content analyzed in *M. nigra* [6,7]. Amongst the Slovenian genotypes, the maxi- mum content was found in genotype SE 5. Screening of phenolics on Slovenian genotypes sampled at the place of their origin revealed chlorogenic acid concentration in the range be- tween 1.80 and 6.89 mg/g DW, whereas the maximum content was determined in genotype from the SM region. The concentrations are in accordance with other authors reporting the concentrations of chlorogenic acid in the range between 3 and 10 mg/g DW [43,49,62].

Generally, the predominant phenolic acids are known to have a stimulative effect on feeding, growth, and development [63]. Chlorogenic acid and other dihydroxybenzoid compounds isolated from mulberry leaves were found to be beneficial for silkworms' growth and development. Chlorogenic acid is sensed by chemosensory organs in the mouthparts and stimulates feeding; hence, high concentrations of chlorogenic acid in mulberry leaves significantly promote feeding and correlate with growth parameters of silkworm larvae [64–68]. Furthermore, Yamagishi et al., identified an additional role of chlorogenic acid in the mid-gut lumen as a cue inducing the tachykinin-related peptide secretion from enteroendocrine cells [63]. These peptide hormones are known to modulate physiological processes such as the release of other hormones, secretion of digestive enzymes, gut motility, feeding behaviour, and energy homeostasis. Thus, the silkworm might use chlorogenic acid in differentially directed functions as a food marker in both the mouthparts and mid-gut.

The beneficial effect also correlates with increased silk production and quality. Naik et al., reported that supplementation of chlorogenic acid increased silk productivity and had a positive influence on different silk parameters, such as silk filament length and weight as well as silk protein fibration [69]. Our results did not show a correlative relationship of silkworm parameters with chlorogenic acid, but out of phenolic acids, there was a negative correlation between 5-coumaroylquinic acid and silkworm weight and p coumaroylquinic acid hexoside and the silk thread length.

In our study, the predominant flavonoids were quercetin and kaempferol glycosides. The main quercetin glycosides were rutin, quercetin malonyl-hexoside, and quercetin-3-glucoside (isoquercetin), whereas the predominant kaempferol glycoside was kaempferol acetyl-hexoside, that were generally recognized as the main flavonols in mulberry leaves [6,7,44,45,70,71]. The nutritional effect of flavonols on the growth and develop- ment of silkworm larvae and cocoon formation has been intensively studied [53–58,71–74]. In low concentrations, they have a beneficial effect on growth and development, whereas high doses might have an antinutrient effect [75]. Quercetin malonyl-hexoside has been recognized as the main quercetin glycoside with antioxidant activities [43,45,49,71,76–79]. In our sampled genotypes the mean concentrations ranged between 0.74 and 2.25 mg/g DW with the highest value in sericultural variety 'Kokusou-20'. A high correlation was found between the length and weight of raw silk.

The second predominant flavonol was rutin, the maximum concentration was 4.31 mg/g DW determined in one Hungarian genotype (TO 1131). We found a medium correlation only with the struse of the raw silk. By reviewing the literature studying the effect of rutin on silkworm larvae, it was reported that rutin has no stimulative effect on the behaviour of silkworm larvae, although it stimulates feeding on many insects. Furthermore, a significant effect of rutin on the growth of silkworm larvae was not confirmed [74]. However, the authors were able to confirm that larvae can differentiate among quercetin glycosides of mulberry leaves. Quercetin-3-glucoside was recognized as a feeding stimulant but the rhamnose conjugate may deter feeding [74]. Furthermore, a positive effect of quercetin-3-glucoside on the growth and development of silkworm larvae was determined [66,74]. In the presented study, the trend towards a weak negative correlation was determined for silkworm weight and silk thread parameters.

The main kaempferol glycoside that was analyzed in leaves of local mulberries was identified as kaempferol acetylhexoside (0.94–4.42 mg/g DW), which reached the highest amount in *M. alba* × *rubra*. Other authors determined kaempferolhexoside in concentrations up to 0.75 mg/g DW, whereas kaempferol malonyl-hexoside was found in traces [71].

Interestingly, we found a negative correlation between kaempherol dirhamnosyl-hexoside and raw silk length and weight, whereas kaempherol rhamnosyl-hexoside correlated positively with struse. The studies on biologically active kaempferol derivatives in mulberry leaves are scarce, and it remains to be seen whether these derivatives may deter feeding.

4.2 The Influence of Macro- and Microelements on Larval, Cocoon and Raw Silk Parameters

Besides phenolics as bioactive compounds, minerals are among the important biochemical components of mulberry leaves, and they may have a high influence on silkworm, cocoon, and raw silk parameters. Previously, positive correlations of nitrogen, phosphorus, potassium, calcium, magnesium, and sulphur were obtained with larval, cocoon, and egg production parameters of mulberry silkworms [80,81]. Shifa et al. considered these macroelements as basic parameters for the evaluation of mulberry varieties for mulberry silkworms rearing in the future [81]. Based on this, we hypothesized that a high amount of these macroelements in mulberry leaves of selected mulberry genotypes will significantly contribute to silkworm and silk thread parameters.

Phosphorus (P) is an important major nutrient in the mulberry plant. It is a component of the complex nucleic acid structure of plants, which regulates protein synthesis. Therefore, it is very important in cell division and the development of new tissue. Phos- phorus is also associated with complex energy transformations such as ATP [80,82]. An inadequate amount of P level affects the uptake of other nutritive elements in mulberry leaves for various other physiological activities, in turn, it hampers the growth and eco- nomic characteristics of silkworms [83,84]. The highest concentration of P was found in Slovenian genotype SE 5 (2.04 g/kg DW), which was superior in silk thread weight and thickness. When compared to other authors, Shifa et al., determined a minimum value of 1.11 g/kg in 'Jimma coll' and a maximum record of 3.22 g/kg in M-4 accession [81]. Similar to Shifa et al., we found no significant correlation between P and silk parameters or with silkworm weight [81].

Sulphur (S) is known to have an important role in the synthesis of proteins, oils, and vitamins [85]. It plays a vital role in the N metabolism and thus proper development of mulberry [86]. It is a constituent of S-containing amino acids, cysteine (contains 27% of S), and methionine (contains 21% of S). Methionine forms one of the ten essential amino acids for silk formation in silkworms. Cystine and cysteine are among the non-essential amino acids, the quantitative presence of which influences the formation of fibroin over sericin [87]. Deficiency of S level leads to low levels of S-containing amino acids, thus reducing protein synthesis. As a result, amino acids without S and amides of nitrate ions accumulate in the plant tissue and lead to a decrease in sugar as well as insoluble N (protein) in plants [86]. Similar to Ca, the highest content of S was found in the Hungarian genotype GMS 2532 (1260 mg/kg DW). The S concentrations of mulberry varieties analysed by Shifa et al., ranged from 0.15 g/kg in K-2 to 0.34 g/kg in M-4 accession [81]. In the presented experiment S ranged between a minimum value of 0.46 g/kg and a maximum value of 1.26 g/kg, which is up to four-fold higher than what was reported by Shifa et al. [81]. These could be due to the different soil conditions and because the mulberry gene bank is on the silicate geological basis of the southern slopes of Pohorje mountain (central alpine region) which might positively affect the uptake of several minerals (Zn, S, Fe), but negatively Ca, Mg, K, P [82]. Shifa et al., found S to have a significant positive correlation with larval weight, cocoon weight, and shell weight. In contrast to these authors, we did not find any correlation with measured parameters [81].

Potassium (K) plays important regulatory roles mostly in cell ion homeostasis, and stomatal conductance, and thus in maintaining water potential on the cell and whole plant level. Furthermore, it is known that the starch synthetase is activated by K. Thus, with inadequate K, the level of starch declines while soluble carbohydrates and N compounds accumulate. Therefore, it also plays a significant role in the high yield and quality of leaves [88]. It is also involved in the translocation of carbohydrates, protein metabolism, and pathogen tolerance in mulberry [80]. In the silkworm body, the strong alkalinity of the gastric juice originates from potassium and sodium compounds present in the haemolymph. The high alkaline condition of digestive fluid has strong germicidal power against pathogens. K is a unique element that contributes to the growth of silkworms to the maximum extent. In addition, K has a stimulating effect on protein synthesis including silk protein in the silk glands [89]. In the presented experiment, K was increased in fruit varieties, although insignificant. The highest concentration of K was present in Hungarian genotype BA 2225 (22.10 g/kg DW), which was also characterized by superior raw silk length, weight, and thickness as well as silk waste parameters. Shifa et al., reported the K contents of mulberry varieties with ranges from 11.35 g/kg in local varieties to 18.61 g/kg in M-4 accession [81].

Calcium (Ca), in the form of calcium pectate, is important for the cell wall structure in plant. Its deficiency causes incomplete cell division or mitosis, without the formation of a new cell wall resulting in multi-nuclear cells. Calcium is also important in activating certain enzymes and to acts as second messengers in cell signalling that coordinates certain cellular activities. Calcium acts as a detoxifying agent by neutralizing organic acids such as oxalic acid which helps in membrane stability and maintenance of chromosome structure, the activity of enzymes, and translocation of carbohydrates. It is also involved in the differential permeability of membranes [82]. Superior Ca contents were found in genotype GMS 2532 (20.00 g/kg DW). Shifa et al., reported calcium concentrations in the range from 13.45 mg/kg (local check) to 20.52 mg/kg (M-4), which is similar to our findings [81].

Micronutrients are needed in small quantities and they play a pivotal role in the enzymatic reactions and thus govern the growth, development, and yield of mulberries. Chloride (Cl) is involved in the hydrolysis of water in photosynthesis, the synthesis of starch, cellulose, and lignin. It influences cell homeostasis (water holding capacity) of plant tissues. It stimulates the activities of some enzymes [82]. Cl correlated with the weight of raw silk. The highest content of Cl was in the Hungarian genotype BE 1264.2.

Manganese (Mn) is essential for the synthesis of chlorophyll and the hydrolysis of water in photosynthesis, and its principal function is to activate some of the enzyme systems in plant physiology and regulation of Fe metabolism. In addition, it has a close relation with N metabolism, assimilation of carbohydrates, and formation of ascorbate. It is involved in redox processes and electron transport systems [82]. Similar to Fe, Mn has the potential to enhance larval development, filament length, cocoon weight, and yield [90]. Mn was the highest in Hungarian genotype BA 2225 (105 mg/kg DW).

Iron (Fe) is present in the chloroplast proteins and several enzymes. It plays a dominant role in protein metabolism and N fixation [82]. Fe has the potential to enhance larval (silkworm) development, filament length of a single cocoon, cocoon weight, and yield [90]. The altered Fe content in mulberry foliage resulted in reduced larval weight, cocoon weight, and silk filament length [89]. In the presented experiment, Fe correlated strongly with telette. The highest content of Fe was found in genotype GMS 2533.

Zinc (Zn) correlates strongly with silk filament length and pupal weight, whereas the excess Zn content in mulberry leaves leads to a reduction in cocoon yield [84,90]. In our experiment, the highest Zn content was found in Slovenian genotype SM6. However, no correlative relationship with Zn was detected in our analysis. We further found a negative correlation between Rb and silkworm weight and a positive correlation between Sr and telette.

In the presented experiment, significant differences between the mulberry groups in the contents of the main elements were determined for S, Mn, Fe, and Sr, which were based on the PCA analysis considered as important markers in the selection of mulberry feed source.

4.3 Correlations between Test Parameters in Multivariate Analysis

The foliar protein, phenolics, and mineral composition of mulberry varieties resulted in significant inter-relationship with larval, cocoon, and silk thread parameters when their leaves served as feeds. This relationship between leaf composition values and important silkworm and raw silk traits has been worked out through correlation and multivariate (PCA) analysis.

Function 1 clearly separated reference varieties, which were characterized by higher larval weight, which coincided with high values of chlorogenic acid, rutin, and other quercetin glycosides, from Slovenian and Hungarian varieties with higher in proteins, length, and weight of raw silk. The second discriminant function was positively associated with cocoon weight, length of raw silk, total phenolics, Mn, and K, and negatively with Ca and quercetin glucoside. Function 2 clearly separated Hungarian mulberries (by over 87%) from Slovenian genotypes.

Positive correlation of coumaroylquinic acid derivatives, certain flavonols, phosphorus, sulphur, Cl, Ca, Mn, Fe, Ni and Rb were obtained with larval, cocoon, and silk thread parameters, whereas caffeoylquinic acid derivatives affected only the length of raw silk. Therefore, it is likely that the levels of these bioactive compounds and elements in mulberry leaves are important feed markers (basis parameters) to the gains on important mulberry silkworm parameters when these leaves served as feed sources.

4.4 Development and Health Status of Larvae

The measurement of body weight gain was used within the experiment to monitor larval development in the different groups. Bodyweight was most successfully increasing in groups of larvae fed with reference varieties of mulberry trees, followed by fruit varieties closely together with the Hungarian genotypes. The slowest gain of bodyweight and lowest weight on day seven of the fifth instar were observed by larvae fed with Slovenian genotypes of mulberry trees. Nonetheless, larvae with lower body weights at this instar started spinning approximately one day later. The additional one-day eating mainly compensated for their backlogs and they have started spinning with approximately the same weight as the other ones.

A correlation was observed between larval body weight in the last days of fifth instar and the weight of the fresh cocoons; however, differences between the averages of sub-groups of cocoons within Slovenian, Hungarian, reference, and fruit variety groups were only 5% (2.14–2.25 g of average fresh cocoon weights). The highest proportion of heavier cocoons was produced in the Hungarian groups. Additionally, the best silk thread parameters (length, weight, and thickness) were also described by cocoons from Hungarian groups, while most cocoon values from reference sericultural varieties and Slovenian groups were positioned within intermediate values. The lowest values of cocoon weight and silk thread parameters were described by fruit groups. Larvae fed with leaves of *M. nigra* showed the worst possible performance, as none of them started spinning (despite rapid larval bodyweight gains)

The general health status of silkworm groups was sufficient, as more than 99% of larvae reached the spinning stage healthy and started to spin (i.e., only three larvae out of 1270 died). By some (about 30) of larvae in the last day or two before spinning some common, general, mild signs of disease were observed: inactivity, cessation of eating, laying on the side of the rearing box, transparent skin, yellow/ivory colour. However, differentia- tion between signs of disease and the physiological changes connected to preparation for spinning was not obvious.

Within larvae that failed spinning BmNPV DNA was not detected, and all other randomly sampled larvae were also negative. However, BmNPV DNA was detected already in a dead and in a cachectic/ill larva in the third instar, though with very low viral DNA content (cT values > 40). BmNPV DNA was also detected in a dead larva after the third moult and in three ill/dead larvae in the fifth instar. Two of these larvae contained high amounts of BmNPV (cT 18–20). These molecular data indicate that BmNPV infection was present in (some of the) the larvae from the beginning of their life. Silkworm rearing has never been performed in the place of the experiment, and there is no known alternative, wild insect hosts of BmNPV, so the infection could not come from the environment, fomites, or contaminated leaves. Nevertheless, the virus amount in the vast majority of the larvae remained under detectable levels, and the larvae stayed healthy.

Whole genome sequencing of silkworm body tissue and 16S rRNA gene sequencing of bacteria in guts did not reveal DNA of BmDNV or relevant amounts of facultatively pathogenic bacteria (e.g., *Enterococcus (Streptococcus) faecalis*, *E. (S.) faecium*, Staphylococcal species, *Serratia marcescens*); however, there were considerable differences in proportions of detected sequences on the level of bacterial classes and orders between the groups (detailed analysis will be described elsewhere). Attempts on the identification of probiotic components of *B. mori* gut microbiota were reviewed by Barretto et al.: studies indicated the impact of Lactobacillus, Enterococcus, and Bacillus spp. as major gut microbiota components [91]. Besides competitive and antimicrobial effects on enteric pathogens, the probiotic effects of Actinobacteria (e.g., Actinomycetales, Bifidobacteriaceae) and Betaproteobacteria (e.g., Burkholderiales) contribute to the digestion with enzymes (e.g., protease, amylase, and lipase production) [91]. In the experimental groups, the highest abundance of DNA sequences from Burkholderiales was detected, followed by Actinomycetales and Bacillales. Lactobacillales (including

Streptococcaceae, Lactobacillaceae, and Enterococcaceae) were found in the lowest amounts. The relative amounts of these detected sequences were higher in the reference, *M. australis* and *M. nigra* groups, however, it was mainly attributed to the high amounts of chloroplast-related sequences in the groups of Slovenian and Hungarian varieties. When chloroplast DNA was excluded from the analysis, the relative abundance of bacterial groups with suspected probiotic effects was similar in the different groups.

Although the trees were cultured in the same collection (i.e., in a common microbial environment), the lower relative amounts of Burkholderiales, and higher relative amounts of Bacillales, Lactobacillales, and Bifidobacteriaceae were detected in the gut microbiome of larvae fed with *M. australis*, compared to the other four groups. No clear connection was identified between the gut microbiome compositions, larval bodyweight gain, cocoon production, and quality in this study.

The mean total haemocyte counts (THC) in the different groups ranged between 45.9 and 74.3 × 104 cells/mL; which is considerably higher than THCs reported by Nematollahian et al. [92]. However, the haemocyte subsets (based on cell morphology) were found similar. No significant differences were found in the haemocyte subset rations between the different feeding groups. Neither was any correlation identified between THCs, haemocyte subset ratios and larval bodyweight gain, cocoon production, and quality in this study.

5. CONCISE SUMMARY

The findings of this experiment indicated that local genetic variations of mulberry had significant qualitative and quantitative variation in chemical properties, including proteins, phenolics, and minerals, when compared to reference sericultural and fruit types. Consequently, positive associations between total proteins and kaempferol derivatives were observed with silk thread metrics (i.e., length, weight). Coumaric acid exhibited a negative correlation with raw silk length, but 5-coumaroylquinic acid associated with larval weight. A positive link was seen between Cl and raw silk weight, while Rb exhibited a negative correlation with larval weight. Iron and Strontium exhibited a favorable correlation with reeling waste. Consequently, this study demonstrates that selecting mulberry varieties from the local gene pool for silkworm rearing, based on foliar protein, specific phenolics, and mineral content, is crucial for optimizing larval development, cocoon yield, and raw silk characteristics. Further research is necessary to substantiate the current findings, taking into account the differing durations of leaf harvesting, nutrient analysis, pruning management, and the agricultural performance of mulberry varieties across various locations when utilizing these types as feed sources. While larvae nourished with reference varieties attained final body weight most rapidly, specific Hungarian genotypes (BA2151, SO 1013,

TO1131) exhibited encouraging outcomes, as their mean larval weight was the highest. Moreover, we identified Slovenian and Hungarian cultivars that exhibited outstanding raw silk characteristics. The minimum was achieved when larvae were nourished with the reference varieties 'Giazzola' and certain fruit genotypes. The highest results in the analysis of reeling wastes were achieved when larvae were fed Hungarian variety.

In addition to the aforementioned results, it is crucial to note that certain reference sericultural and fruit varieties of mulberry trees are beginning to mature earlier in the season compared to the Slovenian and Hungarian genotypes, posing a risk for heightened losses due to early spring frosts, a relatively common climatic phenomenon in these nations. Subsequent spring growth of local genotypes of Slovenian and Hungarian trees, along with their resilience to freezing, can offset the greater leaf yields of reference sericultural variety under field conditions.

Saxena et al. reported that global sericulture incurs annual losses of nearly 20% in potential cocoon production; however, our experiment illustrates that sustainable production of high-quality silk cocoons is feasible in Slovenia and Hungary, contingent upon the selection of superior local genotypes and the implementation of appropriate, locally adapted rearing technology. [93].

REFERENCES

1. Inserco 2022, International Sericultural Commission Web Page. Available online: https://www.inserco.org/en/ (accessed on 13 September 2022).
2. Žontar, J. *Svilogojstvo in Svilarstvo na Slovenskem od 16. do 20. Stoletja*; SAZU, Razred za Zgodovinske in Družbene Vede: Ljubljana, Slovenia, 1957; pp. 1–160.
3. Žontar, J. Gojitev sviloprejk. In *Gospodarska in Družbena Zgodovina Slovencev: Zgodovina Agrarnih Panog; I. Zvezek Agrarno Gospodarstvo*; Blaznik, P., Grafenauer, B., Vilfan, S., Eds.; DZS: Ljubljana, Slovenia, 1970; pp. 409–415.
4. Papp, Z. A selyemhernyó-tenyésztés történeti áttekintése és újrahonosításának ido˝szeru˝ sége. *Erdészeti Lapok* 2000, *85*, 1–12.
5. Tzenov, P.; Cappellozza, S.; Saviane, A. Black, Caspian Seas and Central Asia Silk Association (BACSA) for the Future of Sericulture in Europe and Central Asia. *Insects* 2022, *13*, 44. [CrossRef] [PubMed]
6. Urbanek Krajnc, A.; Ugulin, T.; Paušicˇ, A.; Rabensteiner, J.; Bukovac, V.; Mikulicˇ Petkovšek, M.; Janžekovicˇ, F.; Bakonyi, T.; Bercˇicˇ, R.L.; Felicijan, M. Morphometric and biochemical screening of old mulberry trees (*Morus alba* L.) in the former sericulture region of Slovenia. *Acta Soc. Bot. Pol.* 2019, *88*, 3614. [CrossRef]

7. Šelih, M.; Mikulicˇ Petkovšek, M.; Krajnc, D.; Bercˇicˇ, R.L.; Urbanek Krajnc, A. Screening of leaf metabolites in historical mulberry trees (*Morus alba* L.) from different eco-geographical regions of Slovenia 2020. *Trees* 2020, *34*, 971–986. [CrossRef]

8. Bradford, M.M. A rapid and sensitive method for quantitation of icrogram quantities of protein utilizing the principle of proteindye binding. *Anal. Biochem.* 1976, *72*, 248–254. [CrossRef]

9. Ainsworth, E.A.; Gillespie, K.M. Estimation of total phenolic content and other oxidation substrates in plant tissues using Folin–Ciocalteu reagent. *Nat. Protoc.* 2007, *2*, 875–877. [CrossRef]

10. Necˇemer, M.; Kump, P.; Šcˇancˇar, J.; Jac'imovic', R.; Simcˇicˇ, J.; Pelicon, P.; Budnar, M.; Jeran, Z.; Pongrac, P.; Regvar, M.; et al. Application of X-ray fluorescence analytical techniques in phytoremediation and plant biology studies. *Spectrochim. Acta Part B* 2008, *63*, 1240–1247. [CrossRef]

11. Necˇemer, M.; Kump, P.; Vogel-Mikuš, K. Energy Dispersive X-ray Fluorescence Analysis of Biological Materials. In *X-ray Fluorescence in Biological Sciences: Principles, Instrumentation, and Applications*; Singh, V.K., Kawai, J., Tripathi, D.K., Eds.; John Wiley & Sons: Newark, NJ, USA, 2022; pp. 311–325. [CrossRef]

12. Kump, P.; Necˇemer, M.; Rupnik, Z.; Pelicon, P.; Ponikvar, D.; Vogel-Mikuš, K. Improvement of the XRF quantification and enhancement of the combined applications by EDXRF and micro-PIXE. In *Integration of Nuclear Spectrometry Methods as a New Approach to Material Research*; International Atomic Energy Agency: Vienna, Austria, 2012; pp. 101–109. Available online: https://www.iaea.org/publications/8423/integration-of-nuclear-spectrometry-methods-as-a-new-approach-to-material-research (accessed on 13 September 2022).

13. Brion, H. *Praxishandbuch Seidenraupen-Aufzucht*, 2nd ed.; Vereinigung Schweizer Seidenproduzenten: Wohlen bei Bern, Switzerland, 2016; p. 19.

14. Cappellozza, S. (Council for Agricultural Research and Economics, Research Center for Agriculture and Environment, Padua, Italy). Personal communication, 2015.

15. Gani, M.; Chouhan, S.; Lal, B.; Gupta, R.K.; Khan, G.; Kumar, N.B.; Saini, P.; Ghosh, M.K. *Bombyx mori* nucleopolyhedrovirus (BmBPV): Its impact on silkworm rearing and management strategies. *J. Biol. Control* 2017, *31*, 189–193. [CrossRef]

16. Khurad, A.M.; Mahulikara, A.; Rathod, M.K.; Rai, M.M.; Kanginakudru, S.; Nagaraju, J. Vertical transmission of nucleopolyhe-drovirus in the

silkworm, *Bombyx mori* L. *J. Invertebr. Pathol.* 2004, *87*, 8–15. [CrossRef]

17. Zhang, J.; Kobert, K.; Flouri, T.; Stamatakis, A. PEAR: A fast and accurate Illumina Paired-End reAd mergeR. *Bioinformatics* 2013, *30*, 614–620. [CrossRef] [PubMed]
18. Bolger, A.M.; Lohse, M.; Usadel, B. Trimmomatic: A flexible trimmer for Illumina sequence data. *Bioinformatics* 2014, *30*, 2114–2120. [CrossRef]
19. Rognes, T.; Flouri, T.; Nichols, B.; Quince, C.; Mahé, F. VSEARCH: A versatile open source tool for metagenomics. *PeerJ* 2016, *4*, e2584. [CrossRef] [PubMed]
20. Wood, D.E.; Lu, J.; Langmead, B. Improved metagenomic analysis with Kraken 2. *Genome Biol.* 2019, *20*, 257. [CrossRef] [PubMed]
21. Pruitt, K.D.; Tatusova, T.; Maglott, D.R. NCBI Reference Sequence (RefSeq): A curated non-redundant sequence database of genomes, transcripts and proteins. *Nucleic Acids Res.* 2005, *33*, D501–D504. [CrossRef] [PubMed]
22. McDonald, D.; Price, M.N.; Goodrich, J.; Nawrocki, E.P.; DeSantis, T.Z.; Probst, A.; Andersen, G.L.; Knight, R.; Hugenholtz, P. An improved Greengenes taxonomy with explicit ranks for ecological and evolutionary analyses of bacteria and archaea. *ISME J.* 2012, *6*, 610–618. [CrossRef] [PubMed]
23. R Core Team. *R: A Language and Environment for Statistical Computing*; R Foundation for Statistical Computing: Vienna, Australia, 2021; Available online: http://www.R-project.org/ (accessed on 13 September 2022).
24. McMurdie, P.J.; Holmes, S. phyloseq: An R package for reproducible interactive analysis and graphics of microbiome census data. *PLoS ONE* 2013, *8*, e61217. [CrossRef]
25. Nittono, Y.; Tomabechi, S.; Onodera, N. Formation of haemocytes near the imaginal wingdisc in the silkworm, *Bombyx mori* L. *J. Sericult. Sci. Jpn.* 1964, *33*, 43–45. [CrossRef]
26. Yamashita, M.; Iwabuchi, K. *Bombyx mori* prohemocyte division and differentiation in individual microcultures. *J. Insect Physiol.* 2001, *47*, 325–331. [CrossRef]
27. Lee, Y.W. *Silk Reeling and Testing Manual*; FAO Agricultural Services Bulletin, No. 136; Food and Agriculture Organization of the United Nations: Rome, Italy, 1999; p. 139.
28. Hammer, Ø.; Harper, D.A.T.; Ryan, P.D. PAST: Paleontological Statistics Software Package for Education and Data Analysis. *Palaeontol. Electron.* 2001, *4*, 9. Available online: https://palaeo-electronica.org/2001_1/past/issue1_01.htm (accessed on 13 September 2022).
29. Andadari, L.; Minarningsih, M.; Suwandi. The effect of feeding various

species of mulberry (*Morus* spp.) on the growth of silkworm and quality of cocoon hybrid BS 09. *IOP Conf. Ser. Earth Environ. Sci.* 2021, *914*, 012017. [CrossRef]

30. Kumar, R.V.; Chauhan, S.; Kumar, D.; More, N. Nutritional composition in leaves of some mulberry varieties: A comparative study. In Proceedings of the International Conference of Bioinformatics and Biomedical Technology, Chengdu, China, 16–18 April 2010; pp. 438–442. [CrossRef]
31. Radjabi, R. Effect of mulberry leaves enrichment with amino acid supplementary nutrients on silkworm, *Bombyx mori* L. at north of Iran. *Acad. J. Entomol.* 2010, *3*, 45–51.
32. Wang, C.; Yang, F.; Wang, Q.; Zhou, X.; Xie, M.; Kang, P.; Wang, Y.; Peng, X. Nutritive Value of Mulberry Leaf Meal and its Effect on the Performance of 35-70-Day-Old Geese. *Int. J. Poult. Sci.* 2016, *54*, 41–46. [CrossRef] [PubMed]
33. Rajabi, K.; Ebadi, R.; Mirhosseini, R.; Seidavi, S.Z.; Zolfaghari, M.; Etebari, K. A review on nutritive effect of mulberry leaves enrichment with vitamins on economic traits and biological parameters of silkworm *Bombyx mori* L. *Invert. Surviv. J.* 2007, *4*, 86–91.
34. Mendonça, G.A.; de Marchini, L.C.; Macedo, L.P.M. Cocoon production of the silkworm, *Bombyx mori* L. (Lepidoptera: Bombyci- dae), fed on leaves of mulberry hybrids. *Caatinga* 2010, *23*, 118–122.
35. El-Banna, A.A.; Moustafa, M.N.; Mahmoud, S.M.; El-Shafei, A.M.; Moustafa, A.A. Effect of feeding different mulberry varieties on some biological characteristics of the silkworm, *Bombyx mori* L. *Egypt J. Pure Appl. Sci.* 2013, *51*, 55–60. [CrossRef]
36. Adeduntan, S. Influence of different varieties of mulberry leaves (*Morus alba*) on growth and cocoon performance of biovoltine strain of silkworm (*Bombyx mori*). *Int. J. Biol. Chem. Sci.* 2015, *9*, 751–757. [CrossRef]
37. Radojkovic´, M.M.; Zekovic´, Z.P.; Vidovic´, S.S.; Kocˇar, D.D.; Maškovic´, P.Z. Free radical scavenging activity and total phenolic and flavonoid contents of mulberry (*Morus* spp. L., Moraceae) extracts. *Hem. Ind.* 2012, *66*, 552. [CrossRef]
38. Chung, H.I.; Kim, J.; Kim, J.Y.; Kwon, O. Acute intake of mulberry leaf aqueous extract affects postprandial glucose response after maltose loading: Randomized double-blind placebo-controlled pilot study. *J. Funct. Foods* 2013, *5*, 1502–1506. [CrossRef]
39. Hunyadi, A.; Liktor-Busa, E.; Márki, A.; Martins, A.; Jedlinszki, N.; Hsieh, T.J.; Báthori, M.; Hohmann, J.; Zupkó, I. Metabolic Effects of Mulberry Leaves: Exploring Potential Benefits in Type 2 Diabetes and Hyperuricemia. *Evid.-*

Based Complement. Altern. Med. 2013, *2013*, 948627. [CrossRef]

40. Jeszka-Skowron, M.; Flaczyk, E.; Jeszka, J.; Krejpcio, Z.; Król, E.; Buchowski, M.S. Mulberry leaf extract intake reduces hypergly- caemia in streptozotocin (STZ)-induced diabetic rats fed high-fat diet. *J. Funct. Foods* 2014, *8*, 9–17. [CrossRef]
41. Chen, H.; He, X.; Liu, Y.; Li, J.; He, Q.; Zhang, C.; Wei, B.; Zhang, Y.; Wang, J. Extraction, purification and anti-fatigue activity of γ -aminobutyric acid from mulberry (*Morus alba* L.) leaves. *Sci. Rep.* 2016, *6*, 18933. [CrossRef] [PubMed]
42. Gryn-Rynko, A.; Bazylaka, G.; Olszewska-Slonina, D. New potential phytotherapeutics obtained from white mulberry (*Morus alba* L.) leaves. *Biomed. Pharmacother.* 2016, *84*, 628–636. [CrossRef]
43. Sugiyama, M.; Katsube, T.; Koyama, A.; Itamura, H. Seasonal Changes in Functional Component Contents in Mulberry (*Morus alba* L.) Leaves. *Hortic. J.* 2017, *86*, 534–542. [CrossRef]
44. Ju, W.-T.; Kwon, O.-C.; Kim, H.-B.; Sung, G.-B.; Kim, H.-W.; Kim, Y.-S. Qualitative and quantitative analysis of flavonoids from 12 species of Korean mulberry leaves. *J. Food Sci. Technol.* 2018, *55*, 1789–1796. [CrossRef] [PubMed]
45. Ju, W.-T.; Kwon, O.-C.; Lee, M.-K.; Kim, H.-B.; Sung, G.-B.; Kim, Y.-S. Quali-Quantitative Analysis of Flavonoids for Mulberry Leaf and Fruit of 'Suhyang'. *Korean J. Environ. Agric.* 2017, *36*, 249–255. [CrossRef]
46. Memon, A.A.; Memon, N.; Luthria, D.L.; Bhanger, M.I.; Pitafi, A.A. Phenolic acids profiling and antioxidant potential of mulberry (*Morus laevigata* W., *Morus nigra* L., *Morus alba* L.) leaves and fruits grown in Pakistan. *Pol. J. Food Nutr. Sci.* 2010, *60*, 25–32.
47. Zou, Y.; Liao, S.; Shen, W.; Liu, F.; Tang, C.; Chen, C.-Y.O.; Sun, Y. Phenolics and Antioxidant Activity of Mulberry Leaves Depend on Cultivar and Harvest Month in Southern China. *Int. J. Mol. Sci.* 2012, *13*, 16544–16553. [CrossRef]
48. Iqbal, S.; Younas, U.; Sirajuddin; Chan, K.W.; Sarfraz, R.A.; Uddin, K. Proximate Composition and Antioxidant Potential of Leaves from Three Varieties of Mulberry (*Morus* sp.): A Comparative Study. *Int. J. Mol. Sci.* 2012, *13*, 6651–6664. [CrossRef]
49. Lee, W.J.; Choi, S.W. Quantitative Changes of Polyphenolic Compounds in Mulberry (*Morus alba* L.) Leaves in Relation to Varieties, Harvest Period and Heat Processing. *Prev. Nutr. Food Sci.* 2012, *17*, 280–285. [CrossRef]
50. Thabti, I.; Elfalleh, W.; Hannachi, H.; Ferchichi, A.; Campos, M.D.G. Identification and quantification of phenolic acids and flavonol glycosides in Tunisian Morus species by HPLC-DAD and HPLC–MS. *J. Funct. Foods* 2012,

4, 367–374. [CrossRef]

51. Flaczyk, E.; Kobus-Cisowska, J.; Przeor, M.; Korczak, J.; Remiszewski, M.; Korbas, E.; Buchowski, M. Chemical characterization and antioxidative properties of Polish variety of *Morus alba* L. leaf aqueous extracts from the laboratory and pilot-scale processes. *J. Agric. Sci.* 2013, *4*, 141–147. [CrossRef]

52. Choi, S.W.; Lee, Y.J.; Ha, S.B.; Jeon, Y.H.; Lee, D.H. Evaluation of Biological Activity and Analysis of Functional Constituents from Different Parts of Mulberry (*Morus alba* L.) Tree. *J. Korean Soc. Food Sci. Nutr.* 2015, *6*, 823–831. [CrossRef]

53. Germanò, M.P.; D'Angelo, V.; Catania, S.; Miano, T.C.; Perna, V.; Farago, S.; Cappellozza, L.; Cappellozza, S. Phenolic content of leaf of different mulberry cultivars affects growth in the silkworm. In Proceedings of the 21st International Sericultural Congress of ISC, Athens, Greece, 3–6 November 2008; pp. 25–29.

54. Hirayama, C.; Ono, H.; Tamura, Y.; Konno, K.; Nakamura, M. Regioselective formation of quercetin 5-O-glucoside from orally administered quercetin in the silkworm, *Bombyx mori*. *Phytochemistry* 2008, *69*, 1141–1149. [CrossRef] [PubMed]

55. Daimon, T.; Hirayama, C.; Kanai, M.; Ruike, Y.; Meng, Y.; Kosegawa, E.; Nakamura, M.; Tsujimoto, G.; Katsuma, S.; Shimada, T. The silkworm *Green b* locus encodes a quercetin 5-*O*-glucosyltransferase that produces green cocoons with UV-shielding properties. *Proc. Natl. Acad. Sci. USA* 2010, *107*, 11471–11476. [CrossRef]

56. Vihakas, M. Flavonoid and Other Phenolic Compounds: Characterization and Interactions with Lepidopteran and Sawfly Larvae. Doctoral Thesis, University of Turku, Turku, Finland, 2014.

57. Kusurkar, T.S.; Tandon, I.; Sethy, N.K.; Bhargava, K.; Sarkar, S.; Singh, S.K.; Das, M. Fluorescent silk cocoon creating fluorescent diatom using a "Water glass-fluorophore ferry". *Sci. Rep.* 2013, *3*, 3290. [CrossRef] [PubMed]

58. Shimada, T. Basic and Applied Genetics of Silkworms. In *Silkworm Biofactory; Silk to, Biology*; Maenaka, K., Park, E.Y., Eds.; CRC Press: Boca Raton, FL, USA, 2018; p. 15. [CrossRef]

59. Srivastava, S.; Kapoor, R.; Thathola, A.; Srivastava, R.P. Nutritional quality of leaves of some genotypes of mulberry (*Morus alba*). *Int. J. Food Sci. Nutr.* 2006, *57*, 305–313. [CrossRef]

60. Al-kirshi, R.A.; Alimon, A.R.; Zulkifli, I.; Zahari, M.W.; Sazili, A.Q. The chemical composition and nutritive value of mulberry leaf as a protein source in poultry diets. In *Feed and Nutrition: The 1st International Seminar*

on Animal Industry, Bogor, Indonesia; Bogor Agricultural University: Bogor, Indonesia, 2009; pp. 98–102.

61. Kumar, K.; Mohan, M.; Tiwari, N.; Kumar, S. Production potential and leaf quality evaluation of selected mulberry (*Morus alba*) clones. *J. Pharmacogn. Phytochem.* 2018, *7*, 482–486.

62. Sánchez-Salcedo, E.M.; Mena, P.; García-Viguera, C.; Martínez, J.J.; Hernández, F. Phytochemical evaluation of white (*Morus alba* L.) and black (*Morus nigra* L.) mulberry fruits, a starting point for the assessment of their beneficial properties. *J. Funct. Foods* 2015, *12*, 399–408. [CrossRef]

63. Yamagishi, T.; Endo, H.; Fukumura, K.; Nagata, S.; Hayakawa, T.; Adegawa, S.; Kasubuchi, M.; Sato, R. Glucose, some amino acids and a plant secondary metabolite, chlorogenic acid induce the secretion of a regulatory hormone, tachykinin-related peptide, from the silkworm midgut. *Peptides* 2018, *106*, 21–27. [CrossRef]

64. Kato, M.; Yamada, H. Chlorogenic acid as a growth factor of silkworm. *J. Silkworm* 1964, *16*, 85. [CrossRef]

65. Naito, K.; Hayashiya. Studies on the micro constituents in mulberry leaves. IV. Isolation of chlorogenic acid. *J. Agric. Chem. Soc. Jpn.* 1965, *39*, 237–238.

66. Hamamura, Y.; Kuwata, K.; Masuda, H. Effect of Gallic Acid on the Growth of the Silkworm Larvae *Bombyx mori* L. *Nature* 1966, *212*, 1386–1387. [CrossRef] [PubMed]

67. Yamada, H.; Kato, M. Chlorogenic Acid as an Indispensable Component of the Synthetic Diet for the Silkworm. *Proc. Jpn. Acad.* 1966, *42*, 399–403. [CrossRef]

68. Koike, S.; Iizuka, T.; Mizutani, J. Determination of Caffeic Acid in the Digestive Juice of Silkworm Larvae and Its Antibacterial Activity against the Pathogenic Streptococcus faecalis AD-4. *Agric. Biol. Chem.* 1979, *43*, 1727–1731. [CrossRef]

69. Naik, D.; Patil, G.M.; Biradar, R. Impact of fortification of chlorogenic acid rich botanical on silkworm growth and cocoon parameters of silkworm *Bombyx mori* L. *Plant Arch.* 2017, *17*, 661–665.

70. Jariene˙, E.; Levickiene˙, D.; Danilcˇenko, H.; Vaitkevicˇiene˙, N.; Kulaitiene˙, J.; Jakštas, V.; Ivanauskas, L.; Gajewski, M. Effects of biodynamic preparations on concentration of phenolic compounds in the leaves of two white mulberry cultivars. *Biol. Agric. Hortic.* 2018, *35*, 132–142. [CrossRef]

71. Pothinuch, P.; Tongchitpakdee, S. Phenolic Analysis for Classification of Mulberry (*Morus* spp.) Leaves according to Cultivar and Leaf Age. *J. Food Qual.* 2019, *2019*, 2807690. [CrossRef]

72. Kurioka, A.; Yamazaki, M. Purification and Identification of Flavonoids

from the Yellow Green Cocoon Shell (Sasamayu) of the Silkworm, *Bombyx mori*. *Biosci. Biotechnol. Biochem.* 2002, *66*, 1396–1399. [CrossRef]

73. Kurioka, A.; Yamazaki, M. Antioxidant in the cocoon of the silkworm, *Bombyx mori*. *J. Insect Biotechnol. Sericol.* 2002, *71*, 177–180. [CrossRef]

74. Preiß, S.; Degenhardt, J.; Gershenzon, J. Plant animal dialogues. In *Ecological Biochemistry: Environmental and Interspecies Interactions*; Krauss, G.J., Nies, D.H., Eds.; Wiley-Blackwell: Oxford, UK, 2015; pp. 313–330. ISBN 978-3-527-31650-2.

75. Zhang, Y.-E.; Ma, H.-J.; Feng, D.-D.; Lai, X.-F.; Chen, Z.-M.; Xu, M.-Y.; Yu, Q.-Y.; Zhang, Z. Induction of Detoxification Enzymes by Quercetin in the Silkworm. *J. Econ. Èntomol.* 2012, *105*, 1034–1042. [CrossRef] [PubMed]

76. Enkhmaa, B.; Shiwaku, K.; Katsube, T.; Kitajima, K.; Anuurad, E.; Yamasaki, M.; Yamane, Y. Mulberry (*Morus alba* L.) Leaves and Their Major Flavonol Quercetin- 3-(6-Malonylglucoside) Attenuate Atherosclerotic Lesion Development in LDL Receptor- Deficient Mice. *J. Nutr.* 2005, *135*, 729–734. [CrossRef] [PubMed]

77. Katsube, T.; Imawaka, N.; Kawano, Y.; Yamazaki, Y.; Shiwaku, K.; Yamane, Y. Antioxidant flavonol glycosides in mulberry (*Morus alba* L.) leaves isolated based on LDL antioxidant activity. *Food Chem.* 2006, *97*, 25–31. [CrossRef]

78. Katsube, T.; Tsurunaga, Y.; Sugiyama, M.; Furuno, T.; Yamasaki, Y. Effect of air-drying temperature on antioxidant capacity and stability of polyphenolic compounds in mulberry (*Morus alba* L.) leaves. *Food Chem.* 2009, *113*, 964–969. [CrossRef]

79. Katsube, T.; Yamasaki, M.; Shiwaku, K.; Ishijima, T.; Matsumoto, I.; Abe, K.; Yamasaki, Y. Effect of flavonol glycoside in mulberry (*Morus alba* L.) leaf on glucose metabolism and oxidative stress in liver in diet-induced obese mice. *J. Sci. Food Agric.* 2010, *90*, 2386–2392. [CrossRef] [PubMed]

80. Mahadeva, A. Nutritive Elemental Status in Mulberry (*Morus* sp.) Foliage under Jassids (*Empoasca flavescenes* F.) Infestation. *Indian J. Nat. Sci.* 2016, 7, 11537.

81. Shifa, K.; Terefe, M.; Tilahun, A.; Ibrahim, A.; Nigusu, Y. Leaf macronutrient composition of mulberry (*Morus indica*) varieties and its relationship with productivity of mulberry silkworm (*Bombyx mori* L.). *J. Agric. Sci. Res.* 2020, *8*, 390–398.

82. Biskin, D.P.; Bloom, A. Mineral Nutrition. In *Plant Physiology*, 5th ed.; Taiz, L., Zeiger, P.E.E., Eds.; Sinauer Associatiates Publishers: Munderland, MA, USA, 2010; pp. 107–125.

83. Ito, T.; Nimura, M. Nutrition of the silkworm *Bombyx mori* L., its specific nutrient requirements and its nutrition in relation to the mineral nutrition of its host plant mulberry, *Morus indica* L. *Indian J. Exptl. Biol.* 1966, *4*, 31–36.

84. Chakrabarti, S.; Subramanyam, M.R.; Singhal, B.K.; Datta, R.K. Nutrient deficiency management in mulberry. In *Handbook of Mulberry Nutrition*; Shankar, M.A., Shetty, G.P., Eds.; Multiplex, Karnataka Agro Chemicals: Bangalore, India, 1997; pp. 19–75.

85. Miller, A.J. Plant Mineral Nutrition. In *eLS*; John Wiley & Sons, Ltd.: Chichester, UK, 2014. [CrossRef]

86. Munirathnam Reddy, M.; Subbaswamy, M.R.; Sinha, A.K. Sulphur nutrition in mulberry plants. *Indian Silk* 1990, *29*, 45–46.

87. Mahadevappa, L.; Magadi, S.P.; Prabhuraj, D.K.; Thimmareddy, H.; Bongale, U.D. Available sulphur in some mulberry growing soils of Karnataka. In Nutritional Management and Quality Improvement in Sericulture, Proceedings of the National Seminar on Mulberry Sericulture Research in India, Karnataka, India, 26–28 November 2001; State Sericulture Research and Development Institute: Bangalore, India, 2001; pp. 161–165.

88. Shree, M.P.; Anuradha, R.; Nagaveni, V. Impact of rust disease on the mineral nutrition of mulberry plants. *Sericologia* 2005, *45*, 115–121.

89. Shankar, M.A. *Handbook of Mulberry Nutrition*; Shetty, G.P., Ed.; Multiplex, Karnataka Agro Chemicals: Bangalore, India, 1997.

90. Lokanath, R.; Shivashankar, K.; Kasiviswanathan, K. Effect of foliar application of magnesium and micronutrients to mulberry on the quality and production of cocoons. *Indian J. Seric* 1986, *24*, 40–45.

91. Barretto, D.A.; Gadwala, M.; Vootla, S.K. Chapter 1—The silkworm gut microbiota: A potential source for biotechnological applications. *Methods Microbiol.* 2021, *49*, 1–26. [CrossRef]

92. Nematollahian, S.; Bagheri, M.; Yousefi, Z.; Khezrian, A.; Zahmatkesh, Z. Haematological Changes in *Bombyx mori* (Lepidoptera: Bombycidae) Infected by Nosema bombycis. *J. Agric. Urban Entomol.* 2021, *37*, 38–47. [CrossRef]

93. Saxena, S.; Chandra, R.T.; Kallare, P.S.; Arunkumar, K.P. MicroRNAs in the silkworm-pathogen interactions. *Methods Microbiol.* 2021, *49*, 97–113. [CrossRef]

10 Chapter

COMPREHENSIVE PROFILING OF GENES EXPRESSED IN THE SILK GLANDS OF PHILIPPINE-REARED MULBERRY SILKWORMS (*BOMBYX MORI*)

1. INTRODUCTION: AN OVERVIEW

The Bombyx mori Bombyx mori is a commercially significant insect for silk production, esteemed for its texture, sheen, and status as one of the strongest natural fibers. Silk serves numerous applications, including clothes, furniture, parachute linings, insulation, and surgical sutures. It is regarded as a promising biomaterial in tissue engineering owing to its mechanical strength and biocompatibility. [2].

Sericulture was introduced to the Philippines in the 1970s, with indigenous silkworm stocks derived from Chinese, Japanese, and European strains. Prior research established that temperate strains yield finer and stronger silk fibers, whereas tropical strains are more resilient and disease-resistant but generate coarser and weaker silk fibers [4]. Efforts have been made to develop hybrids that are adapted to the Philippine climate, exhibiting desirable phenotypic features like as elevated hatch rates, cocoon output, cocoon weight, and silk filament quality [5–7]. Local silk production continues to fall short of demand, with an annual requirement of 10 metric tons and a just 1 metric ton of raw silk produced. Inadequate productivity has been ascribed to the quality of indigenous silkworm strains and suboptimal rearing methods [9]. The suboptimal silk output may be enhanced with a comprehensive study of its production process. The molecular mechanisms behind the varying silk yields in silkworm strains require investigation, as they remain inadequately understood [10,11]. Molecular research may thus facilitate the creation of indigenous silkworm strains capable of producing large silk yields. Markers linked to elevated silk output and other commercially significant features may also be established for marker-assisted selection and breeding. Consequently, this undertaking necessitates genetic resources, which can be obtained from genome and transcriptome sequences of foreign B. mori strains available in public databases, including the Silkworm Genomic Database (SilkDB) [12], KAIKObase [13], SilkTransDB (transcriptome) [14], SilkSatDb (microsatellites) [15], and BmTEdb (transposable elements). [16].

While there are already several genomic resources for the silkworm, transcriptome analysis can be used to gain insight into the functional elements of the genome and its biological pathways. One such technique is RNA sequencing (RNA-seq), which utilizes next-generation sequencing to provide high-throughput and high-resolution data to analyze expressed genes in a cost-efficient manner. RNA-seq can be used to identify novel transcripts, splice variants, and single nucleotide polymorphisms [17]. It also allows quan- tification over a dynamic range since it detects both highly and lowly expressed genes, making it ideal for gene expression analysis [18]. Recently, RNA-seq has been used by Zhang et al. to reveal gene expression changes in silkworms exposed to hydrogen sulfide [19]. Yokoi et al. also used RNA-seq to generate reference transcriptome data for major tissues of the silkworm p50T strain [20]. However, available transcriptome resources currently lack representative sequences from Philippine-reared *B. mori* strains; hence, the need to generate such resources is deemed important.

This study sought to produce transcriptome resources for Philippine-reared mulberry silkworms, which may be employed in future research to explore possible industrial applications and to tackle the problem of low silk yield in the Philippines. This study concentrated on the genes present in the silk glands of various strains of silkworms bred in the Philippines.

2. MATERIALS AND METHODS

2.1 Insect and Tissue Collection

Silkworms reared at the Philippine Textile Research Institute Technology Center in Misamis Oriental (PTRI-TCMO, Villanueva, Misamis Oriental) and the Technology Center of the Department of Science and Technology Cordillera Administrative Region (DOST-CAR, Benguet) were collected in 2018. The ambient temperature in TCMO and CAR at the time of collection was 31.7 °C and 23.9°C, respectively. For each site, priority strains (MO204, LAT51, ITA) and a poorly performing strain (DMMMSU119) were used as representative strains. The strains were classified by the rearing facilities as either priority strains or poorly performing strains based on their rearing performance (hatching ratio, larval mortality, and pupation ratio). Economic characteristics such as silk yield and quality were not considered in the classification of priority strains. Three individuals per strain were used as bioreplicates. Silkworms were collected on the 3rd day of the 5th instar larval stage and were characterized morphologically. Larvae were preserved by flash-freezing in liquid nitrogen and stored at − 80°C until ready for dissection. Larvae were dissected in a petri dish on ice to collect whole silk glands. Silk glands were chopped using sterile surgical blades and homogenized mechanically using a mortar and pestle.

2.2 RNA Extraction and Sequencing

Total RNA was extracted from 100 μg of the tissue with the TRIzol reagent (Invitrogen, Life Technologies, Carlsbad, CA, USA). Total RNA was treated with TURBO DNAse (Ambion, Life Technologies, Carlsbad, CA, USA) and purified with the RNA Clean and Concentrator kit (Zymo Research, Irvine, CA, USA). The quality of RNA extracts was visually estimated by running them in 1% agarose gel at 90 V for 40 min. RNA purity was estimated by measuring the 260/280 and 260/230 absorbance ratios in a NanoDrop 2000/2000c Spectrophotometer (Thermo Fisher Scientific, Wilmington, DE, USA), and RNA was quantified by fluorometry using a Qubit RNA BR Assay kit (Thermo Fisher Scientific) according to the manufacturer's instructions. Finally, extracts were analyzed in the Agilent 2200 TapeStation (Agilent Technologies, Santa Clara, CA, USA) to evaluate the RNA Integrity Number (RIN). Samples with RIN $\geq$6 were used for library preparation and sequencing.

The RNA samples were diluted and used as input (0.1–4 μg) for mRNA-enrichment library preparation using the TruSeq Stranded RNA Library Prep Kit (Illumina, San Diego, CA, USA). Twelve libraries were prepared (4 strains × 3 bioreplicates) with fragment sizes ranging 131–279 bp. Libraries were denatured, diluted to 1.8 pM, and combined with a spike-in of 1.0% PhiX sequencing control. The libraries were sequenced in the Philippine Genome Center using a NextSeq 500 (Illumina, San Diego, CA, USA) platform and a NextSeq v2 high output kit (Illumina, San Diego, CA, USA) with 300 cycles (2 × 150 pairedend sequencing).

2.3 Preprocessing, Alignment, and Transcriptome Assembly

Illumina bcl2fastq (v. 2.2.0) was used to convert sequencing base call files to FASTQ format. FastQC was used to evaluate reads and visualize read quality. Raw reads were processed using Trimmomatic (0.39) and fastp (0.20.0) to trim adapter sequences and filter low-quality reads.

The splice-aware aligner STAR (2.7.0f) was used to align reads to the *B. mori* reference genome from NCBI (RefSeq GCF_000151625.1). The BAM file was processed with RNA-SeQC (1.1.9) to evaluate RNA-Seq data bias based on the total read count, coverage, and expression correlation (RPKM-based estimation of expression levels). Another aligner, HISAT (2.1.0), was used for alignment in preparation for generating the StringTie assembly. HISAT alignment files were used for transcriptome assembly evaluation and expression quantification.

Reference-based assembly was performed using Cufflinks (2.2.1) and StringTie (1.3.6). In addition to reference-based assembly, de novo assembly was also performed using Trinity (2.8.6). The percentage of aligned reads was determined using Bowtie.

Contig statistics (number of transcripts, N50, mean, and median contig length) were generated using Trinitystats.pl. Assemblies were evaluated using DETONATE (1.11) to estimate their RSEM- EVAL scores. The completeness of the assemblies was evaluated using BUSCO (v. 3.0.2).

2.4 Expression Quantification and Differential Expression Analysis

FeatureCounts (1.4.6-p5) was used for gene-level quantification and DESeq2 (1.30.1) was used for gene-level differential expression analysis. Differentially expressed genes (DEGs) were filtered based on the false discovery rate (FDR)/adjusted *p*-value < 0.1 and $|\log_2$ fold change$| > 1$. Figures to visualize DEGs were generated from the plotMA function of DESeq2 and R packages EnhancedVolcano (1.8.0) and pheatmap (1.0.12).

The annotation of DEGs was performed against the NCBI nr protein database with an e-value cut-off of 1×10^{-5}. GO terms were mapped to DEGs by matching the protein IDs to the GO annotations in the newest release of the *B. mori* genome annotations

2.5 Quantitative Real-Time PCR (qPCR)

Quantitative real-time PCR assays were carried out using the Bio-Rad CFX96™ Touch system (Bio-Rad Laboratories, Inc., Hercules, CA, USA). Each 10- μL reaction contained 5 μL of 2X iTaq Universal SYBR Green Supermix (Bio-Rad Laboratories, Inc., Hercules, CA, USA), 5 pmol each of forward and reverse primers, and 25 ng of cDNA, following a thermal profile comprising initial denaturation at 95°C for 5 min, 40 cycles of denaturation at 95°C for 10 s, annealing and extension at 60°C for 30 s, and a melt curve assay from 95°C to 65°C in 0.5°C stepwise increments. Primers for select genes differentially expressed between CAR and TCMO were designed using Primer-BLAST and are shown in Table 1.

Table 1. Primers used for qPCR validation of differentially expressed genes in *Bombyx mori* strains.

Primers	Sequences (5› to 3) Forward Reverse	Product Length (bp)	Protein ID
Downregulated genes (CAR vs. TCMO)			
Hsp	GCCAACGATGTCCAGAGATT CTGCCTCTCCTCGTGCTTAC	196	heatshockprotein hsp20.1
odcdl	AAATGTCTGGGTGGAAGCAG GGGTGGAATCATCAGGTTGT	235	oxygen-dependent choline dehydrogenase-like
myr	GGAGACCGAGTGAAGACCTG GCTGTGGCATGAGCTAACAA	152	myrosinase 1
Upregulated genes (CAR vs. TCMO)			

pL	TGTTGTCGCTTTTCATCTGC AGTTTTTGAGCGCCGTATTG	207	peroxidase-like
LOC105842393	AAATGGCGCACAAATAGAGG CCAAGCTCTGCGTAAGGTTC	177	uncharacterized protein
ts	AAACCGAGCGTCCACTTATG TATTGTGATGGCAGCGGTAA	186	uncharacterized protein (tetraspanin family)
Housekeeping gene for normalization			
rp49	CAGGCGGTTCAAGGGTCAATAC TGCTGGGCTCTTTCCACGA	213	ribosomal protein 49

Total RNA was extracted from silkworms reared at the PTRI-TCMO and PTRI-CAR, followed by TURBO DNase treatment as previously described. RNA from two individual silkworms from each location was pooled and used to generate cDNA using the ProtoScript II First Strand cDNA Synthesis Kit (New England Biolabs, Ipswich, MA, USA) according to the manufacturer's recommendations. The cDNA was quantified by fluorometry using the Qubit ssDNA Assay kit (Thermo Fisher Scientific). Serial dilutions (10 ×) of pooled cDNA samples were prepared to generate a standard curve for each primer set used for the qPCR assays, starting at 100 ng of cDNA. The standard curves were used in conjunction with cycle threshold values to calculate starting quantities of the transcripts, and the expression levels and fold changes of each gene were normalized to *B. mori* ribosomal protein 49 [20]. All reactions were performed in triplicate. Calculations were performed using the Bio-Rad CFX Manager (v. 3.1). Statistical significance was determined using Student's t-test with GraphPad Prism 6.

3. RESULTS

3.1 Selection of Representative Silkworm Strains

Representative strains from the two collection sites were selected based on their phenotypic and economic characteristics (Table 2). Each strain can be differentiated by its larval markings. MO204 and DMMMSU119 are milky white, while LAT51 and ITA are beige.

Table 2. Characteristics of silkworm strains used in the study.

Strain	Origin	Strain Classification	Cocoon	Cocoon Shell Percentage	Raw Silk Percentage
MO204	TCMO	Priority	Oval, white	22.92%	22.42%
DMMMSU119	TCMO	Poor performing	Peanut, white	21.81%	21.12%
LAT51	CAR	Priority	Peanut, white	21.57%	17.44%
ITA	CAR	Priority	Oval, yellow gold	20.27%	no data

Cocoon shell percentage: Ratio of the weight of silk shell to the weight of the whole cocoon; Raw silk percentage: Ratio of the weight of raw silk to the weight of the whole cocoon.

The rearing sites had different conditions, particularly temperature and relative humidity (RH), the most important environmental factors in silkworm rearing. During sample collection, the temperature in TCMO was higher by 7.8 ± 0.1°C and RH was also higher by 12%.

3.2 RNA Sequencing and Quality Control

High-quality RNA was extracted (RIN ≥ 6), and the sequencing libraries had fragment sizes ranging from 131 to 279 bp, as determined by the Agilent TapeStation. Sequencing yielded 50.1 Gb of 169.2 M reads passing filter, with 96.1% of reads having > Q30. The number of raw reads from RNA sequencing is shown in Table 3. All ITA bioreplicates yielded lower than the expected number of reads. Because the number of reads was deemed insufficient, ITA was no longer included in downstream analyses.

Table 3. Statistics of raw RNA-seq read sequences of Bombyx mori strains.

Sample	Read Count	% ≥ Q30	Mean Q Score	GC Content (%)
MO204-1	17,531,032	94.11	34.60	41
MO204-3	8,590,562	93.49	34.47	43
MO204-4	13,085,562	94.10	34.59	44
DMMMSU-0	11,406,238	92.68	34.29	46
DMMMSU-2	21,789,720	92.38	34.22	46
DMMMSU-4	15,464,914	91.51	34.04	46
LAT51-1	12,321,992	93.70	34.51	42
LAT51-2	22,767,422	93.91	34.56	41
LAT51-4	8,793,834	94.74	34.73	44
ITA-1	3,621,664	86.80	33.05	48
ITA-2	4,633,324	88.96	33.51	48
ITA-3	3106	65.36	28.55	75

Trimmomatic was used for trimming Illumina adapters that were ligated to tag each sample during sequencing, and for quality trimming on leading and trailing sequences. Sequences that were under 36 bp were also dropped. A sliding window of 4:20 was used, where the sequences will be trimmed if the quality drops below Q20 in a 4-base window. After processing with Trimmomatic, most sequences were still retained, indicating that only a small number of sequences were dropped due to having low-quality bases (Figure 1). After Trimmomatic processing, poly G sequences still appeared in overrepresented sequences in FastQC, particularly in forward reads. Poly G sequences are artifacts in NextSeq sequencing due to its two-color filter that calls for G bases when there is no emission detected. Poly G sequences were filtered out using fastp, after which 77.9–83.1% of the raw reads were retained.

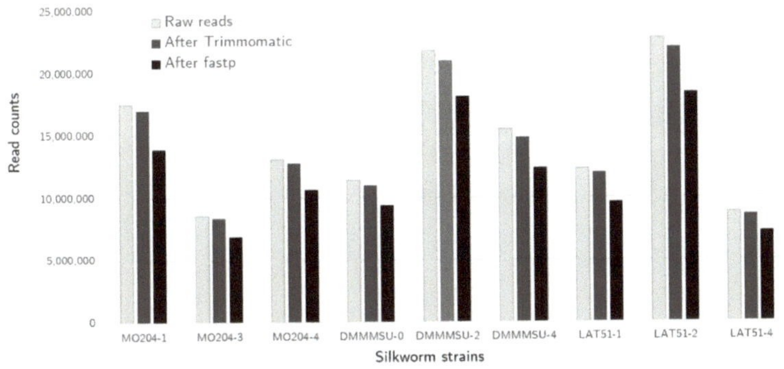

Figure 1. ***Bombyx mori*** **raw read counts and proportion of reads after processing.**

3.3 RNA-seq Read Alignment to the Reference Genome

Two samples per strain with the most read counts were aligned to the *B. mori* reference genome RefSeq GCF_000151625.1 [21] using the splice-aware aligner STAR. Most single and paired reads were uniquely mapped to the reference with only a small percentage (0.11–0.53%) of unmapped reads (Figure 2). The DMMMSU replicates had a higher per- centage of reads that were mapped to multiple loci (22.73% and 14.70% for DMMMSU-2 and DMMMSU-4, respectively), while for LAT51 and MO204 replicates, more than 90% of reads were uniquely mapped.

Moreover, most of the splice sites detected by STAR were in the GTF annotation of the reference genome. It has been reported that STAR generates a large number of putative novel splice sites [22], but in this case, the splice sites were validated by the reference genome annotation.

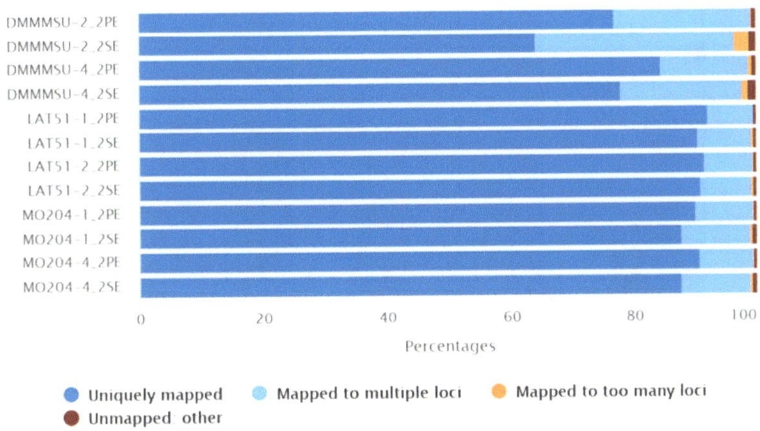

Figure 2. Summary of STAR alignment statistics of Bombyx mori RNA-seq reads to the reference genome. (PE: Paired-end reads, SE: Single-end reads). Generated using STAR.log files and Mul- tiQC [23].

Samtools flagstat was also run to analyze data from the FLAG field of the generated BAM files. According to flagstat, most of the mapped reads were properly paired and mapped with their mate. No duplicates were detected, but secondary alignments were produced (ranging from 9.6 to 31%). This coincides with the STAR statistics indicating that there were several reads mapped to multiple loci (ranging from 7.43 to 22.73%). In the

BAM files of the paired-end reads, less than 1% were singletons, and 0 reads had mates mapped to a different chromosome.

3.4 Transcriptome Assembly and Evaluation Metrics

After assembling the reads using different assemblers, the assemblies were mapped back to the input reads to evaluate their read support. Figure 3 shows that the reads had the highest alignment rates with the Trinity assembly, while the Cufflinks and StringTie assemblies had comparable alignment rates. Since the reads from the DMMMSU samples had a lower percentage of uniquely mapped reads, it could be expected that the reference- based assemblies would have less support from these samples.

Transcriptome assemblies are also evaluated based on the number and length of unigenes (unique sequences reconstructed in the assembly; includes not only genes but all assembled contigs). According to the contig statistics in Table 4, the Trinity assembly had the lowest N50 (1313) and lowest median contig length (468). Cufflinks had the highest contiguity with the highest N50 (3414), highest median contig length (1738), and average contig length (2305.35). However, in a transcriptome assembly, long contig lengths and high N50 are not as important as in genome assemblies. The most highly expressed transcripts are not necessarily the longest ones, and the majority of transcripts are expected to have low expression levels. Transcript lengths are also not indicative of a good transcriptome assembly since transcripts may be present in a wide range of sizes.

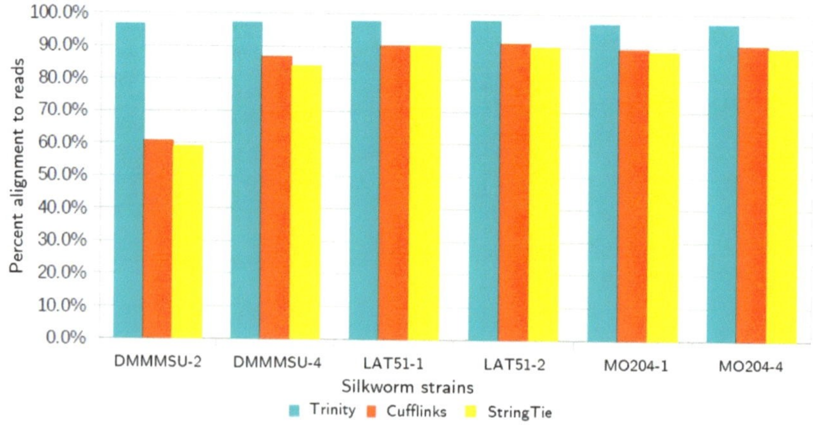

Figure 3. Overall alignment rate of RNA-seq reads to the transcriptome assemblies.

Table 4. Contig statistics of the different transcriptome assemblies from *Bombyx mori* strains. Statistics generated from TrinityStats.pl.

	Trinity	Cufflinks	StringTie
Total Trinity "genes"	47,562	40,593	41,851
Total Trinity transcripts	69,948	40,593	41,851
Percent GC	38.12%	42.08%	41.66%
N10	3257	7944	7333
N20	2487	6108	5595
N30	1992	4987	4491
N40	1635	4126	3640
N50	1313	3414	2981
Median contig length	468	1738	1226
Average contig length	800.21	2305.35	1789.3

Trinity assemblies had the highest overall alignment rates (96.60–98.00%), while the alignment rate for the Cufflinks and StringTie assemblies were comparable (60.5–91.10% and 58.80–90.2%, respectively).

A metric more appropriate for transcriptome assemblies is the ExN50 statistic, which represents N50 based on the most highly expressed transcripts that represent x% of the nor- malized expression data (e.g., E90N50 is the N50 value for 90% of the normalized expressed transcripts, excluding the lowly expressed transcripts) [24]. To obtain the ExN50 scores, the expression data were first normalized using kallisto, then count matrices were generated. Finally, the Trinity scripts contig_ExN50_statistic.pl and plot_ExN50_statistic.Rscript were run to compute and visualize ExN50 contig statistics.

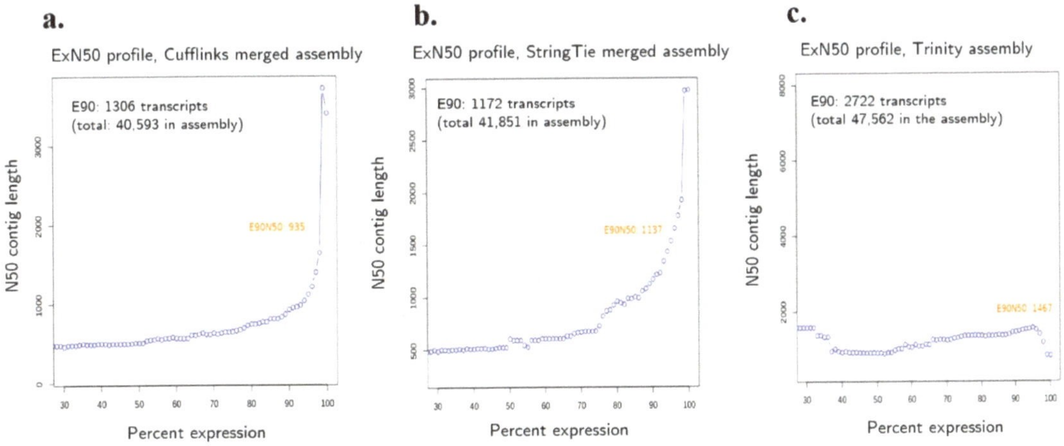

Figure 4. ExN50 profiles of the (a) Cufflinks, (b) StringTie, and (c) Trinity merged *Bombyx mori* transcriptome assemblies.

Comparing the N50 of the assemblies with their E90N50, a reversal is observed. When all contigs were considered (E100), Cufflinks and StringTie had a higher N50 than Trinity, indicating longer contig lengths. However, when N50 is confined to 90% of expression data, the Trinity assembly resulted in a higher value for E90N50. This suggests that lowly expressed genes make up longer contigs in reference-based assemblers but not in the de novo assembler. The ExN50 profiles in Figure 4 show that the peak N50 occurs at higher Ex values, indicating that the sequencing depth was enough to assemble the lowly expressed transcripts.

A reference-free evaluation metric for transcriptome assemblies was developed by Li et al. [25] in the package DETONATE (De novo Transcriptome RNA-seq Assembly with or without the Truth Evaluation). DETONATE has two component packages, RSEM-EVAL and REF-EVAL. RSEM-EVAL is based on the RSEM (RNA-Seq by Expectation Maximization) algorithm. RSEM-EVAL is based on a probabilistic model that combines different factors including the compactness of the assembly and read support [25]. The RSEM-EVAL score is a log joint probability based on three components: Likelihood of the assembly, assembly prior, and a Bayesian information criterion penalty. Penalties are imposed when assemblies have too many bases or contigs or have an unusual distribution of contig lengths relative to the expected read coverage. RSEM-EVAL scores are always reported to be negative, but higher scores are indicative of better assemblies. Out of the three transcriptome assemblies, Trinity obtained the highest RSEM-EVAL score (Table 5). Aside from the RSEM-EVAL scores, REF-EVAL scores are also reported in Table 5. REF-EVAL reports the recall and precision of the assemblies at the contig and nucleotide levels. The recall is the fraction of reference elements (contigs, scaffolds, nucleotides) that are recovered in the assembly,

while precision is the fraction of assembly elements that recover a reference element. The F_1 score represents the harmonic mean of the recall and precision scores:

$$F_1 = \frac{2 \times recall \times precision}{recall + precision}. \qquad (1)$$

Table 5. DETONATE scores of the different *Bombyx mori* transcriptome assemblies.

Assembly	RSEM-Eval	Nucleotide F1 (Unweighted)	Contig F1 (Unweighted)	Weighted k-Mer Recall	k-Mer Compression Score
Trinity	-8.35×10^{9}	0.38019	0.00239295	0.678414	-7.32371
Cufflinks	-1.34×10^{10}	0.682955	0.690038	0.545637	-22.5079
StringTie	-1.55×10^{10}	0.775391	0.677147	0.545727	-17.3472

The k-mer compression (KC) score is computed by the formula:

$$KC = WKR\ (weighted\ kmer\ recall) - ICR\ (inverse\ compression\ rate), \qquad (2)$$

where WKR measures an assembly's recall of the k-mer content in the reference, with each k-mer weighted by relative frequency, and ICR measures the compression of the RNA-seq data.

To evaluate the completeness of assemblies, Benchmarking Universal Single-Copy Orthologs (BUSCO) was employed using the dataset insecta_odb9 (2016). BUSCOs are core genes from OrthoDB that are expected to be present as single copies in orthologous groups. Recovered genes are classified as 'complete' when their lengths are within two standard deviations of the group mean length; 'fragmented' when they are only partially recovered; and 'missing' if they are unrecovered [26]. Both reference-guided assemblies had a high percentage (>95%) of completely recovered BUSCOs, while the Trinity assembly had the lowest number of BUSCOs at 80.10% (Figure 5). The StringTie assembly had more single-copy and fewer duplicate orthologs. Duplicated BUSCOs should be rare and could indicate erroneous assembly of haplotypes [26]. Running BUSCO for unmerged transcriptome assemblies (per sample) resulted in a small percentage of duplicated BUSCOs, so the duplication may be due to heterozygous alleles that were not collapsed when the assemblies were merged.

To summarize the results of the different transcriptome assembly evaluation metrics, the Trinity de novo assembly performed best in terms of alignment rate, assembled the greatest number of transcripts with high contig length represented by E90N50, and had the best RSEM-eval score. However, when it comes to the completeness of assembly based on BUSCO, Cufflinks outperformed the other assemblers.

3.5 Expression Quantification, Normalization, and Differential Expression Analysis

The reads were counted according to gene-level features using FeatureCounts (1.4.6-p5), which is part of the R Subread package [27]. Quantification was run at the meta-feature level (gene level), where reads that overlap multiple exons of the same gene are counted exactly once, provided there is no overlap with another gene (i.e., exon spanning reads will be counted once). Read count normalization was performed with DE analysis using DESeq2's size factor normalization. After running DESeq2, 1647 DEGs were identified within the threshold $p < 0.1$, but after FDR correction, only 476 DEGs were within the adjusted threshold $p_{adj} < 0.1$. Factor levels were set by location, where TCMO is the reference condition and CAR is set as the contrast condition. Among the 476 DEGs, there were 222 upregulated genes and 254 downregulated genes in CAR strains.

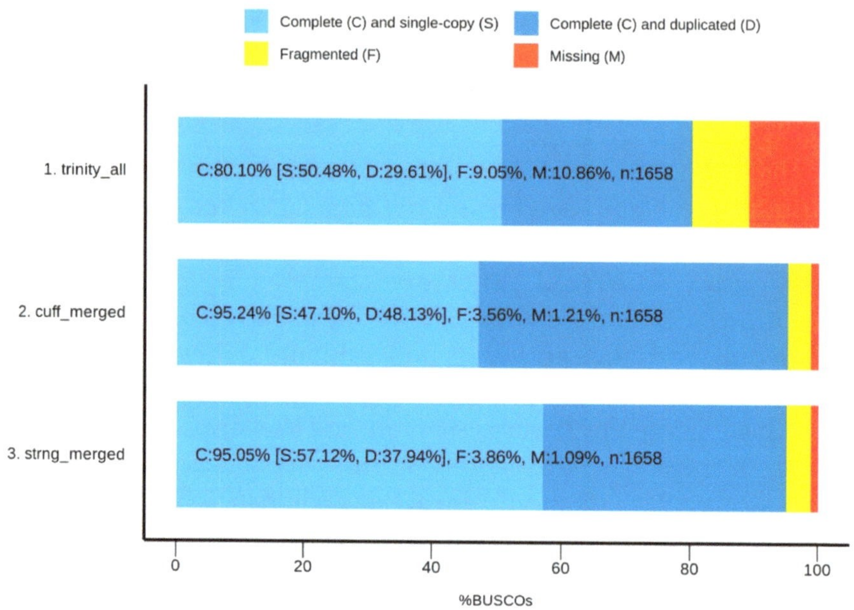

Figure 5. Measure of completeness of transcriptome assemblies from Bombyx mori strains using BUSCO (dataset: insecta_odb9).

A heat map of the 476 DEGs that passed the significance threshold is shown in Figure 6. Clustering of the DEGs by location is apparent in the heat map. The replicates for each strain show similar DEGs, except for a cluster of DEGs that are upregulated in bioreplicate MO204-4 but not in other MO204 replicates. This DEG cluster is composed of loci with uncharacterized proteins (LOC101741578, LOC101739373, LOC101736556, LOC101741319, LOC101738771, LOC101740400, LOC101735788). LOC101741578 has a conserved domain (COG5240 SEC21; vesicle coat complex COPI, gamma subunit) that is found in proteins involved in intracellular traffic and secretion [12]. Since these genes are highly upregulated in MO204-4 but not in other TCMO strains, this may be due to individual differences that were not taken into account (e.g., sex differences).

DEGs were filtered by the $p_{adj} < 0.1$ and $|\log_2 FC| > 1$ to select statistically and biologically significant genes; genes passing this significance threshold are represented by red points in the volcano plot shown in Figure 7. The filtered DEGs with their corresponding gene and protein IDs are listed in Table S1.

3.6 Functional Annotation of Select DEGs

The top DEGs were selected based on statistical significance and high fold changes and are summarized in Table 6. This list includes fold changes, p_{adj} values, as well as the protein names (if available) and the functions derived from Pfam/InterPro annotations available in UniProt [29].

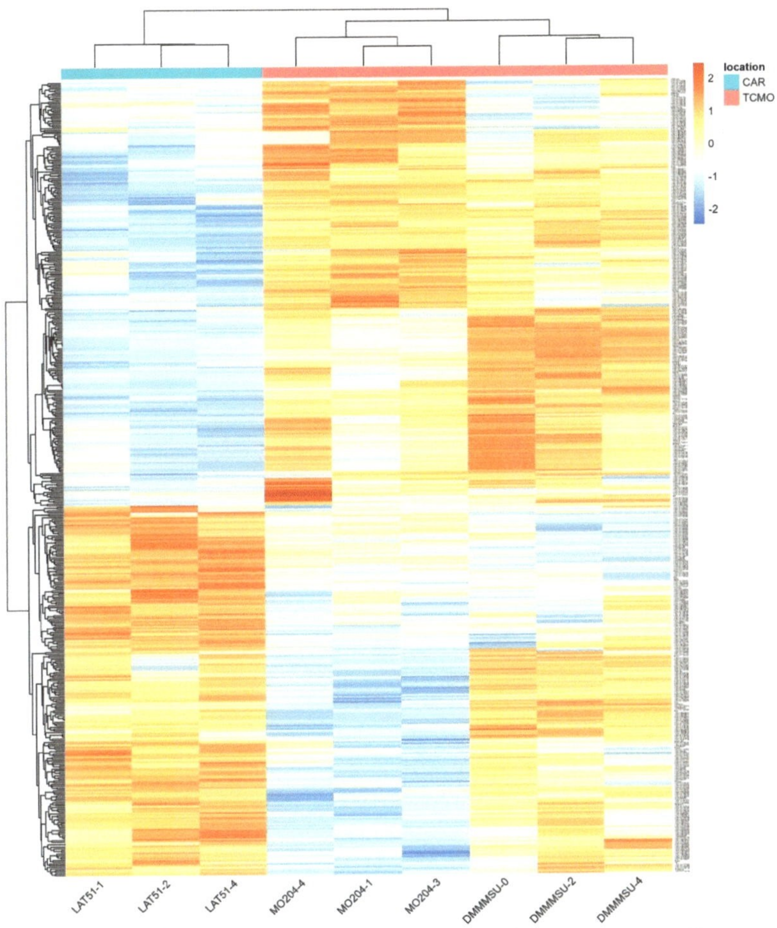

Figure 6. Heat map of the 112 DEGs from Bombyx mori with padj < 0.1. The values shown are scaled to the distances from the average of each row (gene).

Gene Ontology (GO) terms were mapped to the gene sets by matching the protein IDs to the GO annotations in the newest release of *B. mori* genome annotations from KAIKObase (ver. 4.1.0, July 2020). Their annotation contains the newest predicted gene models, descriptions including InterPro IDs, GO terms, and best hit in the NCBI nr database. Out of 14,124 assayed genes, 11,792 were annotated with protein IDs from nr, and 3948 were annotated with GO terms.

GO term enrichment was determined using the R package goseq. Significantly overrepresented GO terms are summarized in Tables S1 and S3 for upregulated and downregulated genes, respectively. Among the upregulated genes, the biological processes that were enriched are related to nucleotide metabolism and biosynthesis, lipid and carbohydrate metabolic processes, and regulation of transcription. Among down-regulated genes, similar biological processes were found to be enriched, but the ontology for carbohydrate metabolic process stands out with a low *p*-value. For cellular components, there is no overlap of GO terms for up- and downregulated genes. For molecular functions, nucleotide, protein, and metal binding are enriched in upregulated genes, while DNA binding, ATP binding, and zinc binding are enriched in downregulated genes. Catalytic activity, oxidoreductase, and hydrolase activities are also enriched molecular functions found in both sets.

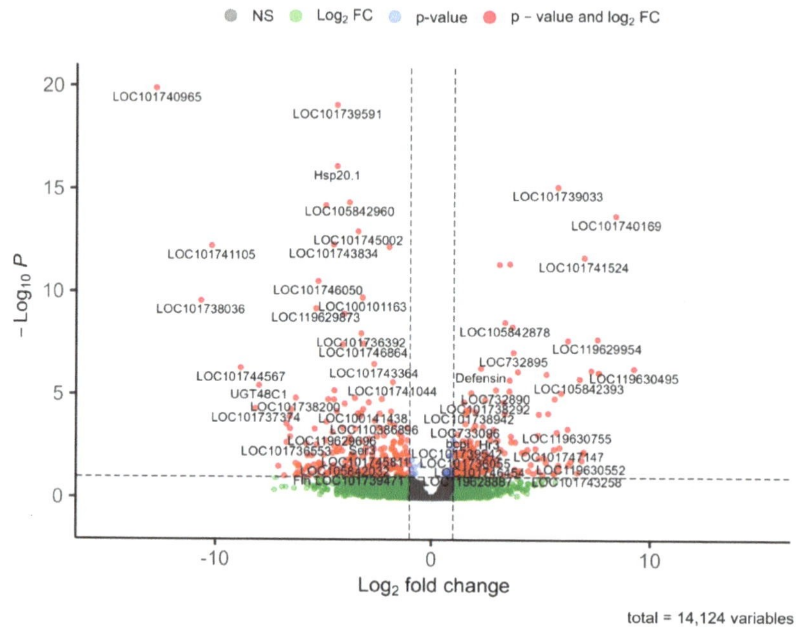

Figure 7. Volcano plot of DEGs from *Bombyx mori*, log fold change vs. -log *P*. Gray: Nonsignificant genes; green: Genes with significant fold changes but insignificant adjusted *p*-values; blue: Genes that passed the significance threshold according to adjusted *p*-value but did not have significant fold changes; red: Genes that have adjusted *p*-value < 0.1 and significant fold change (| $\log_2$ FC| > 1). Graph generated in R package EnhancedVolcano [28]

Table 6. Summary of select top DEGs between different Bombyx mori strains (CAR vs. TCMO).

Downregulated Genes (CAR vs. TCMO)				
Gene ID	**$\log_2$ FC**	**P_{adj}**	**Protein ID**	**Function**
LOC101740965	-12.8030	1.31 × 10-20	myrosinase 1 isoform X2	hydrolysis of glucosinolates, defense enzyme
Hsp20.1	-4.4182	7.45 × 10-17	heat shock protein hsp20.1	chaperone protein
LOC101739591	-4.4249	7.92 × 10-20	transient receptor potential channel pyrexia isoform X1	ion transport, contains ankyrin repeat
LOC101741105	-10.2153	5.85 × 10-13	myrosinase 1	hydrolysis of glucosinolates, defense enzyme

LOC101738036	-10.7102	2.66 × 10-10	serine protease inhibitor swm-1	serine protease inhibitor, trypsin inhibitor
LOC105842960	LOC105842960 -3.8422	4.35 × 10-15	oxygen-dependent choline dehydrogenase-like	oxidoreductase activity, FAD/NAD(P)-binding
Upregulated genes (CAR vs. TCMO)				
Gene ID	**log_2 FC**	**P_{adj}**	**Protein ID**	**Function**
LOC101739033	5.7239	7.09 × 10-16	zinc finger protein 664	dimer formation, metal binding
LOC101740169	8.3899	1.81 × 10-14	uncharacterized protein LOC101740169	tetraspanin family, scaffolding, cell adhesion, proliferation
LOC101741524	6.9289	1.99 × 10-12	synaptic vesicle glycoprotein 2C	sugar transporter
LOC105842878	3.3250	2.85 × 10-9	peroxidase-like	heme peroxidase, redox activity, oxidoreductase activity, defense enzyme
LOC119629954	7.5484	1.91 × 10-8	zinc finger BED domain-containing protein 4-like isoform X2	contains hAT dimerization domain, ribonuclease H-like superfamily: nucleotide metabolism and transport
LOC119630495	9.2422	5.08 × 10-7	uncharacterized protein LOC119630495	contains motifs: peptidase, Hsp70, TetR
LOC105842393	6.7307	1.61 × 10-6	uncharacterized protein LOC105842393	contains motifs: DUF5641 (pfam18701), unknown function, found in retrotransposons
Defensin	2.2251	5.08 × 10-7	defensin like protein 2 precursor	contains signal peptide

3.7 *q*PCR Validation of Select DEGs

qPCR: Heat shock protein 20 (Hsp20), oxygen-dependent choline dehydrogenase-like (odcdl), myrosinase 1 (myr), peroxidase-like (pl), uncharacterized protein (LOC105842393), and uncharacterized protein tetraspanin family (ts). The first three were observed in silico to be downregulated in CAR vs. TCMO silkworms, while the last three were calculated to be upregulated. The results for the qPCR found Hsp20 to be significantly downregulated in CAR vs. TCMO samples ($p < 0.0001$), as were odcdl ($p = 0.0005$) and myr ($p = 0.0008$). LOC105842393 was found to be significantly upregulated in CAR vs. TCMO ($p = 0.028$), as were pl ($p = 0.0001$) and ts ($p = 0.0093$). These data support the results of in silico analysis (Figure 8). Additionally, log_2FC values were calculated and compared with the values produced using DESeq2, as presented in Figure 8. While differences were found between the

log_2FC values between qPCR and DESeq2, these may be due to the individual gene expression differences between samples used for sequencing and samples used for qPCR. Despite these differences, the trends of downregulation or upregulation in CAR vs. TCMO samples observed during qPCR appear to support those found during in silico analysis.

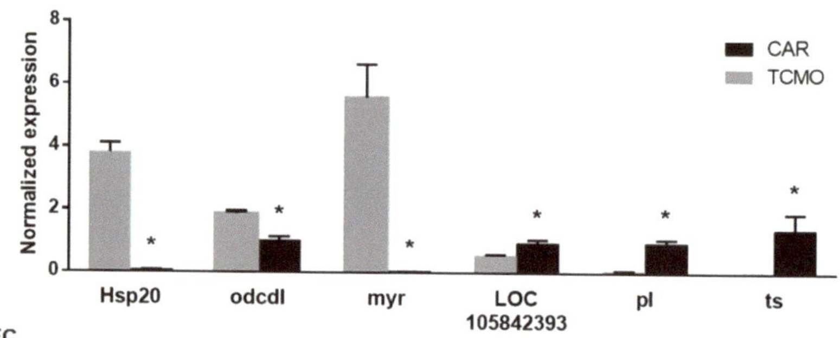

log2FC

	Hsp20	odcdl	myr	LOC 105842393	pl	ts
qRT-PCR	-5.61	-0.88	-8.82	0.77	3.58	7.17
DESeq2	-4.42	-3.84	-12.8	6.73	3.33	8.39

Figure 8. Results of qPCR validation of differentially expressed genes. Heat shock protein 20 (Hsp20), oxygen-dependent choline dehydrogenase-like (odcdl), and myrosinase 1 (myr) were downregulated in CAR vs. TCMO, while uncharacterized protein (LOC105842393), peroxidase-like (pl), and uncharacterized protein tetraspanin family (ts) were upregulated in CAR vs. TCMO. Log_2FC values calculated from qPCR results are compared with values obtained during DESeq2 analysis. Error bars indicate standard deviation; asterisks denote significant change in CAR vs. TCMO (Student's t-test), $p < 0.05$.

4. DISCUSSION

In this study, we compared gene expression in silk glands of B. mori strains reared in two different local facilities with different rearing temperatures. According to previous studies, the optimal temperature for silkworm growth is 20–28°C and 23–28°C for silk productivity [30]. The ambient temperature in the rearing sites during collection was 31.7°C and 23.9°C for TCMO and CAR, respectively. These data are based on bivoltine strains, although polyvoltine strains acclimatized in tropical countries are known to tolerate higher temperatures. Nevertheless, it has been shown that lower temperatures are better for silkworm productivity and larval duration [31].

Closely related to temperature is humidity, thus, an important factor to consider. It influences the withering of leaves in rearing beds, which affects larval feeding. However, there is no limiting range for humidity, and most insects can develop as long as they can control their water balance [30].

Among the downregulated genes in CAR were myrosinase, heat shock proteins, and serine protease inhibitors, which are related to defense and ensuring proper protein folding. These were downregulated in CAR strains likely because chaperone and protease inhibitors are not needed as much in lower temperatures. Aside from Hsp20.1, other heat shock proteins that are significantly downregulated are Hsp 68, Hsp 70, Hsp 23.7 precursor, and Hsp 25.4 precursor.

On the other hand, there were upregulated genes found in CAR related to defense such as peroxidase-like protein, defensin, and other proteins related to oxidoreductase activity. Many of these protein products are yet to be characterized, but sequence motifs provide some hints about their function based on information from protein families and conserved domains.

There were also several dehydrogenases found among the DEGs, such as oxygen-dependent choline dehydrogenase-like protein, L-sorbose 1-dehydrogenase, glucose dehydrogenase [FAD, quinone], 15-hydroxyprostaglandin dehydrogenase, farnesol dehydrogenase, dihydropyrimidine dehydrogenase [NADP(+)], mitochondrial aldehyde dehyxdrogenase, and alcohol dehydrogenases. Dehydrogenases are a type of oxidoreductases that oxidize substrates by reducing an electron acceptor and are important for their role in detoxification of metabolites. Malate dehydrogenase (BmMDH1) and 6-phosphogluconate dehydrogenase (Bm6PGD) were previously found to be differentially expressed in Dazao silkworms exposed to different temperatures, which implies enhancement in energy metabolism and ATPase expression at low temperatures [32].

Another DEG of note is SOSS complex subunit B homolog ($\log_2$FC: −2.0042; p_{adj}: 2.004 × 10^{-13}). The SOSS complex binds to single-stranded DNA and is involved in DNA damage response, cell cycle checkpoint activation, homologous repair of double-stranded breaks, and ATM-dependent signaling pathways [33]. Downregulation of SOSS in CAR strains may indicate that there is less DNA damage response in silkworms reared in low-temperature environments.

Microvitellogenin ($\log_2$FC: 6.2283; p_{adj}: 0.023) was one of the genes found upregulated in CAR strains, which encodes a low molecular weight (30 kDa) lipoprotein (30 K proteins, or 30 KPs). These lipoproteins are synthesized in the *B. mori* fat body and secreted in the hemolymph during the last instar larva stage. They are involved in several physiological processes, such as energy storage, embryonic development, and immune response [34]. Upregulation of microvitellogenin may indicate higher energy storage and enhanced immunoprotection in low-temperature strains [35,36].

Another upregulated DEG was the protein eyes shut ($\log_2$FC: 5.7860; p_{adj}: 0.024), also known as EGF-like protein 10, an essential protein for the formation of photoreceptor cells in *Drosophila* [33]. Most EGF-like domains are found in extracellular domains of membrane-bound proteins or secreted proteins. Aside from its EGF-like domain, *B. mori* protein eyes shut also has a laminin G domain, which is associated with different functions such as cell adhesion, signaling, migration, assembly, and differentiation [37].

Three regulators of the juvenile hormone (JH) were found among the DEGs in this analysis: Juvenile hormone epoxide hydrolase-like protein 2 (Jheh-lp2), juvenile hormone epoxide hydrolase precursor (Jheh2), and cytosolic juvenile hormone binding protein 36 kDa subunit (Cjhbp). JHEH inactivates juvenile hormones by hydrolyzing the epoxide groups in JH. Differential expression of JH regulators could explain the observed differences in larval development between CAR and TCMO strains, wherein CAR strains have a longer larval duration (23 days for TCMO, 27 days for CAR). JH has long been known to affect silk production, and the application of JH analogs has been used to prolong the 5th instar stage to increase larval weight and silk secretion [38]. More recently, the JH regulatory pathway has been demonstrated to influence Fib-H expression through the action of transcription factors *Bmdimm* and *Bmsage* [39]. Zhou et al. [40] showed that JH biosynthesis and sexual maturation are delayed in cotton ballworm (Helicoverpa armigera) reared at 19°C compared to those reared in higher temperatures. Liu et al. [41] showed that JH titers in Formosan termite (*Coptotermes formonasus*) were positively correlated with temperature. In contrast, Geister et al. [42] found no correlation between temperature and JH titers in the tropical butterfly *Bicyclus anynana*.

Caytaxin was downregulated in CAR strains ($\log_2$FC: −2.8609; p_{adj}: 0.032). Caytaxin is a member of the BNIP2 (Bcl2-/adenovirus E1B nineteen kDa-interacting protein 2) protein family (pfam12496). Caytaxins interact with pro- and anti-apoptotic molecules in the cell and are involved in Rho GTPase regulation [43].

Another DEG found was dymeclin ($\log_2$FC: 1.7453; p_{adj}: 0.030), which belongs to the Dymeclin (Dyggve-Melchior-Clausen syndrome protein) protein family (pfam09742) in plants and animals. Dymeclin proteins have a length of approximately 700 residues and contain many leucine and isoleucine in their conserved domain [12]. Mutations in human dymeclin cause Dyggve–Melchior–Clausen syndrome (DMC, MIM 223800), an autosomal recessive disorder. Dymeclin is a peripheral membrane protein dynamically associated with the Golgi apparatus. In *Caenorhabditis elegans*, a member of the Dymeclin protein family is hid1 (high-temperature-induced dauer-formation protein 1), which encodes a highly conserved transmembrane protein that could be involved in vesicle secretion or intercellular signaling [44].

Previous comparative studies found similar enrichment in GO terms as those in this study. In one study comparing wild and domestic silkworms, oxidoreductase activity (GO: 0016491) was the only enriched GO term in four pairwise comparisons [11]. In another study comparing Chinese silkworm strains JingSong and Lan10, which have different rates of silk production, the dominant GO term was membrane-enclosed lumen under the cellular component [10].

CONCISE SUMMARY

This study offers, for the first time, significant insights into the transcriptome of B. mori strains located in the Philippines. The differential expression study conducted on the silk glands of B. mori strains cultivated in various locations provides insight into the processes and roles associated with the down-regulated and up-regulated genes. This molecular data may serve as a repository of phenotypic markers for the enhancement of indigenous silkworm strains to advance the sericulture sector in the Philippines. Future directions include conducting KEGG pathway analysis, in addition to protein annotations and GO keywords, to further clarify the activities of DEGs in silkworms and silk production. Further quantitative PCR experiments are recommended to validate the additional bioinformatically identified DEGs, alongside those verified here. This study could be expanded to include transcriptome analysis of more B. mori strains, organs, and developmental phases.

REFERENCES

Trevisan, A.; Hart, M.; Yuste, V.; Hjoellund, A.; Galle, D.; Bergman, O. Cocoon Silk: A Natural Architecture. Available on- line: https://web.archive.org/web/20120507085636/http://www.senature.com/research/publications/cocoon-silk-a-natural- architecture (accessed on 19 October 2021).

Kunz, R.I.; Brancalhão, R.M.C.; de Ribeiro, L.F.C.; Natali, M.R.M. Silkworm Sericin: Properties and Biomedical Applications. *BioMed Res. Int.* 2016, *2016*, 8175701. [CrossRef] [PubMed]

PhilFIDA. The Negros Silk Project. Available online: http://www.philfida.da.gov.ph/index.php/recent-news-articles/114-the- negros-silk-project (accessed on 4 November 2021).

Zanatta, D.B.; Bravo, J.P.; Barbosa, J.F.; Munhoz, R.E.F.; Fernandez, M.A. Evaluation of Economically Important Traits from Sixteen Parental Strains of the Silkworm Bombyx Mori L (Lepidoptera: Bombycidae). *Neotrop. Entomol.* 2009, *38*, 327–331. [CrossRef] [PubMed]

Reyes, R.C. Silkworm Genetics Research: Developing Silkworm Strains Adapted to Lowland Philippine Conditions. Presented at the Development Fund Professorial Chair Lecture, University of the Philippines Los Baños, College, Laguna. 2003. Available online: https://agris.fao.org/agris-search/search.do?recordID=PH2004000177 (accessed on 4 November 2021).

Supsup, G.E.; Abuan, J.P.; Viduya, M.M.; Sanchez, M.L.; Ulat, M.E. Selection of Superior Bivoltine Silkworm Hybrids through Combining Ability Test and Multiple Traits Evaluation Index. *IAMURE Int. J. Ecol. Conserv.* 2018, *23*, 1.

Velasco, A.B.; Bacuso, P.M.; Cabrito, F.P.; Fernandez, R.W.; Alvarez, V.B.; de Guzman, Z.I.; Villanueva, E.P. New Breeds of Filipino Silkworm Varieties Commercialized. *Philipp. Technol. J.* 1994, *19*. Available online: https://agris.fao.org/agris-search/search.do? recordID=PH9710645 (accessed on 4 November 2021).

Aro, S.C. Silk Production on the Decline. *Sunstar*. 2015. Available online: https://web.archive.org/web/20220722062934/https://www.sunstar.com.ph//article/40526 (accessed on 22 July 2022).

Villanueva, R.C. Problems and Issues Affecting the Pace of Sericulture Development in the Philippines. *Don Marian. Marcos Meml. State Univ. Res. Ext. J. Bacnotan Campus*. 1999. Available online: https://agris.fao.org/agris-search/search.do?recordID=PH200 7M00034 (accessed on 20 June 2022).

Li, J.; Qin, S.; Yu, H.; Zhang, J.; Liu, N.; Yu, Y.; Hou, C.; Li, M. Comparative Transcriptome Analysis Reveals Different Silk Yields of Two Silkworm Strains. *PLoS ONE* 2016, *11*, e0155329. [CrossRef]

Fang, S.-M.; Hu, B.-L.; Zhou, Q.-Z.; Yu, Q.-Y.; Zhang, Z. Comparative Analysis of the Silk Gland Transcriptomes between the Domestic and Wild Silkworms. *BMC Genom.* 2015, *16*, 60. [CrossRef]

Lu, F.; Wei, Z.; Luo, Y.; Guo, H.; Zhang, G.; Xia, Q.; Wang, Y. SilkDB 3.0: Visualizing and Exploring Multiple Levels of Data for Silkworm. *Nucleic Acids Res.* 2020, *48*, D749–D755. [CrossRef]

Shimomura, M.; Minami, H.; Suetsugu, Y.; Ohyanagi, H.; Satoh, C.; Antonio, B.; Nagamura, Y.; Kadono-Okuda, K.; Kajiwara, H.; Sezutsu, H.; et al. KAIKObase: An Integrated Silkworm Genome Database and Data Mining Tool. *BMC Genom.* 2009, *10*, 486. [CrossRef]

Li, Y.; Wang, G.; Tian, J.; Liu, H.; Yang, H.; Yi, Y.; Wang, J.; Shi, X.; Jiang, F.; Yao, B.; et al. Transcriptome Analysis of the Silkworm (Bombyx Mori) by High-Throughput RNA Sequencing. *PLoS ONE* 2012, *7*, e43713. [CrossRef]

Prasad, M.D.; Muthulakshmi, M.; Arunkumar, K.P.; Madhu, M.; Sreenu, V.B.; Pavithra, V.; Bose, B.; Nagarajaram, H.A.; Mita, K.; Shimada, T.; et al. SilkSatDb: A Microsatellite Database of the Silkworm, Bombyx Mori. *Nucleic Acids Res.* 2005, *33*, D403–D406. [CrossRef]

Xu, H.-E.; Zhang, H.-H.; Xia, T.; Han, M.-J.; Shen, Y.-H.; Zhang, Z. BmTEdb: A Collective Database of Transposable Elements in the Silkworm Genome. *Database* 2013, *2013*. [CrossRef] [PubMed]

Xia, Q.; Li, S.; Feng, Q. Advances in Silkworm Studies Accelerated by the Genome Sequencing of Bombyx Mori. *Annu. Rev. Entomol.* 2014, *59*, 513–536. [CrossRef] [PubMed]

Wang, Z.; Gerstein, M.; Snyder, M. RNA-Seq: A Revolutionary Tool for Transcriptomics. *Nat. Rev. Genet.* 2009, *10*, 57–63. [CrossRef]

Zhang, R.; Cao, Y.-Y.; Du, J.; Thakur, K.; Tang, S.-M.; Hu, F.; Wei, Z.-J. Transcriptome Analysis Reveals the Gene Expression Changes in the Silkworm (Bombyx Mori) in Response to Hydrogen Sulfide Exposure. *Insects* 2021, *12*, 1110. [CrossRef] [PubMed]

Yokoi, K.; Tsubota, T.; Jouraku, A.; Sezutsu, H.; Bono, H. Reference Transcriptome Data in Silkworm Bombyx Mori. *Insects* 2021, *12*, 519. [CrossRef]

The International Silkworm Genome Consortium. The Genome of a Lepidopteran Model Insect, the Silkworm Bombyx Mori. *Insect Biochem. Mol. Biol.* 2008, *38*, 1036–1045. [CrossRef]

Dundar, W.F.; Skrabanek, L.; Zumbo, P. Introduction to Differential Gene Expression Analysis Using RNA-Seq 2019. Available online: https://chagall.med.cornell.edu/RNASEQcourse/Intro2RNAseq.pdf (accessed on 2 December 2021).

Ewels, P.; Magnusson, M.; Lundin, S.; Käller, M. MultiQC: Summarize Analysis Results for Multiple Tools and Samples in a Single Report. *Bioinformatics* 2016, *32*, 3047–3048. [CrossRef]

Haas, B. Transcriptome Contig Nx and ExN50 Stats Trinityrnaseq/Trinityrnaseq Wiki. Available online: https://github.com/ trinityrnaseq/trinityrnaseq (accessed on 2 December 2021).

Li, B.; Fillmore, N.; Bai, Y.; Collins, M.; Thomson, J.A.; Stewart, R.; Dewey, C.N. Evaluation of de Novo Transcriptome Assemblies from RNA-Seq Data. *Genome Biol.* 2014, *15*, 553. [CrossRef]

Simão, F.A.; Waterhouse, R.M.; Ioannidis, P.; Kriventseva, E.V.; Zdobnov, E.M. BUSCO: Assessing Genome Assembly and Annotation Completeness with Single-Copy Orthologs. *Bioinformatics* 2015, *31*, 3210–3212. [CrossRef]

Liao, Y.; Smyth, G.K.; Shi, W. FeatureCounts: An Efficient General Purpose Program for Assigning Sequence Reads to Genomic Features. *Bioinformatics* 2014, *30*, 923–930. [CrossRef]

Blighe, K.; Rana, S.; Lewis, M. EnhancedVolcano: Publication-Ready Volcano Plots with Enhanced Colouring and Labeling. 2018. Available online: https://github.com/kevinblighe/EnhancedVolcano (accessed on 2 December 2021).

The UniProt Consortium. UniProt: The Universal Protein Knowledgebase in 2021. *Nucleic Acids Res.* 2021, *49*, D480–D489. [CrossRef] [PubMed]

Rahmathulla, V.K. Management of Climatic Factors for Successful Silkworm (Bombyx Mori L.) Crop and Higher Silk Production: A Review. *Psyche* 2012, *2012*, e121234. [CrossRef]

Kumar, N.S.; Basavaraja, H.K.; Kumar, C.M.K.; Reddy, N.M.; Datta, R.K. On the Breeding of "CSR18xCSR19"- A Robust Bivoltine Hybrid of Silkworm, Bombyx Mori L. for the Tropics. *Int. J. Ind. Entomol.* 2002, *5*, 153–162.

Guo, H.; Huang, C.; Jiang, L.; Cheng, T.; Feng, T.; Xia, Q. Transcriptome Analysis of the Response of Silkworm to Drastic Changes in Ambient Temperature. *Appl. Microbiol. Biotechnol.* 2018, *102*, 10161–10170. [CrossRef]

Zahn-Zabal, M.; Michel, P.-A.; Gateau, A.; Nikitin, F.; Schaeffer, M.; Audot, E.; Gaudet, P.; Duek, P.D.; Teixeira, D.; Rech de Laval, V.; et al. The NeXtProt Knowledgebase in 2020: Data, Tools and Usability Improvements. *Nucleic Acids Res.* 2020, *48*, D328–D334. [CrossRef]

Shi, X.-F.; Li, Y.-N.; Yi, Y.-Z.; Xiao, X.-G.; Zhang, Z.-F. Identification and Characterization of 30 K Protein Genes Found in Bombyx Mori (Lepidoptera: Bombycidae) Transcriptome. *J. Insect Sci.* 2015, *15*, 71. [CrossRef]

Singh, N.K.; Pakkianathan, B.C.; Kumar, M.; Prasad, T.; Kannan, M.; König, S.; Krishnan, M. Vitellogenin from the Silkworm, Bombyx Mori: An Effective Anti-Bacterial Agent. *PLoS ONE* 2013, *8*, e73005. [CrossRef]

Liu, Y.; Chen, M.; Su, J.; Ma, H.; Zheng, X.; Li, Q.; Shi, S.; Qin, L. Identification and Characterization of a Novel Microvitellogenin from the Chinese Oak Silkworm Antheraea Pernyi. *PLoS ONE* 2015, *10*, e0131751. [CrossRef]

Sasaki, T.; Fässler, R.; Hohenester, E. Laminin: The Crux of Basement Membrane Assembly. *J. Cell Biol.* 2004, *164*, 959–963. [CrossRef]

Daillie, J. Juvenile Hormone Modifies Larvae and Silk Gland Development in Bombyx Mori. *Biochimie* 1979, *61*, 275–281. [CrossRef]

Zhao, X.-M.; Liu, C.; Jiang, L.-J.; Li, Q.-Y.; Zhou, M.-T.; Cheng, T.-C.; Mita, K.; Xia, Q.-Y. A Juvenile Hormone Transcription Factor Bmdimm-Fibroin H Chain Pathway Is Involved in the Synthesis of Silk Protein in Silkworm, Bombyx Mori *. *J. Biol. Chem.* 2015, *290*, 972–986. [CrossRef] [PubMed]

Zhou, X.; Coll, M.; Applebaum, S.W. Effect of Temperature and Photoperiod on Juvenile Hormone Biosynthesis and Sexual Maturation in the Cotton Bollworm, Helicoverpa Armigera: Implications for Life History Traits. *Insect Biochem. Mol. Biol.* 2000, *30*, 863–868. [CrossRef]

Liu, Y.; Henderson, G.; Mao, L.; Laine, R.A. Effects of Temperature and Nutrition on Juvenile Hormone Titers of Coptotermes Formosanus (Isoptera: Rhinotermitidae). *Ann. Entomol. Soc. Am.* 2005, *98*, 732–737. [CrossRef]

Geister, T.L.; Lorenz, M.W.; Meyering-Vos, M.; Klaus, H.H.; Fischer, K. Effects of Temperature on Reproductive Output, Egg Provisioning, Juvenile Hormone and Vitellogenin Titres in the Butterfly Bicyclus Anynana. *J. Insect Physiol.* 2008, *54*, 1253–1260. [CrossRef]

Gupta, A.B.; Wee, L.E.; Zhou, Y.T.; Hortsch, M.; Low, B.C. Cross-Species Analyses Identify the BNIP-2 and Cdc42GAP Homology (BCH) Domain as a Distinct Functional Subclass of the CRAL_TRIO/Sec14 Superfamily. *PLoS ONE* 2012, *7*, e33863. [CrossRef]

Ailion, M.; Thomas, J.H. Isolation and Characterization of High-Temperature-Induced Dauer Formation Mutants in Caenorhabdi- tis Elegans. *Genetics* 2003, *165*, 127–144. [CrossRef]

Edgar, R. Gene Expression Omnibus: NCBI Gene Expression and Hybridization Array Data Repository. *Nucleic Acids Res.* 2002, *30*, 207–210. [CrossRef]

11

Chapter

BACTERIAL ANALYSIS IN THE INTESTINAL TRACT OF DOMESTIC SILKWORMS, BOMBYX *MORI L.*

1. INTRODUCTION: AN OVERVIEW

Bombyx mori L., known as the domestic silkworm or mulberry silkworm, solely consumes the leaves of the mulberry plant, Morus alba. The insect, part of the Bombycidae family, is recognized for its capacity to generate silk strands during cocoon formation. Bombyx mori has been cultivated for around 5000 years and has become integral to the silk production business. Bombyx mori serves as the model organism for the order Lepidoptera. It ranks second to Drosophila melanogaster regarding the volume of genetic research performed on insects. [4].

Countries in Asia remain the top producers of natural silk [8], contributing to 98% of the world's supply [2]. High demand for silk in the textile and fashion industries globally prompts sericulturists to improve the production of silk and enhance its quality [9]. Recent efforts to increase the capacity for silk production include breeding silkworms with improved economically important traits. In the Philippines, key qualitative traits used to assess strains for the production of silk include size, weight, and shape of cocoons, hatchability of fertilized eggs, larval development time, larval and pupal weight, moth emergence, reproductive capacity of moths, and mortality rate [10]. Of these, strains exhibiting faster larval development, accompanied by increased cocoon weight and yield are the most preferred. The Philippine silk industry, however, has been facing issues with the quality of silkworms, and these contribute to low cocoon production [11]. The local silk industry, thus, needs silkworms with improved quality. Genetic characterization of the 17 strains that are reared, bred, and maintained in germplasm centers in the country has already been initiated by Alcudia-Catalma et al. 2020 [9], but additional biological information is deemed necessary.

The growth and development of silkworm larvae and the economic characteristics of cocoons are influenced largely by nutritional and climatic factors, rearing techniques, silkworm races, and other factors [12]. Like most insects, silkworms also harbor a consortium of microorganisms that play crucial roles in their survival [13]. The presence of symbiotic bacteria in their gut, for example, allows for the absorption and utilization of nutrients and provides immunity to diseases [14]. These affect silkworm growth, development, and survival, which in turn affect the yield and quality of cocoons. Microbial community investigation can be performed through approaches that utilize high-throughput next generation sequencing techniques. 16S rRNA gene amplicon sequencing is one approach that enables the identification of species of microorganisms present in environmental samples and provides information on the function of these microorganisms through the analysis of genomic data. This approach offers a way to bypass the challenge posed by bacterial species that cannot be cultured because it employs analyses to investigate bacterial diversity, abundance, processes, and possible interactions of an entire bacterial community based on genetic content of samples collected [15]. The process by which 16S rRNA gene amplicon sequencing allows "direct access" to the genetic information of various microorganisms from the environment [16] relies on the 16S rRNA gene conserved in prokaryotic organ- isms [17]. Bacterial species may be identified based on the sequences of this gene [17]. Recently, next-generation sequencing has significantly accelerated this process via various platforms that facilitate the sequencing of target regions across multiple samples, the assembly of numerous reads into contiguous sequences, the alignment of sequences to either a specific target organism or related species, and ultimately the classification of organisms at the chosen taxonomic level [16].

The primary objective of this work was to characterize and compare the bacterial communities present in the gastrointestinal tracts of Philippine silkworm strains and their larval instars by 16S rRNA gene amplicon sequencing. The findings of this study offer information related to the native strains cultivated and preserved in the germplasm centers of the Philippines for decades [9], although biological data are insufficient to improve commercial sustainability. The data produced here can facilitate strain enhancement by traditional hybridization and selective breeding, commonly employed in other silk-producing nations to increase the sustainability of silk production..

2. MATERIALS AND METHODS

2.1 Sample Collection, Preservation, and Selection

Four *B. mori* strains were considered for inclusion in this study. These silkworm strains (K203, MO202, MO204, and DMMMSU 119) were reared at the Philippine

Textile Research Institute Technology Center in Misamis Oriental (PTRI-TCMO) and were chosen for analysis based on the available data describing each strain's larval development, as well as their cocoon and silk characteristics (Table S1). Based on the facility's characterization of strains and on the report of Basaen et al. (1982) [10], high hatch percentage, short larval duration, high cocoon weight, and high silk yield are considered as the most economically important and preferred silkworm traits.

All test strains were reared within the same facility under similar conditions and fed with mulberry leaves collected from the same field. However, early instars (first to third) were fed with young leaves, whereas later instars (fourth and fifth) were fed with fully developed leaves according to the practices of the facility. The second, third, fourth, and fifth instars of each of the silkworm strains were preserved in 95% ethanol and stored at 4°C prior to gut dissection and genomic DNA extraction. Ethanol preservation was adopted based on Hammer et al. (2015) [18].

2.2 Silkworm Gut Dissection and DNA Extraction

One fifth instar larva, one fourth instar larva, three third instar larvae, and five second instar larvae from each strain were first washed with distilled water to remove excess ethanol used for preservation. These were then dissected on the dorsal side to extract the guts, which were washed with distilled water to remove excess plant material within. For the second and third instars, the guts of individuals belonging to the same strain were pooled to obtain 10–30 mg of starting material for DNA extraction. Two biological replicates were made in total using guts of different larvae from the same batch of eggs. Dissection and pooling of silkworm guts was done inside a laminar flow hood. Sterile scalpel blades and forceps used for dissection were also changed in between the dissection of each strain and instar.

DNA was extracted from the guts using the Quick-DNA Tissue/Insect Miniprep Kit (Zymo Research, Irvine, CA, USA) following the manufacturer's instructions. The extracted DNA was quantified using Qubit dsDNA BR assay (Invitrogen, Waltham, MA, USA). The quality and purity of the extracts was assessed using 1% agarose gel electrophoresis and Nanodrop spectrophotometer (Scientific Industries, Bohemia, NY, USA), respectively. DNA extracts were stored at −40°C prior to library preparation.

2.3 Library Preparation and Sequencing on MiSeq FGx

Libraries were prepared and pooled from a total of 32 samples (4 silkworm strains × 4 instar stages × 2 biological replicates) based on the *16S* Metagenomic Sequencing Library Preparation guide by Illumina with a few modifications. DNA extracts were diluted to 10 ng/μL using nuclease-free water. Primers 341F and 805R containing 5′ overhang sequences (Table S2) used for *16S* V3-V4 PCR were from Herlemann et al. (2011) [19]. The amplicon PCR mix contained the following: 2.5 μL DNA template

(15 ng/μL), 5 μL primer 341F (6 μM), 5 μL primer 805R (6 μM), and 12.5 μL MyFi PCR mix (Bioline Reagents Ltd., United Kingdom) for a total reaction volume of 25 μL. PCR was then performed under the following conditions: initial denaturation at 95°C for 3 min; 35 cycles of denaturation at 95°C for 30 s, annealing at 55°C for 30 s; extension at 72°C for 30 s; final extension at 72°C for 5 min; and held at 4°C.

Pairs of different indices (index 1 and index 2) from the Nextera XT Index kit (Illumina, San Diego, CA, USA) were selected such that each sample would have a unique pair of indices, as presented in Table S3. The index PCR mix contained the following: 5 μL amplicon PCR product, 5 μL Nextera XT index primer 1, 5 μL Nextera XT index primer 2, 25 μL MyFi PCR Mix, and 10 μL nuclease-free water, for a total reaction volume of 50 μL. PCR was done under the following conditions: initial denaturation at 95°C for 3 min; 8 cycles of denaturation at 95°C for 30 s; annealing at 55°C for 30 s; extension at 72°C for 30 s; final extension at 72°C for 5 min; and held at 4°C.

Individual library concentrations in nM were calculated based on Qubit concentration (ng/μL) and amplicon size based on TapeStation (Agilent, Santa Clara, CA, USA).

Calculated library concentrations were then normalized, pooled, and denatured. Pre- chilled HT1 hybridization buffer was then added to dilute the denatured library to 20 pM. The PhiX control (Illumina, San Diego, CA, USA) library was prepared following the above procedure. A final concentration of 8 pM was prepared with 15% PhiX spike-in. The combined library was loaded into the MiSeq v3 (600 cycles, 2 × 300 bp reads) reagent cartridge (Illumina, San Diego, CA, USA), and the sequencing run using the MiSeq FGx (Illumina, San Diego, CA, USA) was initiated.

2.4 16S rRNA Amplicon Sequence Analysis

The summary of the sequencing run is detailed in Table S4. The percent Q30 score of the run was 93.0%, which is above the minimum of 70% recommended by Illumina. Raw DNA sequence data generated were deposited in the NCBI Sequence Read Archive (SRA) under the accession numbers listed in Table S5.

Sequence data were then converted to FASTQ format and used for downstream analyses using QIIME 2 version 2021.11 (Available online: https://qiime2.org/ (accessed on 2 January 2022)). The QIIME 2 workflow is summarized in Figure S1. Lengths of 300 bp were obtained for the paired-end reads, and raw sequence counts per sample ranged from 32,620 to 80,412 sequences. Quality control was performed using DADA2 (Available online: https://qiime2.org/ (accessed on 2 January 2022)) included in the QIIME 2 pipeline. The final sequence counts per sample, which ranged from 13,700 to 41,059 after quality filtering, denoising, merging forward and reverse reads, and removing chimeric sequences using DADA2 prior to QIIME 2 analysis, are detailed in Table S6. This allows for replicate sequences to be collapsed into representative sequences (rep-seqs) used for classification. The range of the final sequence counts of this study fell within the range reported by Suenami et al. (2019) [20] (i.e., 12,627 to 56,357 for QIIME 2 analysis of the gut microbiome of hornets).

The representative sequences generated were then used for naïve Bayes taxonomic classification using the Silva 138 classifier, which calculates the probability that a sequence belongs to a certain species instead of sequence alignment [21]. Classification using QIIME 2 in this study, therefore, did not require a percent identity threshold value. Sequences of mitochondria and chloroplasts were then filtered out prior to diversity analyses. A sampling depth of 2420 was used for diversity analyses. This value was used such that all samples and the most sequences possible would be included in the study. This sampling depth would enable the analysis of the majority of the silkworm gut microbiome as the generated alpha-rarefaction plot showed that most sequence counts begin to taper off at this point and sample count declines past this point (Figure S2). Kruskal–Wallis and PERMANOVA tests (999 permutations) were conducted to determine the significance of alpha-diversity and beta-diversity analyses, respectively. A heatmap showing the abundance of bacterial genera across all samples was also generated (Figure S3).

The feature table that was subsequently generated after rep-seqs generation by QIIME 2 was then used to determine the most frequently observed representative sequences across samples. These rep-seqs were used for BLAST queries of bacteria that were associated with each of the genera. NCBI BLAST queries were performed using the nt database on 10 January 2022. The top BLAST hits for the sequences corresponding to the five most abundant bacterial taxa according to QIIME 2 classification were determined.

3. RESULTS

3.1 Analysis of Bacterial Abundance

Bacterial genera were identified based on abundance by QIIME 2 using the Silva classifier. Among the top five most abundant genera, *Pseudomonas* appeared to be the most dominant genus across all silkworm strains and instars (Figure 1). It displayed lower abundance percentages in later instars of DMMMSU 119 (fifth instar) and K201 (fourth and fifth instars), but was still ranked the most abundant. *Sphingomonas* and *Delftia* were typically ranked the second to fourth most abundant in most of the strains and instars. These are followed by *Methylobacterium* which was generally ranked third to fifth most abundant across all instars and strains. *Acinetobacter* was ranked the fourth to fifth most abundant bacterial genus overall.

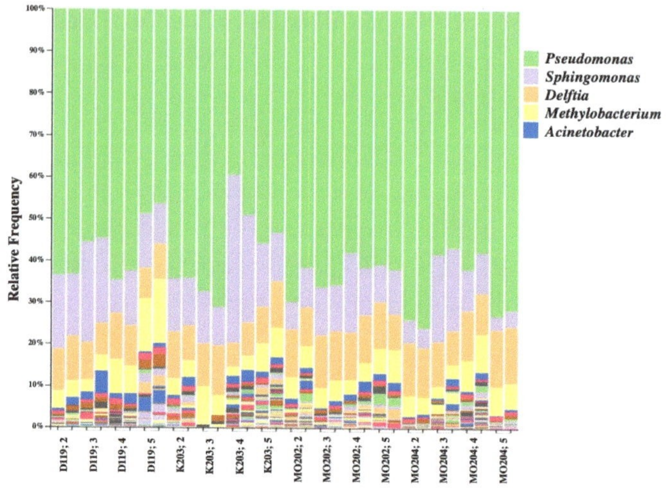

Figure 1. QIIME 2 bar plot of the top five most abundant bacterial genera per instar of each *Bombyx mori* L. strain based on two biological replicates using Silva classifier (top five genera: *Pseudomonas*, *Sphingomonas*, *Delftia*, *Methylobacterium, Acinetobacter*).

NCBI BLAST queries (Table S7) revealed that the most frequent representative sequences under each of the genera were species that have been isolated from various sources, including soil, plant, and insect gut [22]. BLAST queries also returned hits for contami- nating uncultured and cultured bacteria that might have come from the reagents or the laboratory used to prepare the sequencing templates and libraries.

3.2 Alpha-Diversity Analysis

Bar plots detailing Shannon diversity index values per silkworm strain and instar are presented in Figure 2A,B, respectively. Individual Shannon diversity values for each strain and instar are listed in Table S8. Silkworm strains do not seem to differ significantly in terms of diversity based on p-values from Kruskal–Wallis tests for all groups (Table S9). Strain MO202 possesses the closest levels of bacterial diversity among its instars (Figure 2A). This strain, however, also appears to be the second least diverse on average compared to the other strains. DMMMSU 119 appears the most diverse on average compared to other strains, followed by K203 as the second most diverse strain. The individual instars of DMMMSU 119, K203, and MO204 appear to have different diversity values, suggesting that some instars of each strain are rich in the number of bacterial taxa within their guts and some instars have low bacterial diversity. MO204 has the lowest mean Shannon diversity index value, indicating that this strain is the least diverse on average.

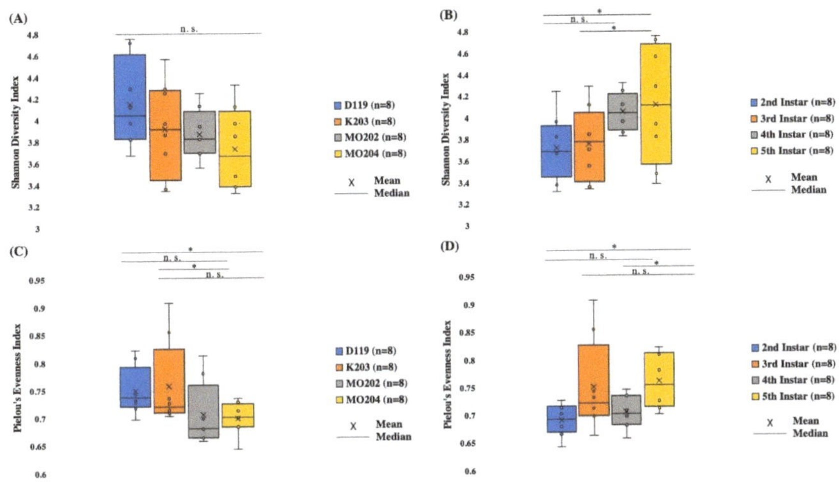

Figure 2. Alpha-diversity plots for Bombyx mori L. strains and instars reared at the Philippine Textile Research Institute Technological Center in Misamis Oriental (PTRI-TCMO). (A) Shannon diversity index boxplot for Bombyx mori L. strains; (B) Shannon diversity index boxplot for Bombyx mori L. instars; (C) Pielou's evenness index boxplot for Bombyx mori L. strains; (D) Pielou's evenness index boxplot for Bombyx mori L. instars. Significance ($p < 0.05$) is indicated with a (*). Comparisons that are not significant ($p > 0.05$) are indicated with "n.s."

Silkworm instars also do not appear significantly different in terms of diversity based on *p*-values from Kruskal–Wallis tests for all groups (Table S9). Among the silkworm instars across the four silkworm strains, the fifth instars seem to be the richest in terms of bacterial taxa, followed by the third instars (Figure 2B). Fourth instars, and subsequently the second instars, appear less diverse on average. Only fourth instars appear to differ significantly from second instars ($p = 0.0117$) and third instars ($p = 0.0460$), however. These significant differences in terms of bacterial diversity in the guts of some silkworm instars may correlate to the increase in mulberry leaf consumption and the potential changes in the gut bacterial community as a result. Chen et al. (2018) [3] also reported a shift in abundance of particular bacteria in the guts of silkworms from early to late instar stages.

Bar plots detailing Pielou's evenness index values per silkworm strain and instar are presented in Figure 2C,D, respectively. Individual Pielou's evenness values for each strain and instar are listed in Table S10. Silkworm strains appear to differ significantly in terms of evenness based on *p*-values from Kruskal–Wallis tests for all groups (Table S11). Strain K203 appears to be the most even in terms of the distribution of bacterial taxa on average (Figure 2C). DMMMSU 119 and MO204 appear as the second and third most even strains in terms of bacterial taxa distribution. DMMMSU 119 was also significantly different from MO204 in terms of evenness (p = 0.0209). The ranking of mean evenness values for DMMMSU 119 and K203 differs from those of their mean diversity values wherein DMMMSU 119 was more diverse than K203 (Figure 2A). MO202 and MO204 appear less even on average (Figure 2C), similar to their lower mean Shannon diversity values (Figure 2A).

Silkworm instars did not seem to differ significantly in terms of evenness based on p-values from Kruskal–Wallis tests for all groups (Table S11). Similar to the trend in diversity, fifth instars appear to be the most even in terms of bacterial taxa distribution (Figure 2D). Fifth instars only appeared significantly different from second instars in terms of evenness (p = 0.0117), however. Unlike the trend in diversity, third instars appeared more diverse than fourth instars (Figure 2D). Second instars were still the least even on average, however (Figure 2D), which is similar to their ranking as the least diverse instar on average (Figure 2B). Second and fourth instars were also more similar in terms of evenness within their respective groups compared to third and fifth instars. The trend of evenness values displayed by the instars matches that of their Shannon diversity index values (Figure 2B), suggesting that evenness possibly contributed to their diversity. This may also suggest that when fewer dominant bacterial taxa are present, there would be more taxa overall due to the absence of certain taxa that monopolize the environment and resources, thus, preventing others from thriving.

3.3 Beta-Diversity Analysis

Bray–Curtis dissimilarity 3D plots sorted by silkworm strain and instars are shown in Figure 3A,B. Silkworm strains appeared to differ significantly in terms of Bray–Curtis dissimilarity based on p-values from PERMANOVA tests (999 permutations) for all groups (Table S12). Figure 3A shows similar bacterial abundance within instars of strain MO202. Instars of strains DMMMSU 119, K203, and MO204 appear more dissimilar to other instars of their respective strain. Strain DMMMSU 119 also appears the most dissimilar compared to the other three strains, but only significantly different from strains MO202 (p = 0.004) and MO204 (p = 0.021). Silkworm instars also appeared significantly different in terms of Bray–Curtis dissimilarity based on p-values from PERMANOVA tests (999 permutations) for all groups (Table S12). Based on the Bray–Curtis dissimilarity values of the instars, diversity in second instars appears to be the most similar group and only varies significantly from the fourth instars (p = 0.029) (Figure 3B). Although there appears to be dissimilarity among the different instars, these are not significant when compared in a pairwise manner.

The Jaccard similarity index 3D plots sorted by silkworm strain and instar stages are shown in Figure 3C,D. Silkworm strains did not appear significantly different from each other in terms of Jaccard similarity based on p-values from PERMANOVA tests (999 permutations) for all groups (Table S13). Figure 3C shows no clustering within strains; however, the values are relatively close to one another. This further suggests that silkworm strains are similar in terms of the bacterial taxa that are present and absent (not considering the abundance of the taxa that are present). Despite the similarities observed, strain DMMMSU 119 appeared significantly different from strains MO202 ($p = 0.018$) and MO204 ($p = 0.002$) once again. Silkworm instars, in contrast, appeared significantly different from one another in terms of Jaccard similarity based on p-values from PERMANOVA tests (999 permutations) for all groups (Table S13). Once again, no clustering was observed within any of the instars. The values all appeared relatively similar to each other. Only the third and fifth instars appeared significantly different from one another ($p = 0.042$). Combined with the generally low Bray–Curtis dissimilarity values for most strains and instars, this suggests that the common bacterial taxa found among silkworm strains and instars possess close abundance values.

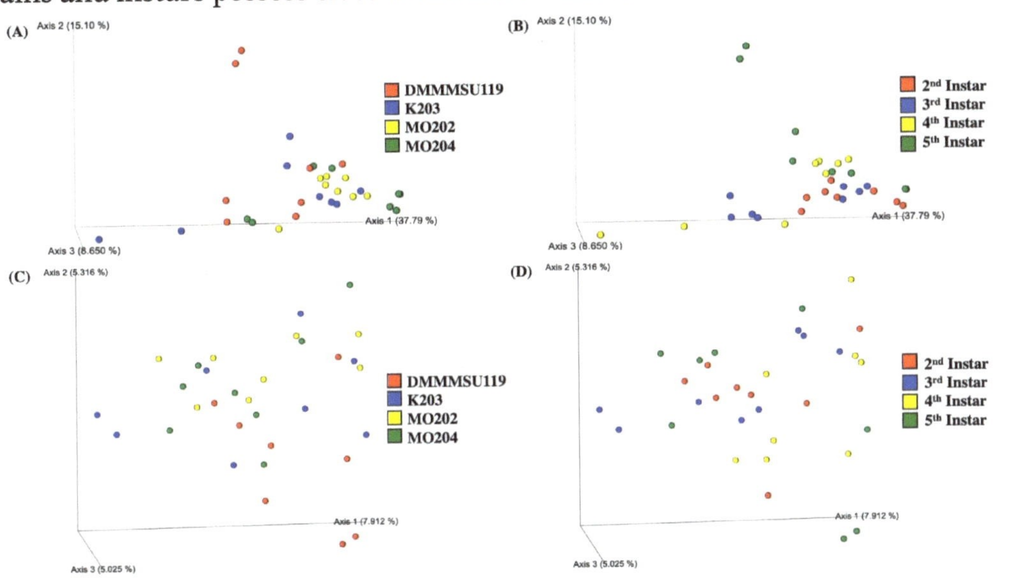

Figure 3. Beta-diversity plots for *Bombyx mori* L. strains and instars reared at the Philippine Textile Research Institute Technological Center in Misamis Oriental (PTRI-TCMO). (A) Bray–Curtis dissim- ilarity sorted by *Bombyx mori* L. strains ($p = 0.03$); (B) Bray–Curtis dissimilarity sorted by *Bombyx mori* L. instars ($p = 0.048$); (C) Jaccard similarity index sorted by *Bombyx mori* L. strains (p = 0.079); (D) Jaccard similarity index sorted by *Bombyx mori* L. instars ($p = 0.024$).

4. DISCUSSION

4.1 Bacterial Abundance

The dominant bacterial communities associated with Philippine *B. mori* belong to the genera *Pseudomonas*, *Sphingomonas*, *Delftia*, *Methylobacterium*, and *Acinetobacter*. These genera have been associated with mulberries [23–25] and were also found prevalent in the guts of other strains of silkworms [3,17]. This indicates shared similarities between silkworms reared in the Philippines and silkworms reared elsewhere, in terms of the bacterial genera present in their guts. However, these bacterial communities still differ in terms of overall genera that are present as well as their relative abundance.

For the species identity of bacteria under each genus, NCBI BLAST queries revealed that the most frequent representative sequences under *Pseudomonas* were mostly identical to *Pseudomonas fluorescens* strains, which were isolated either from sediments or from gut of adult fruit-tree pinhole borer, *Xyleborinus saxesenii,* an ambrosia beetle. Like several bacteria of the genus *Pseudomonas*, the *P. fluorescens* group is found in diverse environments and well-known for its plant-beneficial properties, including pathogen suppression [26]. Thus, the *Pseudomonas* species found in silkworms in this study likely originated from mulberries and from the field where these host plants were cultivated. However, there were recent observations that some strains of this group can also colonize insects and cause severe infections leading to death [27]. *Pseudomonas* strains from other studies also showed amylolytic, cellulolytic, xylanolytic, lipolytic, and esterase activity in silkworms and beetles [28–30]. Hence, the sequence similarity of *Pseudomonas* rep-seqs in this study was notable in consideration of the ability to hydrolyze polymers of mulberries. BLAST queries also returned hits for uncultured bacterium isolated from a human body part, implicating the presence of contaminants. This could have been prevented to a certain extent if a negative control was included in the sequencing and analysis. BLAST also returned hits of most frequent rep-seqs for strains of *Sphingomonas* that were isolated from nectarine tree and paddy soil. Species of this genus have no reported biological function in the other strains of silkworm where they were detected, but some species have been observed to possess the ability to degrade organic matter [31] and also the ability to produce beta-carotene and gellan gum [32,33]. *Sphingomonas* has also been associated with diapause- destined female of the cabbage beetle, *Colaphellus bowringi* [34] and was found to be one of the genera of cellulolytic bacteria in the gut of the Chinese white pine beetle, *Dendroctonus armandi*, larvae [35]. Species of *Sphingomonas* have also been reported to be part of the core members of mulberry endophytes regardless of the season or cultivar [36]. Contaminating *Sphingomonas* species from sterile water was also detected, however, highlighting further the need for controls in this type of study.

Delftia species isolated from the gut of another moth and from soil were also among the BLAST hits for rep-seqs in this study. Besides finding *Delftia* strain in silkworms in other studies, a recent report indicated its presence in diapause-destined *C. bowringi* [34]. A species of *Delftia* (*D. acidovorans*) isolated from the American cockroach, *Periplaneta americana* L., which was found to produce bacteriocins against bacterial and fungal pathogens of hu- mans [37] was also recently reported.

Functions linked to the presence of bacteria under the genus *Methylobacterium* include nitrogen-fixing in silkworm guts [3] and cellulolytic activity in the gut of the Chinese white pine beetle, *D. armandi*, larvae [35]. BLAST hits for *Methylobacterium* in the present study points to strains that might have been involved in nitrogen-fixation and -removal because the strains with hits were isolated from the rhizosphere and sewage sludge undergoing nitrogen removal. The strain similar to that isolated from a fish gut has no associated function, however.

Acinetobacter strains from other studies have shown amylolytic, cellulolytic, xylanolytic, lipolytic, and esterase activity in silkworms and beetles [28–30], but the strains most frequent in the silkworms used in this study were highly similar to those isolated from goose stools and soil. Of interest in the data presented in this study is the relevance of these dominant bacterial communities in Philippine silkworm biology, which remains to be explored further.

4.2 Alpha-Diversity and Beta-Diversity

The increase in average bacterial diversity as the larvae mature suggests a shift in gut bacterial diversity. This occurs either in response to host development or an increase in leaf consumption and the age of mulberry leaves that were used. Strains that possessed high Shannon diversity values also possessed high Pielou's evenness values. This may suggest a potential link between bacterial diversity and evenness among the silkworm strains. This also indicates that bacterial diversity and evenness may be related given that the presence of fewer dominant bacterial taxa may allow more taxa to grow, as it is less likely that a certain taxon would monopolize resources or inhibit the growth of others. It should be noted, however, that neither silkworm strains nor instars appeared significantly different in terms of bacterial diversity overall, and only silkworm strains were significantly different in terms of bacterial evenness. This again relates to the similarity of the strains in terms of phenotypic traits and the most abundant bacterial taxa in their guts. These are also expected as the strains are genetically similar [9].

Quantitative measures of similarity among strains (Bray–Curtis dissimilarity) seem to set strain DMMMSU 119 apart from the other three silkworm strains. The other strains cluster together supporting the observed similar phenotypic traits. Whereas silkworm instars appeared significantly different from one another overall, only second and fourth instars appeared significantly different from each other. Both strains and instars appear similar in terms of qualitative measures of similarity that do not take bacterial abundance into account (Jaccard similarity). DMMMSU 119 was also significantly different from strains MO202 and MO204, and third and fifth instars also appeared significantly different from each other. This indicates that strains and instars are similar in terms of the bacterial taxa present but are less similar in terms of the actual abundance of these bacterial taxa.

5. SUMMARY AND CONCLUSIONS

In this study, the bacterial communities of the four silkworm strains reared in the Philippines were investigated using high-throughput 16S rRNA gene amplicon sequencing and analysis. The results showed the abundance of five bacterial genera that were also found in other silkworm strains. Although the likely association of some of the bacterial species to biological activity in the gut as well as their involvement in the growth, development, and survival of silkworms have been reported, the exact roles of the bacteria within the silkworm strains in the present study and how these roles correlate to the production economically important silk characteristics. remain to be elucidated. Evidently, the re- sults expand the biological data available for the silkworm strains present in the country and will be a useful source of information for future strain improvement to enhance the sustainability of quality silk production.

REFERENCES

1. Mita, K.; Kasahara, M.; Sasaki, S.; Nagayasu, Y.; Yamada, T.; Kanamori, H.; Namiki, N.; Kitagawa, M.; Yamashita, H.; Yasukochi, Y.; et al. The genome sequence of silkworm, *Bombyx mori*. *DNA Res.* **2004**, *11*, 27–35. [CrossRef]
2. Mohanta, M.K.; Saha, A.K.; Saleh, D.K.M.A.; Islam, M.S.; Mannan, K.S.B.; Fakruddin, M. Characterization of *Klebsiella granulomatis* pathogenic to silkworm, *Bombyx mori* L. *Biotech* **2010**, *5*, 577–583. [CrossRef]
3. Chen, B.; Du, K.; Sun, C.; Vimalanathan, A.; Liang, X.; Li, Y.; Wang, B.; Lu, X.; Li, L.; Shao, Y. Gut bacterial and fungal communities of the domesticated silkworm (*Bombyx mori*) and wild mulberry-feeding relatives. *ISME J.* **2018**, *12*, 2252–2262. [CrossRef]
4. Nagaraju, J.; Goldsmith, M.R. Silkworm genomics—Progress and prospects. *Curr. Sci.* **2002**, *83*, 415–425.
5. Xia, Q.Y.; Zhou, Z.Y.; Lu, C.; Cheng, D.; Dai, F.-Y.; Liu, B.; Zhao, P.; Zha, X.; Cheng, T.; Biology Analysis Group; et al. A draft sequence for the genome of the domesticated silkworm (*Bombyx mori*). *Science* **2004**, *306*, 1937–1940. [PubMed]
6. Pereira, N.C.; Munhoz, R.E.F.; Bignotto, T.S.; Bespalhuk, R.; Garay, L.B.; Saez, C.R.N.; Fernandez, M.A. Biological and molecular characterization of silkworm strains from the Brazilian germplasm bank of *Bombyx mori*. *Genet. Mol. Res.* **2013**, *12*, 2138–2147. [CrossRef] [PubMed]
7. Zanatta, D.B.; Bravo, J.P.; Barbosa, J.F.; Munhoz, R.E.F.; Fernandez, M.A. Evaluation of economically important traits from sixteen parental strains of the silkworm *Bombyx mori* L. (Lepidoptera: Bombycidae). *Neotrop. Ento-*

mol. **2009**, *38*, 327–331. [CrossRef]

8. Matienzo, L.H.; Gatica, R.A. Sericulture: Silk Production in the Philippines. Available online: https://aboutphilippines.org/files/ Silk-Production-in-the-Philippines.pdf (accessed on 6 December 2018).
9. Alcudia-Catalma, M.N.; Conde, M.Y.E.D.; Dee Tan, I.Y.; Bautista, M.A.M. First report on the characterization of genetic diversity of Philippine-reared *Bombyx mori* strains based on *COI* and ITS2. *Philipp. J. Sci.* **2020**, *150* (Suppl. S1), 503–517.
10. Basaen, A.M.; Placido, A.; Bacuso, P.; Ladines, A.; Cabrito, F. Silkworm breeding for the development of Philippine purelines. *NSTA Technol. J.* **1982**, *7*, 72–85.
11. Villanueva, R.C. Problems and Issues Affecting the Pace of Sericulture Development in the Philippines. 1999. Available online: https://agris.fao.org/agris-search/search.do?recordID=PH2007M00034 (accessed on 11 January 2022).
12. Sarkar, K. Management of nutritional and climatic factors for silkworm rearing in West Bengal: A review. *Int. J. Agric. Environ. Biotechnol.* **2018**, *11*, 769–780. [CrossRef]
13. Barretto, D.A.; Gadwala, M.; Vootla, S.K. Chapter 1—The silkworm gut microbiota: A potential source for biotechnological applications. *Methods Microbiol.* **2021**, *49*, 1–26.
14. Sun, Z.; Kumar, D.; Cao, G.; Zhu, L.; Liu, B.; Zhu, M.; Liang, Z.; Kuang, S.; Chen, F.; Feng, Y.; et al. Effects of transient high temperature treatment on the intestinal flora of the silkworm *Bombyx mori*. *Sci. Rep.* **2017**, *7*, 3349. [CrossRef]
15. National Research Council (US). Committee on Metagenomics: Challenges and Functional Applications. In *The New Science of Metagenomics: Revealing Secrets of Our Microbial Planet*; National Academies Press (US): Washington, DC, USA, 2007. Available online: https://www.ncbi.nlm.nih.gov/books/NBK54011/ (accessed on 7 May 2019).
16. Thomas, T.; Gilbert, J.; Meyer, F. Metagenomics—A guide from sampling to data analysis. *Microb. Inform. Exp.* **2021**, *2*, 3. [CrossRef]
17. Sun, Z.; Lu, Y.; Zhang, H.; Kumar, D.; Liu, B.; Gong, Y.; Zhu, M.; Zhu, L.; Liang, Z.; Kuang, S.; et al. Effects of BmCPV infection on silkworm *Bombyx mori* intestinal bacteria. *PLoS ONE* **2016**, *11*, e0146313. [CrossRef]
18. Hammer, T.; Dickerson, J.; Fierer, N. Evidence-based recommendations on storing and handling specimens for analyses of insect microbiota. *PeerJ* **2015**, *3*, e1190. [CrossRef]
19. Herlemann, D.P.R.; Labrenz, M.; Jürgens, K.; Bertilsson, S.; Waniek, J.J.; Andersson, A.F. Transitions in bacterial communities along the 2000 km

salinity gradient of the Baltic Sea. *ISME J.* **2011**, *5*, 1571–1579. [CrossRef]

20. Suenami, S.; Nobu, M.K.; Miyazaki, R. Community analysis of gut microbiota in hornets, the largest eusocial wasps, *Vespa mandarinia* and *V. simillima*. *Sci. Rep.* **2019**, *9*, 9830. [CrossRef]
21. Wang, Q.; Garrity, G.; Tiedje, J.; Cole, J. Naïve Bayesian classifier for rapid assignment of rRNA sequences into the new bacterial taxonomy. *Appl. Environ. Microbiol.* **2007**, *73*, 5261–5267. [CrossRef]
22. Schoch, C.L.; Ciufo, S.; Domrachev, M.; Hotton, C.L.; Kannan, S.; Khovanskaya, R.; Leipe, D.; McVeigh, R.; O'Neill, K.; Robbertse, B.; et al. NCBI Taxonomy: A comprehensive update on curation, resources and tools. *Database* **2020**, *2020*, 1–21. [CrossRef]
23. Ou, T.; Xu, W.-F.; Wang, F.; Strobel, G.; Zhou, Z.-Y.; Xiang, Z.-H.; Liu, J.; Xie, J. A microbiome study reveals seasonal variation in endophytic bacteria among different mulberry cultivars. *Comput. Struct. Biotechnol. J.* **2019**, *17*, 1091–1100. [CrossRef]
24. Lukša, J.; Servienė, E. White mulberry (*Morus alba* L.) fruit-associated bacterial and fungal microbiota. *J. Environ. Eng. Landsc. Manag.* **2020**, *28*, 183–191. [CrossRef]
25. Dong, H.L.; Zhang, S.X.; Chen, Z.H.; Tao, H.; Li, X.; Qiu, J.F.; Xu, S.Q. Differences in gut microbiota between silkworms (*Bombyx mori*) reared on fresh mulberry (*Morus alba* var. *multicaulis) leaves or an artificial diet. RSC Adv.* **2018**, *8*, 26188–26200.
26. Flury, P.; Aellen, N.; Ruffner, B.; Péchy-Tarr, M.; Fataar, S.; Metla, Z.; Maurhofer, M. Insect pathogenicity in plant-beneficial pseudomonads: Phylogenetic distribution and comparative genomics. *ISME J.* **2016**, *10*, 2527–2542. [CrossRef] [PubMed]
27. Flury, P.; Vesga, P.; Péchy-Tarr, M.; Aellen, N.; Dennert, F.; Hofer, N.; Kupferschmied, K.P.; Kupferschmied, P.; Metla, Z.; Ma, Z.; et al. Antimicrobial and insecticidal: Cyclic lipopeptides and hydrogen cyanide produced by plant-beneficial *Pseudomonas* strains CHA0, CMR12a, and PCL1391 contribute to insect killing. *Front. Microbiol.* **2017**, *8*, 100. [CrossRef]
28. Prem Anand, A.A.; Vennison, S.J.; Sankar, S.G.; Gilwax Prabhu, D.I.; Vasan, P.T.; Raghuraman, T.; Vendan, S.E. Isolation and characterization of bacteria from the gut of *Bombyx mori* that degrade cellulose, xylan, pectin and starch and their impact on digestion. *J. Insect. Sci.* **2010**, *10*, 107.
29. Briones-Roblero, C.I.; Rodríguez-Díaz, R.; Santiago-Cruz, J.A.; Zúñiga, G.; Rivera-Orduña, F.N. Degradation capacities of bacteria and yeasts isolated from the gut of *Dendroctonus rhizophagus* (Curculionidae: Scolytinae). *Folia Microbiol. (Praha)* **2017**, *62*, 1–9. [CrossRef]
30. Delalibera, I., Jr.; Handelsman, J.; Raffa, K.F. Contrasts in cellulolytic ac-

tivities of gut microorganisms between the wood borer, *Saperda vestita* (Coleoptera: Cerambycidae), and the bark beetles, *Ips pini* and *Dendroctonus frontalis* (Coleoptera: Curculionidae). *Environ. Entomol.* **2005**, *34*, 541–547. [CrossRef]

31. Gong, B.; Wu, P.; Huang, Z.; Li, Y.; Dang, Z.; Ruan, B.; Kang, C.; Zhu, N. Enhanced degradation of phenol by *Sphingomonas* sp. GY2B with resistance towards suboptimal environment through adsorption on kaolinite. *Chemosphere* **2016**, *148*, 388–394. [CrossRef]
32. Silva, C.; Cabral, J.M.S.; van Keulen, F. Isolation of a beta-carotene over-producing soil bacterium, *Sphingomonas* sp. *Biotechnol. Lett.* **2004**, *26*, 257–262. [CrossRef]
33. Wang, X.; Xu, P.; Yuan, Y.; Liu, C.; Zhang, D.; Yang, Z.; Yang, C.; Ma, C. Modeling for gellan gum production by *Sphingomonas paucimobilis* ATCC 31461 in a simplified medium. *Appl. Environ. Microbiol.* **2006**, *72*, 3367–3374. [CrossRef]
34. Liu, W.; Li, Y.; Guo, S.; Yin, H.; Lei, C.; Wang, X. Association between gut microbiota and diapause preparation in the cabbage beetle: A new perspective for studying insect diapause. *Sci. Rep.* **2016**, *6*, 38900. [CrossRef]
35. Hu, X.; Yu, J.; Wang, C.; Chen, H. Cellulolytic bacteria associated with the gut of *Dendroctonus armandi* larvae (Coleoptera: Curculionidae: Scolytinae). *Forests* **2014**, *5*, 455–465. [CrossRef]
36. Xu, W.; Wang, F.; Zhang, M.; Ou, T.; Wang, R.; Strobel, G.; Xiang, Z.; Zhou, Z.; Xie, J. Diversity of cultivable endophytic bacteria in mulberry and their potential for antimicrobial and plant growth-promoting activities. *Microbiol. Res.* 2019, *229*, 126328. [CrossRef] [PubMed]
37. Amer, A.; Hamdy, B.; Mahmoud, D.; Elanany, M.; Rady, M.; Alahmadi, T.; Alharbi, S.; AlAshaal, S. Antagonistic activity of bacteria isolated from the *Periplaneta americana* L. gut against some multidrug-resistant human pathogens. *Antibiotics* 2021, *10*, 294. [CrossRef] [PubMed]

12
Chapter

ENTEROCOCCI AS PREDOMINANT XYLOSE-UTILIZING LACTIC ACID BACTERIA IN THE MIDGUT OF ERI SILKWORMS AND THE PROSPECTIVE APPLICATION OF ENTEROCOCCUS HIRAE AS A PROBIOTIC FOR ERI CULTIVATION.

1. INTRODUCTION: AN OVERVIEW

Eri silk is a type of non-mulberry silk that is gaining popularity due to advancements in silkworm domestication and its multivoltine characteristics. Eri silk is produced from the Eri silkworm (Samia ricini), which is thought to have originated in northeast India, particularly in Assam and Meghalaya, before being disseminated to other Asian nations. The production of Eri silkworm is advantageous and pertinent to rural on-farm and off-farm activities, regarded as a beneficial strategy for socio-economic development, especially in certain Asian emerging nations. Moreover, Eri silk textile goods are marketed as environmentally sustainable due to their natural production methods and eco-friendly characteristics [3]. The typical diet of the Eri silkworm consists of various plant leaves, which are primarily composed of cellulose, emicelluloses like xylan and mannan, pectin, lignin, and trace amounts of protein and fat. Consequently, the gut of this Lepidopteran plays a crucial role in the digestion of these leaf components for nutritional purposes. The microbial community within the insect gut has been thoroughly examined concerning its symbiotic traits and function as advantageous gut microbiota, where its metabolically significant enzymes are essential for substrate digestion and the release of nutritionally vital compounds [5]. The bacterial community in the midgut of this Lepidopteran insect has been examined for its production of digestive enzymes necessary for the breakdown of leaf components [6]. Numerous reports have determined that gut microbiota affects insects in various manners, including facilitating nutrient digestion and detoxification [7], enhancing innate immunity [8], supplying nutrients and growth-promoting compounds [9], and safeguarding against infectious pathogens and parasites [10]. Although numerous research have examined the gut microbiota associated with insect metabolism and developmental development, the connection between the gut microbiota community and its potential implications for enhancing Eri silk production remains inadequately understood.

Microbial biodiversity is increasingly recognized as crucial for the metabolism of higher eukaryotic systems; thus, comprehending the pertinent microbial community in the gut of Eri silkworms could enhance these microbial resources in Eri silkworm production as a source of advantageous probiotic lactic acid bacteria. This has been specifically identified as a source of microbes that produce optically pure L- or D-lactic acid through biological processes, particularly utilizing pentose substrates derived from lignocellulose, which are essential for the production of bioplastics like polylactic acid (PLA) [11,12]. In addition to serving as biological catalysts for lactic acid synthesis, certain lactic acid bacteria (LAB), including Lactobacillus sp. and Bifidobacterium sp., have been studied and employed as probiotics to confer health benefits to both people and animals. The influence of probiotic bacteria on the enhancement of growth and quality in silkworms has also been documented [13–16]. Nevertheless, the majority of prior research has concentrated on the mulberry silkworm (Bombyx mori), with no studies addressing the Eri silkworm. The generation of antimicrobial metabolites by LAB probiotics, which counteract pathogenic or undesirable microbes in the host gastrointestinal tract, is a significant characteristic. Consequently, the antimicrobial activity against insect pathogens in response to LAB is also regarded as a crucial attribute of probiotics for silkworm cultivation. Furthermore, the utilization of live probiotics is more environmentally sustainable for the control of silkworm infections compared to the prevalent usage of antibiotics in the sericulture business [18].

This work explored the isolation of xylose-utilizing lactic acid bacteria (LAB) and their molecular identification, as well as their antibacterial efficacy against insect pathogenic bacteria. Additional examination of the application of certain xylose-utilizing lactic acid bacteria as probiotics in Eri culture was also progressed.

2. MATERIALS AND METHODS

2.1 Isolation and Screening of Xylose Utilizing LAB

Ten mature Eri silkworm larvae (5th instar day 5) were collected from a local farm in Sankampang district, Chiang Mai province, Thailand. The larvae were rinsed twice with sterile water and surface sterilized in ethanol (70%, *v*/*v*) for 60 s, followed by a final rinsing in sterile distilled water before dissection. Then, the worm's gut tissue was aseptically dissected and immediately transferred to a sterile microcentrifuge tube containing sterile 0.85% (*w/v*) NaCl solution in a ratio of 1:10. A 10-fold serial dilution was carried out by pipetting 0.5 mL of homogenized sample into 4.5 mL of 0.85% (*w/v*) NaCl solution. The mixture was then mixed well, and 100 μL of each dilution was transferred to nutrient agar (NA) containing 10 g/L glucose and 0.04% (*w/v*) bromocresol purple using the spread plate technique. After 24 h incubation at 37°C, the acid-forming bacterial isolates observed from the yellow clear zone around their colonies were then transferred into NA containing 10 g/L xylose and 0.04% (*w/v*) bromocresol purple. The viable bacterial cells were enumerated as colony-forming units (CFU) per gram of sample.

All acid-forming bacterial colonies capable of growth on NA xylose were transferred into modified de Man, Rogosa and Sharpe (MRS) agar containing 10 g/L beef extract, 10 g/L peptone, 5 g/L yeast extract, 5 g/L sodium acetate, 2 g/L K_2HPO_4, 2 g/L triammonium citrate, 0.2 g/L $MgSO_4$, 0.2 g/L $MnSO_4$, 0.1% (*v/v*) Tween80, and 10 g/L xylose supplemented with 0.04% (*w/v*) bromocresol purple and then incubated at 37°C for 24 h. The colonies that exhibited a yellow zone around their respective colony were presumptively considered to be LAB and were individually streaked on MRS-xylose agar plate to obtain single colonies. The cultures of pure isolates were stored at – 80°C in MRS broth mixed with 25% (*v/v*) glycerol.

2.2 Identification of LAB by 16S rDNA Sequence Analysis

The molecular identification of LAB isolates was carried out using the method described by Unban et al. [19] with some modification. Briefly, the bacterial cells were collected from 10 mL of an overnight bacterial culture by centrifugation at 10,000 × *g* for 10 min, and the genomic DNA was extracted by using the Wizard Genomic DNA purification kit (Promega Corp., Madison, WI, USA), according to the manufacturer's instructions. The 16S rRNA gene was amplified by using genomic DNA as a template with a standard poly- merase chain reaction (PCR). Forward primers 27F, 5′-AGAGTTTGATCMTGGCTCAG-3′, and reverse primers 1525R, 5′-AAGGAGGTGWTCCARCC-3′, were used for 16S rRNA gene amplification. The amplification cycles were initially performed at 94°C for 5 min of denaturation, followed by 30 cycles of denaturation at 94°C for 20 s, annealing at 55°C for 20 s, extension at 72°C for 1 min 30 s, and additional final extension at 72°C for 5 min. All PCR reactions were performed using Phusion High-Fidelity PCR Master Mix (New England Biolabs, MA, USA). The PCR products were purified and sequenced by a sequencing service provider (1st BASE Laboratory Company, Singapore). The comparison of 16S rRNA gene sequences was performed with the genetic database from National Center for Biotechnology Information (NCBI) GenBank and EzBioCloud databases. The multiple sequence alignment was performed using

BioEdit 7.0 software tool and MEGA 4.0 [20]. The phylogenetic tree was constructed by the neighbor-joining methods with 1000 replications bootstrap analysis. All 16S rRNA gene sequences generated in this study have been deposited in the NCBI GenBank database under accession number MZ127632–MZ127643 and OM090176–OM090214.

2.3 Growth and Acid Formation on Polysaccharides

The selected strains of xylose utilizing LAB from Eri silkworm midgut were investi- gated for their abilities to utilize the selected polysaccharides as the sole carbon source. Briefly, individual LAB isolate was spiked on MRS agar containing 0.5% (*w*/*v*) of single carbon source of individual polysaccharides including starch (Fisher Scientific, Loughborough, Leicestershire, UK), pectin (Fisher Scientific, Loughborough, Leicestershire, UK), beechwood xylan (Megazyme International, Bray, Co., Wicklow, Ireland), locust bean gum (Sigma Aldrich, St. Louis, MO, USA) or carboxymethyl cellulose (Sigma Aldrich, St. Louis, MO, USA) supplemented with bromocresol purple (0.04%, *w*/*v*). After incubation at 37°C for 24 h, the bacterial growth and the yellow halo-formed surrounding colonies were observed. Further duplicated experiment was also carried out, but the bromocresol purple was replaced with trypan blue (0.01%, *w*/*v*) for detection of extracellular activities of amylase, pectinase, xylanase, β-mannanase, and cellulase, respectively. The clear zone formed surrounding colonies were observed after incubation at 37°C for 24 h.

2.4 Tannin-Tolerance Test

The ability of LAB isolates to tolerate tannins was evaluated on MRS-xylose agar supplemented with tannin. To prepare tannin solution, 10 g of tannic acid (LOBA Chemie, Mumbai, India) was dissolved in 100 mL of 0.1 M sodium phosphate buffer (pH 7.0) and sterilized by filtering through a 0.2 μm filter cartridge (Millipore, Billerica, MA, USA). The tannin solution was aseptically added into the sterilized MRS-xylose agar to achieve the final concentrations of 1 and 5 g/L. A single colony of each LAB isolate was picked up and spiked on the supplemented MRS-xylose agar. Growth of bacterial colony and the yellow clear zone formation of the isolates were observed after incubation at 37°C for 24 h.

2.5 Antimicrobial Activity against Insect Pathogens

The antimicrobial activity against insect pathogenic bacteria was investigated by independent cultivation of all selected LAB isolates in MRS-xylose broth at 37°C for 24 h. Then, cell-free culture supernatants (CFCS) were collected by centrifugation at 10,000 × *g* for 15 min. The CFCS were then neutralized to pH 6.5 with 5 N NaOH and filtered through a 0.22 μm filter cartridge (Millipore, Billerica, MA, USA). The antimicrobial activity was investigated by the disc-agar diffusion method following the method of Fehlberg et al. [21]. Briefly, 100 μL of the overnight grown pathogen including *B. cereus* TISTR 747, *S. aureus* TISTR 746, and *P. vulgaris* TISTR 100 cultures (10^6–10^7 CFU/mL) was spread onto NA agar plate, and paper discs (diameter 6 mm, Macherey-Nagel, Duren, Germany) were placed on each plate. Then, 50 μL of CFCS from LAB isolates were transferred into paper discs. The diameter of the inhibition zone against pathogenic bacteria was measured after incubation at 37°C for 24 h.

2.6 Effect of E. hirae SX2 Supplementation on Eri Silkworm Growth

A pure culture of *E. hirae* SX2 was inoculated in MRS-xylose broth and incubated at 37°C under static conditions for 24 h. The bacterial cells were collected by centrifugation at 8000 × *g*,4°C for 10 min and cell pellets was washed twice with sterile 0.85% (*w*/*v*) NaCl solution, and the cell pellets were resuspended in NaCl solution and adjusted to an optical density (OD600) of around 0.5 corresponding to approximately 10^8 CFU/mL. The freshly picked castor leaves were cleaned by excess volume of tap water and finally washed with sterile water. The cleaned leaves targeted for feeding were then sprayed with a cell suspension of *E. hirae* SX2 on both sides of leaves and shade dried before feeding. The 1st instar larvae of Eri silkworm were divided into two groups for the treatment, with each group consisting of 50 larvae; one group was reared on castor leaves served as control, while the other was fed on probiotics treated leaves. Three replications were maintained for each treatment. The treatment was given the first feed on the first day of 1st instar larvae up to 5th instar stage (3 weeks) under laboratory conditions at 25 1°C, the humidity of about 75 ± 5%, and photoperiod of 16 h of light and 8 h of dark. The treatment was fed 2–3 times a day, and the unfed leaves were removed from the trays daily. The growth of larva was monitored (10 larvae per treatment were selected randomly) by recording the larva weight once a week. At the end of rearing (3 weeks), 10 larvae per treatment were randomly selected as the representative of each treatment for dissection, and the midgut content were collected and mixed well for use as the representative of each group for determination of total bacterial count on NA agar supplemented with 10 g/L glucose and 0.04% (w/v) bromocresol purple as an indicator. The viable cell number of acid-forming bacteria was also enumerated and observed from the yellow clear zone formed surrounding the colony. The viable cell number of E. hirae SX2 was enumerated by spread plating on MRS-xylose agar supplemented with 0.04% bromocresol purple.

2.7 Statistical Analysis

The collected data were analyzed using the statistical program SPSS/PC version 17.0 (SPSS Inc. Chicago, IL, USA). The data were analyzed for achieving both normality and homoscedasticity, and the statistical significance of the differences among the treatments was evaluated by one-way ANOVA followed by Tukey's multiple range test. Values not sharing a common letter are significantly different from each other at $p < 0.05$.

3. RESULTS AND DISCUSSION

3.1 Xylose Utilizing LAB Isolation

The healthy mature fifth instar Eri silkworm larvae cultivated at the local farm in Sankampang district, Chiang Mai, Thailand (Figure 1a) was used as the source of midgut content. The numbers of acid-forming bacteria that grew on different kinds of media are presented in Figure 1b. The total viable bacterial population of the Eri silkworm midgut observed on NA glucose were found to be 3.95×10^7 CFU/g of larval gut, while acid- forming bacteria were found at a high number of 3.76×10^7 CFU/g, which represented 95.4% of the total number of bacteria. Thereafter, 1000 isolates of these acid-forming bacteria were randomly replicated into NA-xylose and incubated at 37°C for 24 h. Only 608 isolates (60.8%) showed the ability to grow and produce acid around the colonies. Of all isolates that were able to grow and produce acid in NA-xylose, only 51 isolates (10.2%) showed an acid-forming capability in the selective MRS-xylose agar and were presumptively identified as LAB. These results indicated that even xylose utilizing bacteria were found in significant numbers among acid-forming bacteria capable of growth in NA, but LAB was a minor component representing only 10.2% of the total number when tested on the LAB selecting medium such as MRS agar. These results were in agreement with the previous study, which suggested that the main population of culturable bacteria in Eri silkworm gut were non-LAB, including Enterobacter sp., Pseudomonas sp., Citrobacter sp., and Bacillus sp. [22]. This might be due to the pH condition of the insect gut, which is reported to be alkaline [10], which may limit the growth and number of LAB.

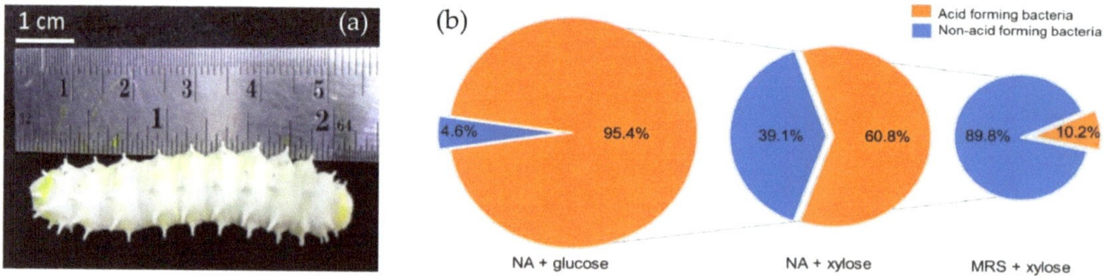

Figure 1. The mature Eri silkworm larvae cultivated at local farm in Chiang Mai, Thailand (a) and the pie diagram presenting the percentage of acid-forming bacteria from Eri silkworm midgut found in different agar medium (b).

3.2 Xylose Utilizing LAB Identification and Phylogenetic Analysis

The morphological study of 51 LAB isolates found that 47 isolates were Gram-positive with cocci shape, while the last four isolates were Gram-positive with irregular rod shape. All the xylose utilizing LAB isolates were identified using the sequence analysis of the 16S rRNA gene. The sequence was aligned with the related type strains, and the construction of phylogenetic trees was created and compared with the sequences of their closest relatives (Figure 2a).

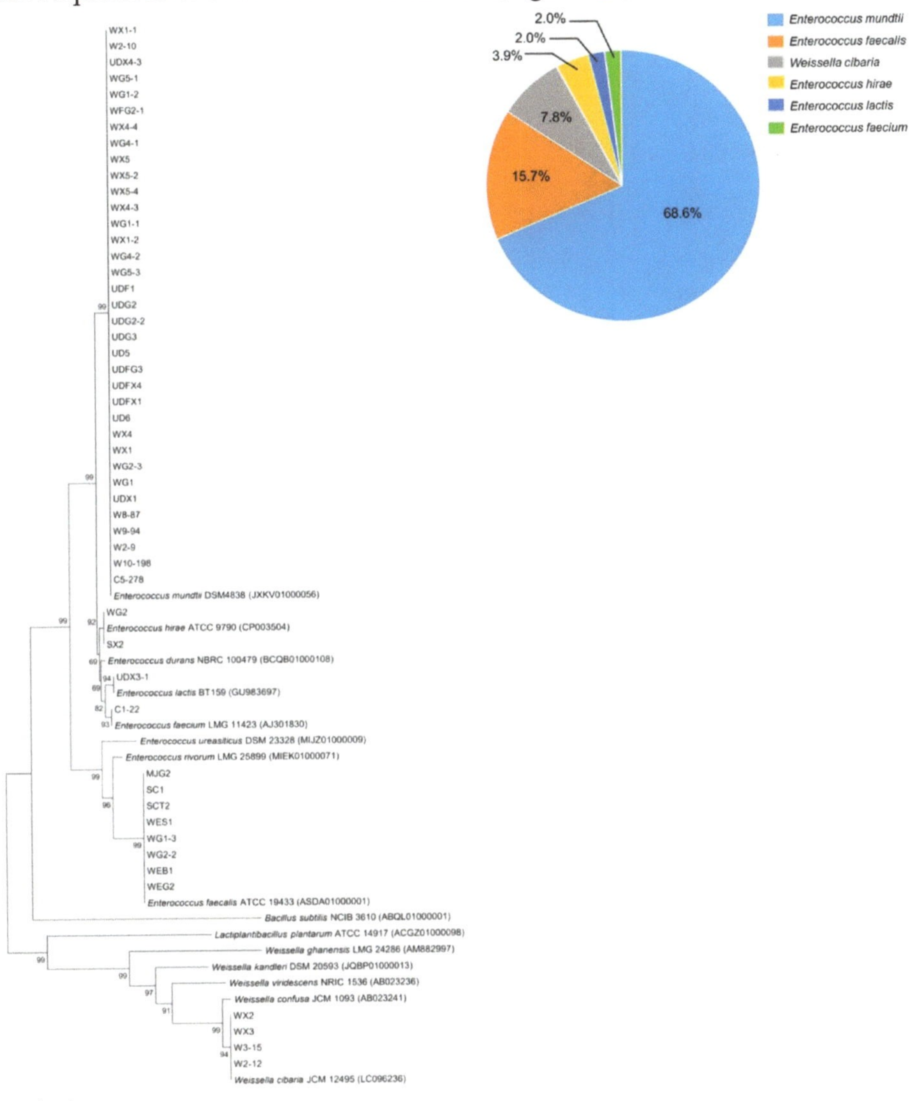

(a) (b)

Figure 2. Phylogenetic tree based on the 16S rRNA gene sequence analysis (a) and the proportion percentage of the identified species (b) of 51 selected LAB from Eri silkworm midgut.

The full-length sequence of 16S rRNA gene of LAB iso- lates shared similarity to genera Enterococcus and Weissella with higher than 99%. Of 51 isolates, 35 were identified to be Enterococcus mundtii, which represented 68.6% of total isolates, followed by Entercoccus faecalis (15.7%), Weissella cibaria (7.8%), Enterococcus hirae (3.9%), Enterococcus lactis (2%), and Enterococcus faecium (2%) (Figure 2b). Enterococcus and Lactobacillus have been reported to be the predominant LAB genera in mulberry silkworm midgut, where the relative abundance varied depending on silkworm species and their physiological activities [23]. In another study, the predominant gut microbiota from the healthy silkworm larvae were Delfitia sp., Ralstonia sp., Enterococcus sp., Staphylococcus sp., and Pelomonas sp.; nevertheless, the observation of relative abundance change depended on the larval stage [24]. According to Chen et al. [25], approximately 10^7 CFU of bacteria were found in the whole gut of each sample. The identification using 16S rDNA sequence found that Enterococcus (62.1%) and Clostridium (35.4%) were the two main genera in each sample. Furthermore, the effect of bacteria including Lactobacillus spp., Bifidobacterium, B. licheniformis, and B. niabensis reflected in the improvement of growth and productivity characteristics of silkworm Bombyx mori L. [15,26,27]. Several studies recently have reported microbial abundance and diversity of Lepidopteran insect guts [22,23,28,29], and some xylose utilizing bacteria in Lepidoptera were also reported [16,30].

3.3 Growth and Acid Formation on Polysaccharide Substrates

Silkworm does not code for cellulolytic genes; therefore, some of the cellulolytic-degrading enzymes might be produced by gut microbiota [31]. All selected xylose utilizing LAB do not show the extracellular polysaccharide degrading enzymes production on MRS agar supplemented with various polysaccharide substrates. However, all selected LAB showed positive growth, but the acid formation was found in varying levels when cultured on MRS agar supplemented with individual polysaccharides as the sole carbon source (data not shown). This means that these LAB may possess membrane-bound enzymes with the capability to convert these polysaccharides to free simple sugars and assimilate them into their metabolism via the lactic acid fermentation pathway. The results of this study revealed that some xylose utilizing LAB isolated in this study did not utilize various polysaccharide substrates by directly secreting the extracellular polysaccharide degrading enzymes, which is different from previous research where Enterococcus sp. was the host gut microbe with the capability of cellulolytic enzyme activity [14,32]. This evidence leads to a furthersuggestion that the main role of polysaccharides degradation in Eri silkworm may be undertaken by other non-LAB microbes such as the well-recognized polysaccharides degrading bacteria Bacillus spp., which is also reported to be the dominant bacterial population in the midgut of Bombyx mori L. silkworm [14,15].

3.4 Tannin-Tolerance Test

The results from these experiments found that all 51 LAB selected isolates showed tolerance against tannins supplemented in MRS-xylose agar at the final concentration of 1 g/L tannin with variation observed based on the extent of yellow clear zones around their colonies (Figure 3). The clear zones formed by 51 isolates on MRS-xylose with 1 g/L of tannin were found in the range of 4–15 mm, which was comparable to a lesser degree in size to their clear zone on the MRS-xylose agar without tannin supplementation. Interestingly, only three isolates (*E. faecalis* SC1, *E. faecalis* SCT2, and *E. hirae* SX2) were able to grow and form yellow clear zones on MRS-xylose supplemented with 5 g/L of tannin (Figure 3c), whereas the other isolates could not grow under this condition. Phenolic compounds of plants are secondary metabolites, including such as flavonoids, tannins, and phenolic acids, which exhibit many biological activities associated with plant growth and development [33]. Tannin is the polyphenolic compound that is commonly found in plant leaves, and this phenolic compound has been recognized for its antimicrobial activity effect as a result of binding either with the cellular enzymes in the cytoplasm or cell surface receptor proteins in the cell wall [34]. Thus, it seems likely that a tannin-rich environment provides a selective condition to the LAB that can survive in tannin-rich conditions. The capability of growth of *E. faecalis* SC1, *E. faecalis* SCT2, and *E. hirae* SX2 indicates the special beneficial characteristic for active survival of these LAB in the tannin-rich condition in Eri silkworm gut.

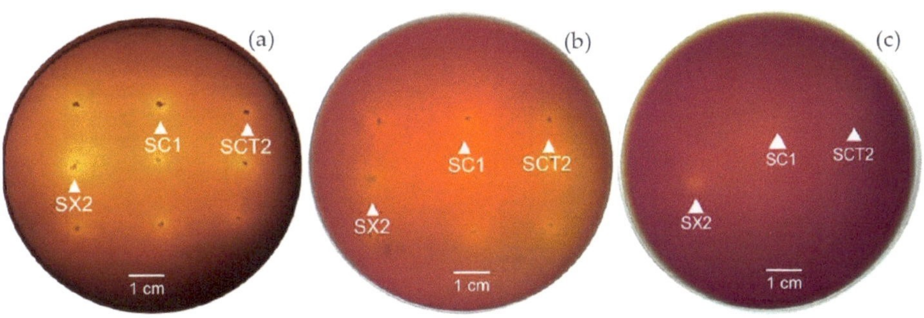

Figure 3. Growth and clear zone formation of xylose-utilizing LAB, E. faecalis SC1, E. faecalis SCT2, and E. hirae SX2, on MRS-xylose agar (a), MRS-xylose agar supplemented with 1 g/L tannin (b), and 5 g/L tannin (c), when cultivated at 37° C for 24 h.

3.5 Antimicrobial Activity

One of the common benefits provided by probiotics is the ability to combat infection of pathogenic bacteria. In this study, *B. cereus*, *S. aureus*, and *P. vulgaris* were used as Gram-positive and Gram-negative pathogenic indicators, and the result is presented in Table 1. Overall, all 51 isolates exhibited inhibitory effects against at least one of the tested pathogens. Five isolates with the most excellent antagonistic activity against all three tested insect pathogenic bacteria were *E. faecalis* SCT2, *E.*

hirae SX2, *E. mundtii* UDF1, *E. mundtii* UDFX1, and *E. mundtii* UDG3. *P. vulgaris* was found to be the most sensitive pathogenic strain based on the CFCS inhibitory effect produced by LAB isolates. The highest inhibitory effect was observed in *E. hirae* SX2 against all tested pathogenic bacteria (Figure 4).

Table 1. Antimicrobial activity of LAB isolates against some pathogenic bacteria.

Inhibition Zones against Pathogenic Bacteria (mm)						
Isolates	*Proteus vulgaris*	*Bacillus cereus*	*Strepto-coccus aureus*			
	Unneu-tralized	Neutral-ized	Unneu-tralized	Neutral-ized	Unneu-tralized	Neutral-ized
	CFCS	CFCS	CFCS	CFCS	CFCS	CFCS
C1-22	++	++	+	+	ND	ND
C5-278	++	++	+	+	ND	ND
MJG2	++	++	+	+	ND	ND
SC1	++	++	+	+	ND	ND
SCT2	+++	+++	++	+	++	+
SX2	+++	+++	++	++	++	+
UD5	+++	+++	+	+	ND	ND
UD6	++	++	+	+	ND	ND
UDF1	+++	+++	++	+	+	ND
UDFG3	+	+	+	+	ND	ND
UDFX1	+++	+++	++	+	+	ND
UDFX4	++	++	+	+	ND	ND
UDG2	++	++	+	+	ND	ND
UDG2-2	++	++	+	+	ND	ND
UDG3	+++	+++	++	+	+	ND
UDX1	++	++	+	+	ND	ND
UDX3-1	++	++	+	+	ND	ND
UDX4-3	++	++	+	+	ND	ND
W10-198	++	++	+	+	ND	ND
W2-10	++	++	+	+	ND	ND
W2-12	++	++	+	+	ND	ND
W2-9	++	++	+	+	ND	ND
W3-15	++	++	+	+	ND	ND
W8-87	++	++	+	+	ND	ND

W9-94	++	++	+	+	ND	ND
WEB1	++	++	+	+	ND	ND
WEG2	++	++	+	+	ND	ND
WES1	+++	+++	+	+	ND	ND
WFG2-1	+	+	+	+	ND	ND
WG1	++	++	+	+	ND	ND
WG1-1	++	++	+	+	ND	ND
WG1-2	+	+	+	+	ND	ND
WG1-3	+	+	+	+	ND	ND
WG2	++	++	+	+	ND	ND
WG2-2	++	++	+	+	ND	ND
WG2-3	+	+	+	+	ND	ND
WG4-1	+	+	+	+	ND	ND
WG4-2	++	++	+	+	ND	ND
WG5-1	+	+	+	+	ND	ND
WG5-3	+	+	+	+	ND	ND
WX1	+	+	+	+	ND	ND
WX1-1	+	+	+	+	ND	ND
WX1-2	++	++	+	+	ND	ND
WX2	+++	+++	++	+	ND	ND
WX3	+++	+++	++	+	ND	ND
WX4	+++	+++	+	+	ND	ND
WX4-3	+++	+++	+	+	ND	ND
WX4-4	++	++	+	+	ND	ND
WX5	++	++	+	+	ND	ND
WX5-2	++	++	+	+	ND	ND
WX5-4	++	++	+	+	ND	ND

Note: clear zone around disc; +: 1–3 mm; ++: 3–5 mm; +++: >5 mm; ND: no inhibition zone was detected.

Some previous reports confirmed that the silkworm gut is colonized with many bacte- ria [22,35]. *Streptococcus* sp. was found to be the most abundant pathogen in *Bombyx mori* larvae, whereas LAB from genus *Pediococcus*, *Leuconostoc*, and *Lactobacillus* did not cause any infection to silkworm [36]. However, the beneficial effect on silkworm host or in- teraction among the different microflora and the precise mechanism of action is unclear. Even though microbial supplementation as probiotics is commercially used for humans, ruminants, poultry, and fisheries, there are only few options of a probiotic formulation designed for silkworms [36]. Meanwhile, applying the commercial probiotic formulation was found to improve the larval body weight, effective rate of rearing, cocoon weight, pupal weight, shell ratio, and silk productivity [37–40]. Since synthetic antibiotics are widely applied in sericulture for disease control and some previous studies showed the antibiotic potential of LAB from silkworm against a range of both Gram-positive and Gram-negative pathogenic bacteria, considering the use of LAB isolated from silkworm gut with the antimicrobial activity has been proposed to be an alternative way for ecofriendly management of silkworm diseases [18]. Furthermore, some enterococci strains showed their ability to produce several bacteriocins [41], which confirm a competitive advantage toward the pathogenic microbes, and some *Enterococcus* spp. have been reported to be used as probiotics, such as *E. faecium*, *E. faecalis*, *E. lactis*, *E. hirae*, and *E. durans* [15,42,43]. However, the potential of enterococci used as probiotics supplement in Eri silkworm has not yet been reported. The dominant strains of xylose utilizing *Enterococcus* spp. isolated from the healthy silkworm of this study, and particularly *E. hirae* SX2 has the potential to contribute beneficial effects to the silkworm, and it is expected to be used as probiotic for Eri silkworm culture.

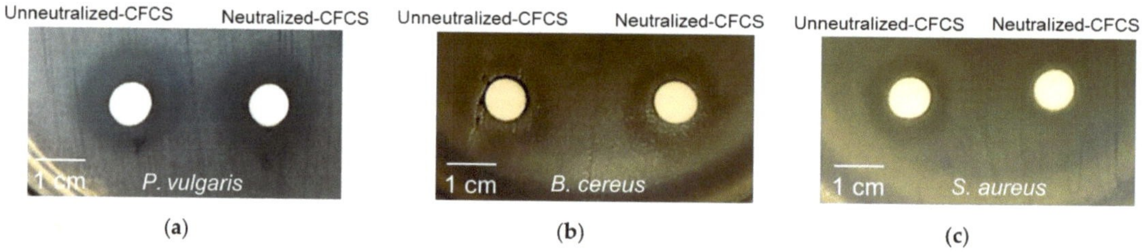

Figure 4. Antimicrobial activity of cell-free culture supernatant (CFCS) of E. hirae SX2 cultivated in MRS xylose against Proteus vulgaris (a), Bacillus cereus (b), and Streptococcus aureus (c).

3.6 Effect of E. hirae SX2 Supplementation on Eri Silkworm Growth

The effects of presumptive probiotic *E. hirae* SX2 on the growth of Eri silkworm are shown in Table 2. The results revealed that the larvae fed on probiotics treated castor leaves exhibited a significant difference in weight when compared to the naturally fed control group. After 3 weeks of cultivation (fifth instar), larval weight of Eri silkworm reared with castor leaves supplemented with *E. hirae* SX2 were 4.53 ± 0.07 g, whereas larval weights of the control group were 4.20 ± 0.04 g. The results showed that larval weight of Eri silkworm reared with probiotic showed a significant increase of 7.85% ($p < 0.05$) when compared to the control leaves. In addition, the mortality rate of 3.3% was observed from the Eri silkworm reared with castor leaves supplemented with *E. hirae* SX2, which was 9.3% lower than that of the control. The culturable bacteria of Eri silkworm midgut reared on probiotic after 3 weeks of cultivation were enumerated (Figure 5). The results revealed that probiotic supplementation significantly affected total viable cells in the Eri silkworm gut. The high numbers of total viable cells and acid-forming bacteria in probiotic-treated larvae were detected when compared with untreated. Furthermore, the viable cells of LAB significantly increased when compared to the control treatment of castor leaves. These results indicated that probiotic *E. hirae* SX2 might be able to survive and grow in larval gut and could provide some benefits to the host larvae quality leading to the increase of the larval weight.

Table 2. Weight (g) of Eri silkworm reared with castor leaves and *E. hirae* SX2 supplemented castor leaves during 3 weeks of cultivation.

Treatments	0 Week (1st Instar Larva)	1 Week	2 Weeks	3 Weeks
Castor	0.03± 0.04	0.48± 0.03^{a}	1.53± 0.04^{b}	4.20± 0.04^{b}
Castor+SX2	0.02± 0.02	0.47± 0.01^{a}	1.69± 0.09^{a}	4.53± 0.07^{a}

Note: means in columns with different superscripts are statistically different at $p < 0.05$.

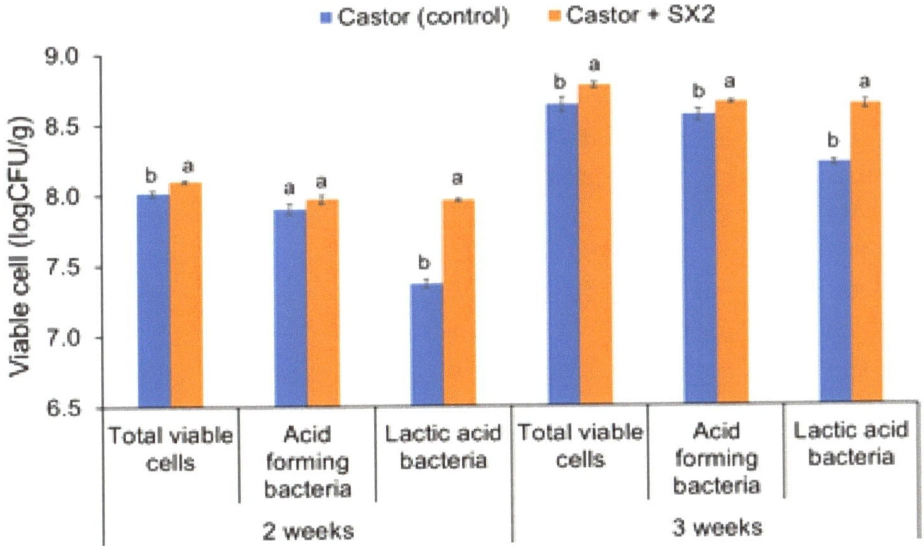

Figure 5. Total viable cells, acid-forming cells, and lactic acid-forming cells of probiotic-supplemented Eri silkworm midgut. Different small letters indicate significant differences in the viable cell (logCFU/g) in midgut between Eri silkworm reared with castor leaves and E. hirae SX2 treated castor ($p < 0.05$).

The probiotic microbial flora activity in the host gut might increase the efficiency of digestion and food assimilation, potentially leading to increase protein synthesis and assim- ilation in the host [44]. Moreover, probiotics have the ability to produce some vitamins and rapidly degrade the digestible compounds, potentially leading to nutritional assimilation. Based on the theory that beneficial microbial ecology is essential for eukaryotic metabolism, insects need gut microbial flora to provide a diversity of enzymes required for digestion of the nutritional components, which can contribute to the release of amino acids, fermentable sugars, and other molecules, which are beneficial to the growth of the insect [45]. Furthermore, the effect of probiotic bacteria including Lactobacillus spp., Bifidobacterium, B. licheniformis, and B. niabensis exhibited improvement in growth and quality characteristics of silkworm Bombyx mori L. [15,16,26,27]. Even some enterococci strains produce several bacteriocins, which have been reported with reference to the use of probiotics as mentioned previously, but there are no reports of the use of enterococci as probiotics supplement in Eri silkworm. The present study is the first report that indicates the potent beneficial effects of probiotic Enterococcus sp. on the Eri silkworm larval growth. However, more details concerning the influencing factors affecting the efficacy of probiotics, such as the probiotic dosage and probiotic formulation, are essential for advancing practical uses in Eri culture.

4. CONCISE SUMMARY

This study involved the isolation and characterization of xylose-utilizing lactic acid bacteria (LAB) from the midgut of Eri silkworm larvae to assess probiotic traits, including tannin tolerance, cellulolytic enzyme synthesis, and antibacterial efficacy against insect pathogens microorganisms that may assist the Eri silkworm. Additionally, from 51 LAB isolated from Eri silkworm, E. hirae SX2 was identified as the promising probiotic LAB due to its advantageous traits, including tolerance to 0.5% (w/v) tannin and antimicrobial efficacy against insect pathogenic bacteria. The preliminary assessment of E. hirae SX2 treatment in Eri silkworm rearing shown a beneficial impact on larval quality, leading to the proposal of this enterococci LAB as a novel probiotic for Eri silk cultivation. Furthermore, these findings will enhance the economic features of Eri silkworms and may serve as a paradigm for the future development of innovative probiotics for Bombyx mori or sericulture research.

REFERENCES

1. Pallavi, S.B. Commercial characters of selected eco-races of eri silkworm (*Samia cynthia ricini* Boisduval) reared on castor hybrid/variety. *Inter. J. Res. Anal. Rev.* **2018**, *5*, 513–518.

2. Charungkiattikul, S.; Joneurairatana, E. A revival for thailand's textile traditions: New value for local materials (eri silk) through art practice. *Textile* **2021**, 1–14. [CrossRef]

3. Leamkaew, V.; Jitjankarn, P.; Chairat, M. Polydopamine-dyed eri silk yarn for the improvement of wash and light fastness properties. *J. Text. Inst.* **2021**, *112*, 553–560. [CrossRef]

4. Prabhakar, C.S.; Sood, P.; Kanwar, S.S.; Sharma, P.N.; Kumar, A.; Mehta, P.K. Isolation and characterization of gut bacteria of fruit fly, *Bactrocera tau* (Walker). *Phytoparasitica* **2013**, *41*, 193–201. [CrossRef]

5. Paniagua Voirol, L.R.; Frago, E.; Kaltenpoth, M.; Hilker, M.; Fatouros, N.E. Bacterial symbionts in Lepidoptera: Their diversity, transmission, and impact on the host. *Front. Microbiol.* **2018**, *9*, 556. [CrossRef] [PubMed]

6. Nandy, G.; Chakraborti, M.; Shee, A.; Aditya, G.; Acharya, K. Gut microbiota from lower groups of animals: An upcoming source for cellulolytic enzymes with industrial potentials. *Biointerface Res. Appl. Chem.* **2021**, *11*, 13614–13637.

7. Ceja-Navarro, J.A.; Vega, F.E.; Karaoz, U.; Hao, Z.; Jenkins, S.; Lim, H.C.; Kosina, P.; Infante, F.; Northen, T.R.; Brodie, E.L. Gut microbiota mediate caffeine detoxification in the primary insect pest of coffee. *Nat. Commun.* **2015**, *6*, 7618. [CrossRef]

8. Kwong, W.K.; Moran, N.A. Gut microbial communities of social bees. *Nat. Rev. Microbiol.* **2016**, *14*, 374–384. [CrossRef]

9. Storelli, G.; Defaye, A.; Erkosar, B.; Hols, P.; Royet, J.; Leulier, F. *Lactobacillus plantarum* promotes Drosophila systemic growth by modulating hormonal signals through TOR-dependent nutrient sensing. *Cell Metab.* **2011**, *14*, 403–414. [CrossRef]

10. Engel, P.; Moran, N.A. The gut microbiota of insects—Diversity in structure and function. *FEMS Microbiol. Rev.* **2013**, *37*, 699–735. [CrossRef]

11. Naser, A.Z.; Deiab, I.; Darras, B.M. Polylactic acid (PLA) and polyhydroxyalkanoates (PHAs), green alternatives to petroleum- based plastics: A review. *RSC Adv.* **2021**, *11*, 17151–17196. [CrossRef]

12. Kim, J.-H.; Block, D.E.; Mills, D.A. Simultaneous consumption of pentose and hexose sugars: An optimal microbial phenotype for efficient fermentation of lignocellulosic biomass. *Appl. Microbiol. Biotechnol.* **2010**, *88*, 1077–1085. [CrossRef] [PubMed]

13. Bai, P.; Bai, M. Studies on the effect of aprobiotic and a neutraceutical agent on growth, development and commercial characteris- tics of silkworm, *Bombyx mori* L. *Indian J. Seric.* **2012**, *51*, 37–42.

14. Dantur, K.I.; Enrique, R.; Welin, B.; Castagnaro, A.P. Isolation of cellulolytic bacteria from the intestine of *Diatraea saccharalis* larvae and evaluation of their capacity to degrade sugarcane biomass. *AMB Express* **2015**, *5*, 1–11. [CrossRef] [PubMed]

15. Mala, M.; Vijila, K. Beneficial effects of *Bacillus licheniformis* and *Bacillus niabensis* on growth and economic characteristics of silkworm, *Bombyx mori* L. *Int. J. Chem. Stud.* **2018**, *6*, 1750–1754.

16. Prem Anand, A.A.; Vennison, S.J.; Sankar, S.G.; Gilwax Prabhu, D.I.; Vasan, P.T.; Raghuraman, T.; Jerome Geoffrey, C.; Vendan, S.E. Isolation and characterization of bacteria from the gut of *Bombyx mori* that degrade cellulose, xylan, pectin and starch and their impact on digestion. *J. Insect Sci.* **2010**, *10*, 107.

17. Quinto, E.J.; Jiménez, P.; Caro, I.; Tejero, J.; Mateo, J.; Girbés, T. Probiotic lactic acid bacteria: A review. *Food Nutr. Sci.* **2014**, *5*, 1765. [CrossRef]

18. Sonnenburg, J.L.; Chen, C.T.L.; Gordon, J.I. Genomic and metabolic studies of the impact of probiotics on a model gut symbiont and host. *PLoS Biol.* **2006**, *4*, e413. [CrossRef]

19. Unban, K.; Kodchasee, P.; Shetty, K.; Khanongnuch, C. Tannin-tolerant and extracellular tannase producing *Bacillus* isolated from traditional fermented tea leaves and their probiotic functional properties. *Foods* **2020**, *9*, 490. [CrossRef]

20. Tamura, K.; Dudley, J.; Nei, M.; Kumar, S. MEGA4: Molecular evolutionary genetics analysis (MEGA) software version 4.0. *Mol. Biol. Evol.* **2007**, *24*, 1596–1599. [CrossRef]

21. Fehlberg, L.C.C.; Nicoletti, A.G.; Ramos, A.C.; Rodrigues-Costa, F.; de Matos, A.P.; Girardello, R.; Marques, E.A.; Gales, A.C. *In vitro* susceptibility of *Burkholderia cepacia* complex isolates: Comparison of disk diffusion, Etest®, agar dilution, and broth microdilution methods. *Diagn. Microbiol. Infect. Dis.* 2016, *86*, 422–427. [CrossRef] [PubMed]

22. MsangoSoko, K.; Gandotra, S.; Chandel, R.K.; Sharma, K.; Ramakrishinan, B.; Subramanian, S. Composition and diversity of gut bacteria associated with the eri silk moth, *Samia ricini*, (Lepidoptera: Saturniidae) as revealed by culture-dependent and metagenomics analysis. *J. Microbiol. Biotechnol.* **2020**, *30*, 1367–1378. [CrossRef] [PubMed]

23. Yeruva, T.; Vankadara, S.; Ramasamy, S.; Lingaiah, K. Identification of potential probiotics in the midgut of mulberry silkworm, *Bombyx mori* through metagenomic approach. *Probiotics Antimicro. Prot.* **2020**, *12*, 635–640. [CrossRef] [PubMed]

24. Sun, Z.; Lu, Y.; Zhang, H.; Kumar, D.; Liu, B.; Gong, Y.; Zhu, M.; Zhu, L.; Liang, Z.; Kuang, S. Effects of BmCPV infection on silkworm *Bombyx mori* intestinal bacteria. *PLoS ONE* **2016**, *11*, e0146313. [CrossRef]

25. Chen, B.; Teh, B.-S.; Sun, C.; Hu, S.; Lu, X.; Boland, W.; Shao, Y. Biodiversity and activity of the gut microbiota across the life history of the insect herbivore *Spodoptera littoralis*. *Sci. Rep.* **2016**, *6*, 29505. [CrossRef]

26. Masthan, K.; Rajkumar, T.; Narasimha Murthy, C.V. Studies on fortification of mulberry leaves with probiotics for improvement of silk quality. *Int. J. Biotechlol. Biochem.* **2017**, *13*, 73–80.

27. Moustafa, M.N.; Soliman, S. Nutritional efficiency and economic traits of silkworm *Bombyx mori*, L. reared on mulberry leaves fortified with synbiotics. *J. Plant Prot. Pathol.* **2019**, *10*, 671–675. [CrossRef]

28. Chen, B.; Yu, T.; Xie, S.; Du, K.; Liang, X.; Lan, Y.; Sun, C.; Lu, X.; Shao, Y. Comparative shotgun metagenomic data of the silkworm *Bombyx mori* gut microbiome. *Sci. Data* **2018**, *5*, 1–10. [CrossRef]

29. Chen, B.; Du, K.; Sun, C.; Vimalanathan, A.; Liang, X.; Li, Y.; Wang, B.; Lu, X.; Li, L.; Shao, Y. Gut bacterial and fungal communities of the domesticated silkworm (*Bombyx mori*) and wild mulberry-feeding relatives. *ISME J.* **2018**, *12*, 2252–2262. [CrossRef]

30. Klongklaew, A.; Unban, K.; Kanpiengjai, A.; Wongputtisin, P.; Pamueangmun, P.; Shetty, K.; Khanongnuch, C. Improvement of enantiomeric L-lactic acid production from mixed hexose-pentose sugars by coculture of *Enterococcus mundtii* WX1 and *Lactobacillus rhamnosus* SCJ9. *Fermentation* **2021**, *7*, 95. [CrossRef]

31. Pandiarajan, J.; Revathy, K. Cellulolytic potential of gut bacterial biomass in silkworm *Bombyx mori*. L. *Ecol. Genet. Genom.* **2020**, *14*, 100045. [CrossRef]

32. Shil, R.K.; Mojumder, S.; Sadida, F.F.; Uddin, M.; Sikdar, D. Isolation and identification of cellulolytic bacteria from the gut of three phytophagus insect species. *Braz. Arch. Biol. Technol.* **2014**, *57*, 927–932. [CrossRef]

33. Despres, L.; David, J.-P.; Gallet, C. The evolutionary ecology of insect resistance to plant chemicals. *Trends Ecol. Evol.* **2007**, *22*, 298–307. [CrossRef]

34. Buzzini, P.; Arapitsas, P.; Goretti, M.; Branda, E.; Turchetti, B.; Pinelli, P.; Ieri, F.; Romani, A. Antimicrobial and antiviral activity of hydrolysable tannins. *Mini-Rev. Med. Chem.* **2008**, *8*, 1179. [CrossRef] [PubMed]

35. Gandotra, S.; Kumar, A.; Naga, K.; Bhuyan, P.; Gogoi, D.; Sharma, K.; Subramanian, S. Bacterial community structure and diversity in the gut of the muga silkworm, *Antheraea assamensis* (Lepidoptera: Saturniidae), from India. *Insect Mol. Biol.* **2018**, *27*, 603–619. [CrossRef]

36. Bhalchandra, P.M.; Pathade, G. Antibacterial and cholesterol reducing lactic acid bacteria from silk worm (*Bombyx mori*) gut environment—A review. *Nat. Environ. Pollut. Technol.* **2011**, *10*, 319–326.

37. Anitha, J.; Sathish, J.; Sujatha, K. Nutritional efficiency and economic traits in *Samia ricini* donovan reared on castor leaves fortified with probiotic agent. *Int. J. Inn. Res. Sci. Eng. Tech* **2015**, *4*, 2319–8753.
38. Shruti, A.; Hadimani, D.; Sreenivas, A.; Beladhadi, R. Effect of probiotic feed supplements to mulberry silkworm, *Bombyx mori* L. for larval growth and development parameters. *Int. J. Chem. Stud.* **2019**, *7*, 3914–3919.
39. Singh, K.; Chauhan, R.; Pande, A.; Gokhale, S.; Hegde, N. Effect of use of *Lactobacillus plantarum* as a probiotics to improve cocoon production of mulberry silkworm *Bombyx mori* (L). *J. Basic Appl. Sci.* **2005**, *1*, 1–8.
40. Suraporn, S.; Sangsuk, W.; Chanhan, P.; Promma, S. Effects of probiotic bacteria on the growth parameters of the Thai silkworm, *Bombyx mori*. *Thai J. Agric. Sci.* **2015**, *48*, 29–33.
41. Shao, Y.; Chen, B.; Sun, C.; Ishida, K.; Hertweck, C.; Boland, W. Symbiont-derived antimicrobials contribute to the control of the Lepidopteran gut microbiota. *Cell Chem. Biol.* **2017**, *24*, 66–75. [CrossRef] [PubMed]
42. Li, B.; Zhan, M.; Evivie, S.E.; Jin, D.; Zhao, L.; Chowdhury, S.; Sarker, S.K.; Huo, G.; Liu, F. Evaluating the safety of potential probiotic *Enterococcus durans* KLDS6. 0930 using whole genome sequencing and oral toxicity study. *Front. Microbiol.* **2018**, *9*, 1943. [CrossRef] [PubMed]
43. Baccouri, O.; Boukerb, A.M.; Farhat, L.B.; Zébré, A.; Zimmermann, K.; Domann, E.; Cambronel, M.; Barreau, M.; Maillot, O.; Rincé, I. Probiotic potential and safety evaluation of *Enterococcus faecalis* OB14 and OB15, isolated from traditional tunisian testouri cheese and rigouta, using physiological and genomic analysis. *Front. Microbiol.* **2019**, *10*, 881. [CrossRef] [PubMed]
44. Rowland, I.; Gibson, G.; Heinken, A.; Scott, K.; Swann, J.; Thiele, I.; Tuohy, K. Gut microbiota functions: Metabolism of nutrients and other food components. *Eur. J. Nutr.* **2018**, *57*, 1–24. [CrossRef]
45. Liang, X.; Fu, Y.; Liu, H. Isolation and characterization of enzyme-producing bacteria of the silkworm larval gut in bioregenerative life support system. *Acta Astronaut.* 2015, *116*, 247–253. [CrossRef]

13 Chapter

A PATHWAY FOR TRANSLATING A SILK-BASED MEDICAL DEVICE FROM LABORATORY TO CLINICAL APPLICATION: THE EXPERIENCE OF SILK BIOMATERIALS SRL.

1. INTRODUCTION: AN OVERVIEW

Silk is a fibrous protein produced by certain species (such as insects and spiders) for reasons of shelter, defense, or predation. It is produced within the cells of specific glands, stored in the lumen, and ultimately expelled to form extracorporeal filamentous constructions (nets, cocoons, nests, etc.) [1,2]. The silks generated by Lepidoptera (genus Bombyx, Antheraea, Phylosamia) serve as a significant supply of raw materials for the textile industry [3]. Domesticated or mulberry silk, derived from cultivating Bombyx mori silkworms and unwinding their cocoons, has been utilized for millennia to create exquisite fabrics renowned for their sheen, softness, and sophistication. Silk manufacturing accounts for less than 0.1% of the world fiber market, predicted at approximately 110 million tons in 2020; nonetheless, it remains a substantial billion-dollar sector due to silk's prominence in fashion design. (https://fashionunited.com, accessed on 15 November 2021).

The first documented medical use of silk dates back to about 2200 years, when Claudius Galenus proposed to use silk as suture thread [4]. Other similar applications appeared throughout the centuries, but only in the 20th century have silk sutures become commercially available for routine clinical use. A key step for the development of a wider range of medical applications has been the dissolution of native silk fibroin fibers and the obtainment of a liquid feedstock from which different regenerated silk-based materials such as films, hydrogels, porous scaffolds, nanofibers, particles, etc., could be prepared [5]. Of the two main components of silk, fibroin and sericin, fibroin is the material that has attracted the most attention, although sericin has recently been carving out a significant space in the biomaterial field [6]. However, every time silk is mentioned in this article, reference will be made solely to the fibroin component.

From the 1980s onwards, the scientific community's interest in silk as a biomaterial has dramatically increased. Just to give an idea of the development of scientific interest in this area, a quick search in the Scopus database using "silk fibroin" as keyword returned about 50 papers published in the year 2000 and more than 800 in 2020, with an annual increase trend that is expected to continue in the future. While Japanese, European, and American research teams dominated the scientific production about silk at the beginning of the 21st century, the current panorama of the most prolific countries is led by China, whose scientific production is now three-times larger than that of America. It seems, therefore, that the millennial history of silk, which began in China and spread throughout the world via the Silk Road, has made a U-turn to return to where it originated, coupling the more traditional Chinese monopoly in silk production to the newly acquired scientific leadership. The range of biomedical applications targeted by silk-based scaffolds is very wide, spanning from implantable devices for engineering and/or regeneration of soft and hard tissues to the development of tissue models, carriers for drug release, diagnostic devices, etc. [4,7,8]. The reason for such a wide interest in silk lies in its intrinsic properties, including biocompatibility, controllable degradability, absence of toxicity of degradation products, structural stability, mechanical performance, wide choice of manufacturing options which makes it possible to design scaffolds with desirable features for specific applications [5]. Different silk-based products have reached the pre-clinical stage and have been tested in small and large animal models, and in clinical trials with promising results [4]. Even if the number of silk-based devices approved for clinical use is still very low, the number of companies involved in the development of silk-based medical products has been increasing over the last decade; thus, it is likely that, quite soon, many of them will undertake the regulatory path that leads to the approval of devices for use in the clinical setting [4,9].

The translational process that allows a silk-based material to go through the regulatory pathway that transforms the initial idea and the starting material into a product to be used in the therapeutic field is particularly long and complex. Although there might be differences depending on the different rules put in place by European (CE mark) or American (Food and Drug Administration-FDA approval) regulatory agencies, many basic steps are common. These include the acquisition of raw materials in accordance with well- established quality assurance programs, the rigorous control of the manufacturing process that must be robust and compliant in terms of product specification, the implementation and execution of on-bench and in vitro testing programs, the thorough biocompatibility evaluation of the product according to guidelines such as ISO 10993-1 (International Organization for Standardization), the execution of relevant in vivo functional tests using appropriate animal models, the access to clinical trials to evaluate safety and performance of the device and the submission of the dossier to a regulatory body for device approval [9]. The aim of this paper is to disclose how the clinical translation route works using the experience of the Italian start up Silk Biomaterials srl in the development of a silk-based medical device, the SILKBridge® nerve conduit, as case study. The device is designed for the surgical repair of peripheral nerve discontinuities to support the regeneration of injured nerves [10]. It is intended to act as a bridge to guide and structurally support axonal growth across the gap during the healing process. The conduit is entirely made of silk fibroin obtained from *Bombyx mori* silkworms. During the manufacturing process, sericin is completely removed by degumming to leave pure silk fibroin. The conduit has a tubular structure with a three-layered wall comprising an intermediate textile layer intimately and firmly coupled with two electrospun layers, one in the inner and the other in the outer parts of the device wall [10]. This design allowed for optimizing both mechanical and biological properties, because the electrospun layers have biomimetic characteristics that enhance cell's adhesion and integration with the surrounding tissues, while the textile layer was designed to provide the required mechanical resistance during implantation (suture retention strength) and in vivo functioning (compression strength) [11]. The conduit displayed a slow degradation rate both in vitro [12] and in vivo [11]. As deterioration progressed, the load-bearing responsibility shifted to the newly formed tissue ingrowth, therefore preserving mechanical integrity at the implantation site and facilitating a comprehensive morphological and functional regeneration of the transected nerve [11].

The findings presented herein augment previously published results [10–12] regarding the synergistic effectiveness of silk as a biomaterial and the design of a three-layered wall architecture for nerve conduit fabrication; furthermore, they specifically address the subsequent aspects of the regulatory process: Demonstration of the production process's robustness via the assessment of the device's functional parameters established during the design phase, verification of potential risk sources arising from the chemicals utilized in production through toxicological analysis, evaluation of biological safety in accordance with current regulations for medical devices, and initiation of the first phase of clinical trials.

2. MATERIALS AND METHODS

2.1 Starting Materials and Device Manufacturing

The wall of the tubular device consists of three layers: two electrospun layers (ES, inner and outer) and an intermediate textile layer (TEX) [10]. Two different raw materials were used to produce the device: (i) a degummed silk yarn (organzine, 40 dtex), with which a tubular braid was manufactured by warp needle braiding technology; the braid forms the intermediate TEX layer; (ii) silk cocoons, which were degummed in an autoclave to remove sericin; the pure silk fibroin fibers were used as starting material to produce the inner and outer ES layers. Both silk cocoons and yarn were acquired from external suppliers, in compliance with traceability requirement. The incoming raw materials entered the device's manufacturing cycle after having passed control tests aimed at addressing quality and purity requirements.

The manufacturing process has been described in detail elsewhere [10]. Briefly, the TEX and ES layers comprising the wall of the device were coupled during electrospinning according to a patented process [13]. Two ES layers were assembled onto the inner and outer faces of a TEX textile braid. Coupling of the TEX layer with the two ES layers was made by means of two different welding media: (i) a solution of ionic liquid (1-ethyl-3-methylimidazolium acetate; EMIMAc; #51053, Sigma-Aldrich, Merk Life Science S.r.l., Milano, Italy) in water (EMIMAc/water 80/20% *v/v*); (ii) a solution of 15% *w/w* SF in EMIMAc. After electrospinning, the three-layered tubular structure was consolidated by immersion in 80% *v/v* ethanol for 30 min at room temperature, followed by overnight washing with distilled water and drying. Finally, the tubular structure was purified by extraction with ethanol (EtOH) in a microwave assisted extractor, washed with ultrapure water, dried, packaged in double pouches under laminar flow cabinet, and, finally, sterilized with ethylene oxide.

2.2 On-Bench Testing

2.2.1 Relaxed Internal Diameter

The cross-sections of the final finished devices were mounted on aluminum stubs and sputter coated with Au/Pd (Desk V, Denton Vacuum, LLC, Moorestown, NJ, USA). A Zeiss EVO MA10 scanning electron microscope (SEM) operating at 10 kV acceleration voltage, 100 μA beam current, and 35 mm working distance was used to acquire the image of the tubular cross-section of the device. The internal diameter was determined from the SEM image using the SEM software measuring tools.

2.2.2 Wall Thickness

Wall thickness was measured according to the ISO 7198:2016 standard method. The tubular device was cut longitudinally, flattened, and measured with a thickness tester MarCator 1075R (Mahr) equipped with a constant load thickness gauge of 0.3 cm^2 foot area that exerts a pressure of 1 kPa.

2.2.3 Linear Density (Mass Per Unit Length)

The length (L) of the sample was measured with a digital Vernier caliper (Metrica, Brighton, East Sussex, UK; Carbon Fiber Composite Digital Caliper), with 0.01 cm resolution. The mass (M) of the samples was measured with an analytical balance (Mettler Toledo, Milano, Italy, XRS105 Dual Range), with 0.01 mg resolution. The linear density (mass per unit length) of the sample, expressed in mg/cm, was calculated from the M/L ratio.

2.2.4 Porosity

Porosity (%P) was determined according to ISO 7198:2016, using the gravimetric test procedure, according to the following formula:

$$\%P = 100 \times [1 - (M/V\rho)]$$

where:

- M is the mass of the device (see: Section 2.2.3);
- V is the volume of the device, calculated from the values of wall thickness (see: Section 2.2.2) and relaxed internal diameter (see: Section 2.2.1);
- ρ is the density of silk fibroin (1.35 g/cm^3) [14].

2.2.5 Degree of Crystallinity

Attenuated total reflectance Fourier transform infrared spectroscopy (ATR-FTIR) was used to determine the degree of crystallinity. ATR-FTIR spectra were acquired with an AL- PHA FTIR spectrometer (Bruker) equipped with an ATR platinum diamond accessory, at a resolution of 4 cm^{-1}, in the infrared range of 4000–400 cm^{-1}. The tubular specimen was cut longitudinally and the surface of the inner electrospun layer was pressed against the ATR crystal. The spectra recorded were corrected with a linear baseline and normalized to the CH_2 bending peak at about 1445 cm^{-1}. The crystallinity index was calculated by ratioing the intensity of the Amide III bands at 1260 cm^{-1} and 1230 cm^{-1} ($CI = A_{1260}/A_{1230}$) [15].

2.2.6 Thermal Properties

Thermal properties were determined by differential scanning calorimetry (DSC) using a calorimeter DSC 3500 Sirius (Netzsch, Selb, Germany). Samples (3–5 mg) were sealed in aluminum pans and subjected to a heating cycle from 50°C to 400°C, at a heating rate of 10°C/min, under an N_2 atmosphere (flow rate: 20 mL/min). The following parameters were determined from the DSC curves: peak temperature of the endotherms associated with melting/degradation of the ES (T_{ES})and TEX (T_{TEX}) layers; cumulative enthalpy ($D_{HES/TEX}$) of the ES and TEX endotherms; ES:TEX weight ratio, calculated from the relative intensity of the respective DSC endotherms, using the DH values of – 402 J/g and –307 J/g for the pure TEX and ES layers, respectively, to normalize the contribution of the two materials.

2.2.7 Compression Strength

Compression tests were performed under wet conditions, using an all-electric dynamic test instrument ElectroPuls E3000 (Instron, Norwood, MA, USA), equipped with a load cell of 10 N, a thermostatic bath (BioPuls), and appropriate grips submersed in water at 37°C [10]. The sample of 10 mm length was mounted between the upper and lower plates and tested at a crosshead speed of 1.0 mm/min, with a preload of 0.1 N. The sample was conditioned in the thermostatic bath for 5 min before starting the test. The force was applied perpendicular to the longitudinal axis of the tubular sample. Compressive load values were recorded at strains of 20% and 40%. The modulus was calculated from the slope of the compression curve in the 0–40% load range.

2.2.8 Suture Retention Strength

The suture retention strength is the force necessary to pull a suture from the device. The ElectroPuls E3000 (Instron) tester equipped with a load cell of 10 N was used. The conduit was cut normal to the long axis and a suture was inserted 2 mm from the end of the sample through the wall to form a half loop. The sample was clamped in the lower fixed grip and the suture thread in the upper moving grip, which was pulled at the rate of 50 mm/min. The test was performed in the wet state with the sample submersed at 37°C. The sample was conditioned in the thermostatic bath for 5 min before starting the test. The force required to pull the suture through the device was recorded.

2.2.9 Tensile Properties

Breaking load and elongation at break in uniaxial tensile test were determined under submersed conditions (in water at 37°C), with the ElectroPuls E3000 (Instron) tester, equipped with a 250 N load cell. The test was performed in accordance with the provisions of ISO 7198:2016. The sample was conditioned in the thermostatic bath for 5 min before starting the test. The gauge length was 30 mm; a preload of 0.5 N was applied; the tests were run at 30 mm/min crossbar rate.

2.2.10 Statistical Analysis

Data from different production batches were compared using the one-way ANOVA statistical test to confirm the statistical equivalence between different production lots. Statistical analyses were conducted using Minitab statistical software.

2.3 Chemical Analysis

2.3.1 Determination of Bromide Ion

The sample was weighed into a crucible and reduced to ash. Ashes were recovered with water and the solution was filtered with a 0.45 μm filter before analysis by ion chromatography, with a Dionex™ ICS-4000 Integrated Capillary HPIC™ System, equipped with a Dionex IonPac AS11-HC-4 mm column, using a KOH solution as mobile phase in the gradient mode (0.02–80 mM in 15 min), at a flow rate of 15 mL/min. The injection volume was 0.4 mL; detection was made by a conductivity and QD charge detector. The results were expressed in mg/kg.

2.3.2 Determination of Lithium Ion

The sample was digested in a microwave oven (Anton Paar Multiwave 3000) with an acid mixture containing 4 mL HNO_3/2 mL H_2O_2/0.25 mLHF, with a power ramp from 0 to 1200 W in 25 min, a 15 min hold step at 1200 W and a cooling step for 20 min.

The analysis was performed by ICP-MS (inductively coupled plasma—mass spectrometry) employing a cross flow nebulizer with a Scott spray chamber (ICP-MS Perkin Elmer SCIEX mod ELAN 9000, autosampler AS90, Waltham, MA, USA). The element was quantified with an external calibration (AccuStandard ICP-MS-CAL2-1 multielement, New Haven, CT, USA), employing Germanium as the internal standard to compensate plasma fluctuations. The results were expressed in mg/kg.

2.3.3 Determination of Methyl and Ethyl Alcohols

The analysis was performed by head-space gas chromatography/mass spectrometry (HS-GC/MS) with a Shimadzu, mod. 2010 Plus, equipped with a MS detector QP2010 Ultra and an autosampler Shimadzu AOC 5000 Plus. Head space conditions were as follows: temperature 80°C, time 60 min, injection volume 500 μL. A capillary column Agilent HP5-ms (30 m length, 0.25 mm inner diameter, 0.25 μm film thickness) was used for the separation. Other analytical conditions were as follows: carrier gas Helium at 1 mL/min; incubation temperature 80°C; incubation time 60 min; splitless injector temperature 200°C; temperature program hold 40°C for 3 min—from 40°C to 260°C at 40°C/min; injection volume 500 μL; transfer line temperature 280°C; source temperature 200°C; E 70 eV; SIM acquisition mode. Target ions were m/z 32 methyl alcohol (qualifier ion 31), m/z 31 ethyl alcohol (qualifier ion 45), and m/z 31 for 1-propanol (internal standard). The results were expressed in mg/kg.

2.3.4 Determination of 1-ethyl-3-methyl Imidazolium Acetate

The samples were finely chopped and extracted with methanol in an ultrasonic bath. The supernatant was filtered on a 0.22 μm PVDF syringe filter, dried, and recov- ered with 300 μL of eluent A. The analysis was performed by ultra high-pressure liquid chromatographyion trap-mass spectrometry (UHPLC-IT-LTQ-MS/MS, FinniganMAT), with an Alltima HP C18 (3 μm, 150 mm × 2.0 mm) column, in the gradient mode (Eluent A: H_2O + 0.1% CH_3COOH; Eluent B: CH_3OH + 0.1% CH_3COOH; from 90% A to 100% B in 17 min), at a flow rate of 0.2 mL/min. The injection volume was 10 μL, the operating mode of the MS detector H-ESI(+), and the mass range m/z 50–300 uma. The results were expressed in mg/kg.

2.3.5 Determination of Formic Acid

The analysis was performed by head-space gas chromatography/mass spectrometry (HS-GC/MS), by using the same apparatus described in paragraph 2.3.3. for the determina- tion of alcohols. Analytical conditions were the same, with exception of incubation time (60 min) and injection volume (500 μL). Target ions were formic acid (m/z 46) and acetic acid (m/z 60; internal standard). The results were expressed in mg/kg.

2.4 Biological Safety Evaluation

The ISO 10993 standard was used as benchmark for the biological safety evaluation of the nerve conduit. The following tests were performed: cytotoxicity (ISO 10993-5:2009); delayed hypersensitivity (ISO 10993-10:2010); irritation/intracutaneous reactivity (ISO 10993-10:2010); pyrogenicity (ISO 10993-11:2017; US Pharmacopoeia-USP 40<151>, and European Pharmacopoeia-EU 8.0 §2.6.8.); LAL test (USP <85>; USP <161>; EU par. 2.6.14 e par. 5.1.10; AAMI ST72); acute systemic toxicity (ISO 10993-11:2017); genotoxicity (ISO 10993- 3:2014). Tests were carried out at Eurofins Biolab S.r.l. test facility (Good Laboratory Practice-GLP cert. n. 2017/16), Vimodrone (Milan), Italy.

3. RESULTS

3.1 On-Bench Testing and Property/Performance Evaluation

Sources of possible variability may emerge at different points of the manufacturing process of a medical device and must be carefully monitored to avoid any negative impact on the properties and performance of the final device. Therefore, the design controls requirements according to USA (FDA, 21 CFR 820.30) and EU (MDR, 2017/745) directives were put in place for the design and manufacturing of the SILKBridge® nerve conduit, and an onbench testing plan has been specifically developed for an in-depth property/performance evaluation. Three test categories, each one addressing specific functional endpoint, were defined. In particular:

1. the morphological and geometrical characteristics, which may affect the usability of the device during surgery, the performance at the site of implantation, and the biological response of surrounding tissues;
2. the physical and structural characteristics, which may impact the biological interaction with surrounding tissues and the rate and extent of degradation after implantation (the device is entirely made from a biodegradable natural protein polymer);
3. the mechanical characteristics, which must achieve a level high enough to withstand the mechanical stresses caused by suturing during implantation, as well as tensile, compression, and bending stresses caused by the surrounding tissues during in vivo functioning.

At the end of the manufacturing process, each device was visually inspected, cut to its final length (30 mm), and packaged for sterilization. The portions of the device exceeding the length of 30 mm were collected and analyzed for the determination of the morphological, geometrical, physical, and structural properties. This means that each device has undergone testing for the characteristics listed under points 1 and 2. On the other hand, the mechanical properties listed under point 3 were analyzed by sampling (1 device every 50 devices produced) because the tests require an entire device to be executed and are destructive. Figure 1 shows some representative experimental data for the three test categories analyzed on the finished device.

Tables 1–3 list the testing results obtained from the analysis of samples representative of three different production batches. For each batch, five samples were analyzed, and the results were reported as mean ± standard deviation. No statistically significant differences were observed among the production batches for each measured characteristic, indicating that the manufacturing process is under control and that the devices will perform as expected.

Table 1. On-bench testing results (mean ± Std. Dev.): morphological and geometrical characteristics (n = 5).

	Batch A	Batch B	Batch C	p Value[1]
Internal diameter (mm)	1.78 ± 0.10	1.76 ± 0.04	1.82 ± 0.07	>0.05
Wall thickness (mm)	0.43 ± 0.05	0.47 ± 0.04	0.47 ± 0.04	>0.05
Linear density (mg/cm)	5.7 ± 0.4	5.8 ± 0.2	6.2 ± 0.3	>0.05
Porosity (%)	86 ±2	87 ±2	86 ± 2	>0.05

[1]One-way ANOVA (confidence interval of the mean 95%). All data followed a normal distribution.

Table 2. On-bench testing results (mean ± Std. Dev.): physical and structural characteristics (*n* = 5).

	Batch A	Batch B	Batch C	p Value[1]
Degree of crystallinity	0.66± 0.03	0.66± 0.03	0.67± 0.01	>0.05
Thermal properties:				
TES (°C)	291 ± 1	291 ± 2	290 ± 1	>0.05
TTEX (°C)	321 ± 1	322 ± 1	321 ± 2	>0.05
ΔHES/TEX (J/g)	411 ± 21	402 ± 32	410 ± 31	>0.05
ES:TEX ratio (%)	53 ±1	53 ±3	54 ± 2	>0.05

[1]One-way ANOVA (confidence interval of the mean 95%). All data followed a normal distribution.

Table 3. On-bench testing results (mean ± Std. Dev.): mechanical characteristics (*n* = 5).

	Batch A	Batch B	Batch C	p Value 1
Suture retention (N)	4.6± 0.4	4.5± 0.8	4.7± 0.8	>0.05
Compression strength: Load at 20% strain (cN) Load at 40% strain (cN) Modulus (kPa)	4.6± 1.2 93± 12	4.2± 1.0 84± 17	4.6± 1.0 97± 14	>0.05 >0.05 >0.05
Tensile strength: Breaking load (N) Elongation at break (%)	24.8± 0.5	23.3± 0.9	23.1 ±1.8	>0.05 >0.05

[1]One-way ANOVA (confidence interval of the mean 95%). All data followed a normal

distribution.

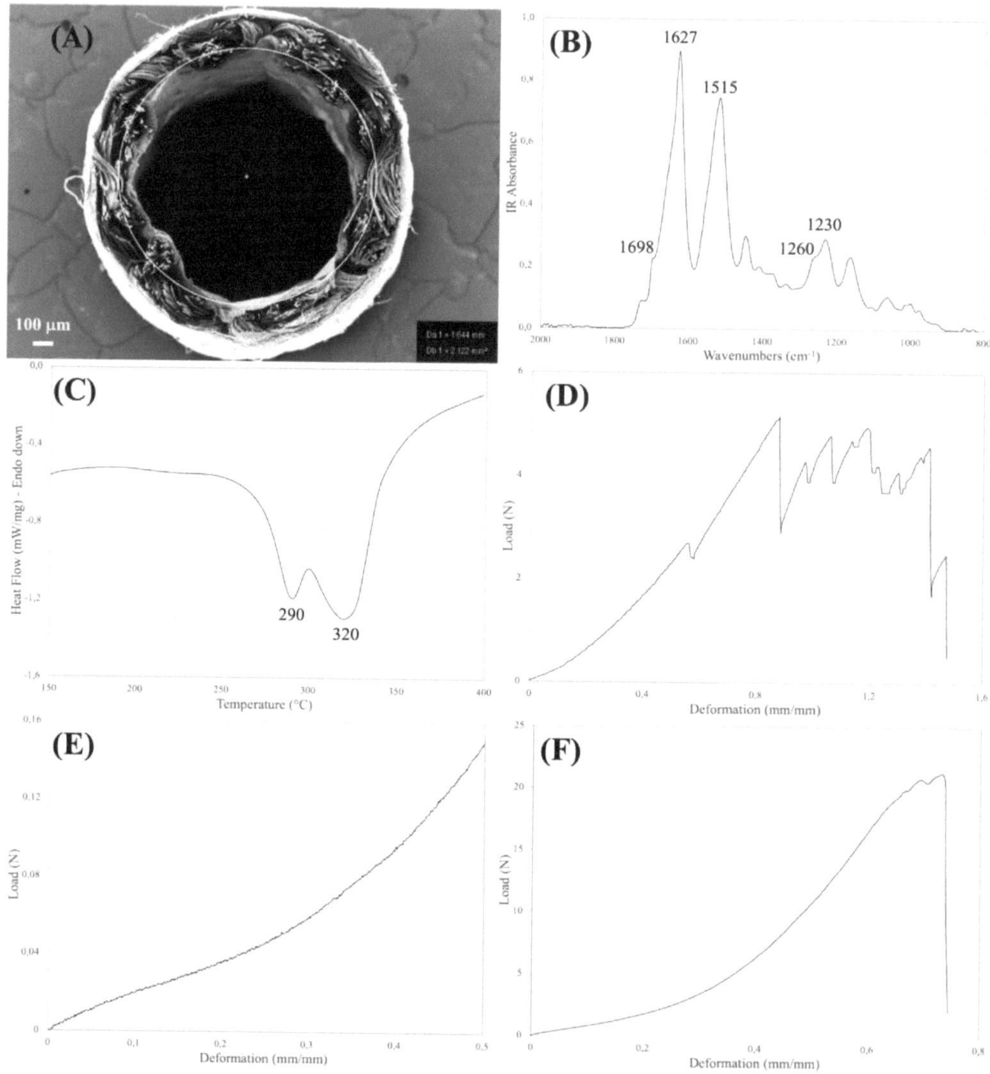

Figure 1. (A). SEM cross-section of the SILKBridge® nerve conduit showing the two inner and outer electrospun layers that enclose the intermediate textile layer. (B). ATR-FTIR spectrum of the inner electrospun layer in the 2000–800 cm^{-1} range showing the Amide bands characteristic of silk fibroin: Amide I, peak at 1627 cm^{-1} and shoulder at 1698 cm^{-1}; Amide II: peak at 1515 cm^{-1}; Amide III: peak at 1230 cm^{-1} and shoulder at 1260 cm^{-1}. The Degree of crystallinity is calculated from the I1260/I1230 intensity ratio. (C). DSC thermograms of the device in the 150–400 ° C temperature range. Peak at 290° C: thermal degradation of the electrospun component. Peak at 320° C: thermal degradation of the textile component. (D). Typical load/deformation curve of the suture retention test. The series of peaks recorded before ultimate failure represent the breakage of the yarns forming the textile layer. (E). Typical load/deformation curve of the compression strength test. (F). Typical load/deformation curve of the device analyzed in the uniaxial

tensile mode.

3.2 Chemical Analysis and Toxicological Evaluation

For the manufacturing of medical devices, chemicals could be used as processing aids. The safety profile and the toxicological risks related to their use must be assessed. In the case of SILKBridge®, the processing aids used during manufacturing are: (i) methanol (MeOH), to clean starting materials; (ii) ethanol (EtOH), to consolidate the electrospun layers by solvent-induced random coil $\rightarrow \beta$-sheet conformational transition and to clean the final device; (iii) lithium bromide (LiBr), to dissolve degummed cocoon silk fibroin fibers to produce SF films as a preparatory step of the electrospinning process; (iv) formic acid (FA), to dissolve SF films and prepare the electrospinning dope; (v) 1-ethyl-3-methyl imidazolium acetate (EMIMAc), the welding agent used to couple the TEX and ES layers. In terms of hazard classification (Regulation EC No 1272/2008), all these substances share an acute toxicity designation (oral, dermal, or by inhalation); some may cause eye/skin irritation (LiBr, EtOH, EMIMAc); EtOH and MeOH may show specific target organ toxicity, but none of them are reported as carcinogenic, mutagenic, toxic to reproduction or with endocrine-disrupting activity. Despite the thorough cleaning procedures adopted, residues of these chemicals may remain on the device and leach in the surrounding tissues during in vivo implantation. This might impact on the biocompatibility of the device and interfere with the smooth progress of tissue regeneration.

To evaluate the amount of residual processing aids that may remain in the final device, sterilized samples of nerve conduits were extracted, and the extracts were analyzed (Table 4, line 1). These results represent the starting point to perform the toxicological evaluation following the provisions of ISO 10993-17. As a first step, the concentration of each chemical expressed in mg/kg was transformed in μg/device to obtain the total amount of each compound released by one device (line 2). For this calculation, an average weight of 25 mg/device was considered. Then, assuming a worst-case scenario of multiple devices implanted at the same time in one patient (in this case: 10 devices), the value was multiplied by 10 to calculate the potential patient daily intake (PPDI, expressed in μg/day) (line 3), i.e., the total amount of chemical to which the patient is potentially exposed during the lifetime. A further worst case assumption was that the total amount of the chemical expressed by the PDDI value is released just in one day.

Table 4. Chemical and toxicological analysis of leachable processing aids.

		LiBr					
		Li^+	Br^-	EtOH	MeOH	EMIMAc	FA
1	Concentration (mg/kg)	4.0	324	2.7	1	4.1	200

2	Total Released Amount (μg/device)[1]	0.1	8.0	0.07	0.03	0.1	5.0
3	Potential Patient Daily Intake (PPDI; μg/day)	1.0	80	0.7	0.3	1.0	50
4	Tolerable Exposure (TE; μg/day)	250	1750	50,000	9000	102	35,400
5	Margin of Safety (MoS)	0.25×10^3	22	71×10^3	30×10^3	10	0.71 × 103

1 An average weight of 25 mg has been considered for the device. 2 The TTC (threshold of toxicological concern) is used for EMIMAc instead of TE.

For each compound, the toxicological data were retrieved from specialized databases (TOXNET Databases, Pubmed, Pubchem, ECHA—European Chemical Agency, FDA—Food and Drug Administration, etc.). They allowed setting the values of tolerable exposure (TE, expressed in μ g/day), i.e., the amount of chemical agent that doesn't pose concerns for human health (line 4). Since no relevant toxicological data were available for EMIMAc, the TE could not be calculated as recommended by ISO 10993-17. Instead, the value of TTC (threshold of toxicological concern) was estimated according to ISO/TS 21726. It is important to note that the TTC value assigned to EMIMAc (10 μ g/day) derives from an extremely conservative and protective classification based on individual mutagenic impurities in pharmaceuticals and is presumed to be protective for both the potential carcinogenic and non-cancer effects that would occur following patient exposure to the medical device. Finally, the TE/PPDI ratio was calculated to obtain the margin of safety (MoS, line 5). For values of MoS 1, the compound is considered to pose a toxicological concern. For values of MoS > 1, the compound does not raise risks for human health. We adopted a more conservative approach, setting a threshold for MoS $\geq$ 10. As can be observed in Table 4, despite the worst-case assumptions made at different levels of the toxicological evaluation, i.e., total amount released in one day, multiple devices, MoS threshold higher than recommended by the reference standard, none of the compounds tested can be held responsible for causing toxicological concerns.

3.3 Biological Safety Evaluation

The ISO 10993-Part 1 provides a guidance for the assessment of medical devices according to the nature and duration of their contact with human tissues, and the other parts of the standard give indications on how to perform the tests required. SILKBridge® is categorized as an implant device in long term contact with tissue/bone (>30 days). Based on this categorization, as well as on the analysis of the risks associated with the constituent materials, their source, the manufacturing process, and the morphological, physical, chemical, and mechanical characteristics, the endpoints of the biological safety evaluation were established according to the whole ISO 10993 standard. Sterilized devices were then tested for each biological endpoint; the results are summarized in Table 5.

Table 5. Results of the biological safety evaluation.

Test	Evaluation	Final Score
Cytotoxicity	After 24 h of contact with the test sample extracts, discrete intracytoplasmic granules, no cell lysis, and no reduction of cell growth were observed (reactivity grade 0). The cell viability reduction was 4%.	Not cytotoxic
Delayed hypersensitivity	No abnormalities were observed in treated and control animals duringthechallengephase(grade0).	Notsensitizing
Intracutaneous reactivity	No abnormalities were observed immediately after injection. With polar extract all treated and control sites did not show signs of erythema, eschar or edema. With apolar extract, all treated and control sites showed a slight erythema, no eschar or edema. The Primary irritation index was 0.	Not irritating
Pyrogen test	The summed temperature rise of rabbits was +0.3°C and the summed response did not exceed 2.8°C. Only an individual temperature rise of one rabbit was higher than 0.5°C and the summed temperature rise was lower than 3.3°C.	Not pyrogenic
LAL test	The test sample solution had an endotoxin content <20 EU/device.	Not endotoxic
Acute systemic toxicity	None of the treated and control animals showed toxic signs, symptoms, mortality, or weight loss.	Not toxic
Genotoxicity	Bacterial Reverse Mutation Test: no toxic effects of the extracts were noted in any of the tester strains used; no biologically relevant increases in revertant colony numbers of any of the tester strains were observed following treatment with extracts. Mouse Lymphoma Test: no growth inhibition was observed for the polar and lipophilic extracts; no biologically relevant increase of mutants was found for the polar and lipophilic; the global evaluation factor (GEF) was not exceeded by the induced mutant frequency at any concentration; no dose-response relationship was observed.	Not mutagenic

4. DISCUSSION

The development of an implantable medical device as the SILKBridge® nerve conduit must undergo a comprehensive series of control tests before being released for use to ensure that the manufacturing process is under control and that the device performs as expected. The on-bench testing framework complies with the design controls requirements set out according to FDA and EU/MDR regulations, and comprises a list of functional parameters and test methods inspired by the ISO 10993-18 standard, which can be used for the identification and evaluation of the physical, chemical, morphological, and topographical properties of materials in finished medical devices, all having a strong impact in terms of verification and validation of the target performance characteristics.

The results of the tests reported in Table 1 (morphological and geometrical characteristics), Table 2 (physical and structural characteristics), and Table 3 (mechanical characteristics) demonstrate that the comparison of three production lots gave satisfactory results. All the measured characteristics appeared highly reproducible, with a very low level of batch-to-batch variability, thus, ensuring the robustness of the manufacturing process. With reference to the suitability of the device for the intended use, the following discussion will provide deeper insights into the set of functional specifications that were defined during the design controls to comply with end-use requirements.

The inner diameter of the device is a geometrical characteristic determined by the foreseen clinical use. The transversal dimension of the conduit must be large enough to accommodate the distal and proximal extremities of the severed nerve and to allow their fixation by suturing during surgery. The size of the nerve conduits currently produced is 1.5–2.0 mm inner diameter, which is optimized for repairing transected digital nerves [16]. To repair larger nerves located in other anatomical positions, larger devices will be required. Usually, marketed nerve conduits cover a range of discrete sizes from 1 mm to 10 mm.

The wall thickness can be a more critical characteristic. A relationship between the formation of neuromas in regenerated nerves tissues and the thickness of the conduit wall has been reported [17]. The wall-thickness of SILKBridge® has been carefully designed considering the properties of the selected starting materials and the requirements in terms of mechanical performance, suturability, targeted degradation rate, and the permeability specifications [18]. In general, wall thickness values lower that 1 mm are highly suggested for nerve guide application, with 0.6 mm as the optimum target for the maximum value [19]. As shown in Table 1, the final device falls within this optimal range.

Linear density, i.e., the weight per unit length, allows for evaluating the amount of material implanted into the body. Silk, as any other non-autologous biomaterial, is likely to elicit a foreign body response at the site of implantation [7]. Moreover, it has been reported that silk-based devices implanted in soft tissues or with longer degradation times tend to induce a long-lasting inflammatory response than those with shorter degradation times or those located within hard tissues [8]. On this basis, it has been decided to keep the linear density as low as possible to reduce the impact of the device in terms of local reactions. The choice of a macro-porous texture of the textile braid for the central layer of the wall allowed for keeping the linear density of the final device largely below the threshold of 10 mg/cm, as defined in the design stage (Table 1).

The porosity of the wall of a nerve conduit is a key functional parameter. It must be optimized to maximize the influx of oxygen and nutrients from the interstitial fluid and to avoid the loss of neurotrophic factors secreted by Schwann cells at the distal stump. A wall porosity of about 80% or higher is considered ideal for peripheral nerve repair [17–19]. This target has been achieved with SILKBridge®, whose wall porosity higher than 80% results from the combination of the macro- and micro-porosity of the TEX and ES layers, respectively (Table 1). With reference to the pore size, this characteristic is entirely determined by the two ES layers and does not exceed the 5 μm threshold on average (data not shown). This allows for the permeation of small nutrient molecules, while preventing the migration of inflammatory cells and the infiltration of fibrous tissue inside the lumen of the conduit [18].

The physical and structural features of the device were explored by FTIR and DSC analyses (Table 2). Both the spectroscopic and thermal properties of silk fibroin-based materials have been extensively investigated [20,21]. The type, position, and intensity of the bands in the IR spectra and of the thermal transitions in the DSC thermograms not only reflect the chemical structure of the silk materials but also provide information about the molecular conformation taken by the silk fibroin chains in native and regenerated formats. The aim of the FTIR analysis was to verify that the manufacturing process did not alter the chemical integrity of the constituent material, did not leave foreign matters or contaminants on the device, and allowed for achieving the desired degree of crystallinity of the regenerated electrospun layers. Loss of integrity and changes in the chemical and conformational structure of the constituent silk material may have a strong impact on the biodegradation rate of the device upon implantation. Moreover, the presence of contaminants may impact on the biocompatibility of the device. For the purpose of integrity and purity checks, the position and intensity of the IR bands found in the spectrum of the sample were compared with reference spectra, while the crystallinity index was calculated from the intensity ratio of the Amide III bands at 1260 cm^{-1} (crystalline phase) and 1230 cm^{-1} (amorphous phase) [15].

The major thermal events characteristics of silk fibroin fall in the 150–400°C temperature range (glass transition and melting/degradation). As with FTIR, the aim of the DSC analysis was to verify the structural integrity of the constituent silk fibroin materials to avoid any impact on the performance of the device. The DSC analysis allowed us to also calculate the mass ratio between the electrospun and textile layers, which present distinct endothermic peaks [10]. This parameter is a very useful tool to further assess the robustness of the manufacturing process. The DSC and FTIR results listed in Table 2 allowed for confirming the structural integrity of the silk fibroin materials comprising the device.

Medical devices must be designed with an optimal set of mechanical properties fitting the requirements of the specific clinical need. Nerve conduits must provide a balanced combination of strength and elasticity to withstand clinical operation stresses, such as manipulation and suturing during implantation, to resist deformation caused by the biomechanical stresses generated in vivo, and to avoid channel collapse, since compression can result in damage to the growing axon [10,11]. To verify that the manufacturing process allows for reaching the desired mechanical characteristics, the devices were analyzed for determining the tensile, suture retention, and compression properties (Table 3).

To withstand physiological loads, the tensile properties of nerve conduits should at least approach the ones of natural nerves [18,22]. The device must be elastic enough to match the deformation of the natural nerve and strong enough to protect the growing axons from breaking. From the data reported in the literature, the stress and strain values of human nerves are about 7 MPa and 60%, respectively. Considering the geometry of the device, the stress value corresponds to a breaking load of about 15 N. As shown in Table 3, the SILKBridge® nerve conduit largely met the target values.

The suture retention test is intended to verify whether the device withstands the mechanical stresses applied by the surgeon during implantation and later during in vivo functioning. Suture failure may cause dramatic consequences to the patients, due to the loss of stability of the implant, lack of support for tissue regeneration, and need to re-operate. No threshold values for suture retention strength of nerve conduits have been reported in literature. On the contrary, limits were imposed for vascular grafts, a much more demanding application in terms of anastomotic strength, which requires that grafts are capable to exceed 2 N as suture retention strength [23–25]. Thanks to the strength of the textile braid, which is the load-bearing element of the wall of the device, the target threshold of suture retention strength was achieved by the final device (Table 3).

Finally, nerve conduits must withstand the mechanical compression stresses of surrounding tissues until complete nerve regeneration, avoiding collapse that may hinder the healing and cause pain to the patient [18]. To set acceptable threshold values for the compression performance, reference was made to the limit pressure values reported for the human median nerve exposed to carpal tunnel pressure [26]. In healthy individuals, carpal tunnel pressures are typically well below 10 mmHg, corresponding to a stress 1.33 kPa.

Considering the geometry of the SILKBridge®, this stress corresponds to a compression load of about 1.1 cN. The results listed in Table 3 indicate that the compression resistance threshold was easily achieved, and that the device can withstand compression stresses and remain open to allow for a smooth progression of the nerve regeneration.

The evaluation of the potential toxicological risks associated with leachable substances released by a medical device in the surrounding tissues is another important step for the identification and quantification of the biological hazards related to its use. The ISO 10993-12 standard provides provisions for the preparation of samples for analysis. The ISO 10993-18 standard specifies a framework for the identification of leachable substances and for the quantitative determination of their potential release in the human body. The ISO 10993-17 standard guides the manufacturer through the process of estimation and control of the toxicological risks associated with the medical device use. The chemical analysis addressed all the processing aids used along the manufacturing process. Their amount was determined by applying suitable laboratory extraction conditions and advanced analytical methods.

With reference to the mechanism by which the chemical compounds may be released into the tissues surrounding the implant area, a combination of contributions can play a role. Simple diffusion from the device to the tissue may occur in the first period after implantation, when the device is still intact. Afterwards, when the device starts to degrade, swelling and fragmentation of the polymer texture may open new ways for the chemical to leach outside. However, it must be considered that the release will never occur in a bursting way, but gradually over time and in a more physiologically compliant manner, so that the local load is diluted over time. This assumption is supported by the fact that SILKBridge® is made of silk fibroin materials characterized by a slow rate of degradation, i.e., from months to years for regenerated electrospun fibers and native microfibers, respectively [12].

As shown in Table 4, the results of the toxicological evaluation have demonstrated that the leachable substances coming from the manufacturing process were under control and that the cleaning procedures were effective in removing the greatest part of the processing aids from the final device. The residual amounts still present were largely below the threshold for toxicological concern, also considering the worst-case scenarios taken into account during the toxicological evaluation, i.e., the possibility of multiple implants (up to 10, one device for each finger), the tenfold increase of the acceptable value of the Margin of Safety, and the assumption that the total amount is released in one day in a bursting way. The targeted toxicological analysis here reported has been complemented with an untargeted one, which allowed to identify possible leachable compounds beyond those used for manufacturing, to include also unpredictable contaminants of environmental source, including laboratory materials, packaging, etc. (data not shown). It is worth noting that also this additional approach did not reveal the presence of unexpected contaminants likely to pose a health risk.

The evaluation of the toxicity profile of a degradable medical device such as SILKBridge® cannot disregard the possible toxicity of the degradation products. This issue has been addressed in a previous study where the in vitro degradation profile and the cytotoxicity of the degradation products were reported [12]. The bacterial protease type XIV from *Streptomyces griseus* was used as hydrolytic agent at three different enzyme/substrate ratios and for incubation times as long as 91 days. Degradation of the device occurred by surface erosion. The mass spectrometry analysis of the degradation products showed that the silk fibroin polypeptides recovered in the incubation buffers were representative of the aminoacidic sequence of the fibroin light and heavy chains, indicating that virtually the entire sequence of the fibroin protein was degraded. More important, the incubation buffers containing the soluble degradation products were tested with human HEK293 cells and mouse neuroblastoma N2a cells to assess cytotoxicity. No detrimental effects on cell viability were observed, suggesting that the degradation products, consisting of amino acids and small peptides, did not show any toxic property and their more likely fate was to enter the metabolic pathways of the host tissue cells [7]. Therefore, it is possible to conclude that the SILKBridge® nerve conduits will not elicit any toxic effect due to the constituent materials and/or to the manufacturing process.

Biocompatibility is an essential requirement for the medical devices and the large amount of scientific data on the safety of silk is not sufficient on a regulatory prospective to allow the use of silk-based device since the manufacturing process can change the characteristics of the raw material and have an impact on the biocompatibility of the finished device [8]. This perspective imposes that a comprehensive evaluation of the device biocompatibility is carried out to prevent any possible adverse effect for the patient's health. Different parts of the ISO 10993 standard provide a framework for the evaluation of the biological safety of the device. A plan has been designed and a set of tests, including cytotoxicity, delayed hypersensitivity, intracutaneous reactivity, pyrogen test, LAL test, acute systemic toxicity, and genotoxicity has been carried out. The overall results of the biological response allowed confirming that SILKBridge® is fully biocompatible (Table 5). This is in good agreement with the numerous data reported in the scientific literature about the biocompatibility of silk materials [8].

Finally, the evaluation of a medical device can't exempt from carrying out performance tests to verify that the product achieve it's intended use. In a previously published article [11] the implantation of the SILKBridge® in the median nerve of rats was discussed to demonstrate the regeneration of myelinated fibers along the conduit filling the gap previously created and leading to an effective morphological and functional recovery of the median nerve, like that observed with the reference autograft nerve reconstruction.

Altogether, the results reported here, and others previously published [10–12] represent an important achievement towards the implementation of a clinical study aimed at investigating the safety and efficacy of the SILKBridge® nerve conduit. The device has demonstrated an optimized balance of biomechanical and biological properties; it is intended as an "off-the-shelf" product ready to be used in the operating room as it is, without the need of adding neurotrophic and/or angiogenic factors or cells. The encouraging on-bench and pre-clinical results allowed us to proceed quickly towards the submission of a first in-human clinical study aimed at evaluating the reconstruction of digital nerve defects in humans (ClinicalTrials.gov identifier: NCT03673449). The study has already started at the Department of Plastic Surgery and Hand Surgery of the University Hospital of Zurich. Four out of 15 patients have been enrolled so far and implanted with SILKBridge® nerve conduit to repair a digital nerve gap, with very satisfactory outcome after one year in terms of functional recovery.

5. CONCISE SUMMARY

TThe journey from silk-derived raw materials (cocoons, threads) to the fabrication of a medical device and its application in people is extensive and complex. Patient safety and health are key objectives that must not be compromised in any strategy designed to address clinical issues. It is unsurprising that regulatory bodies have established comprehensive directions to objectively evaluate all risks related with the use of implantable medical devices.

This study demonstrates that converting silk material into a medical device and advancing it to clinical application necessitates a network of skills encompassing polymer science, chemistry, toxicity, biomedical engineering, medicine, biology, regulatory affairs, and quality assurance. It is essential to establish that all production processes culminating in the final device are scalable from laboratory or pilot scale to industrial scale. The support of doctors is crucial from the outset, beginning with the identification of the clinical need to be addressed and continuing through to the evaluation of clinical outcomes and the facilitation of product marketing.

REFERENCES

1. Vollrath, F.; Knight, D.P. Liquid crystalline spinning of spider silk. Nature **2001**, 410, 541–548. [CrossRef]
2. Fedic, R.; Zurovec, M.; Sehnal, F. The Silk of Lepidoptera. J. Insect Biotechnol. Sericol. **2002**, 71, 1–15. [CrossRef]
3. Babu, K.M. Silk: Processing, Properties and Applications; Woodhead Publishing: Sawston, UK, 2013; pp. 1–55.
4. Holland, C.; Numata, K.; Rnjak-Kovacina, J.; Seib, F.P. The Biomedical Use of Silk: Past, Present, Future. Adv. Healthc. Mater. **2019**, 8, 1800465. [CrossRef]
5. Rockwood, D.N.; Preda, R.C.; Yücel, T.; Wang, X.; Lovett, M.L.; Kaplan, D.L. Materials fabrication from Bombyx mori silk fibroin. Nat. Protoc. **2011**, 6, 1612–1631. [CrossRef]
6. Jo, Y.Y.; Kweon, H.Y.; Oh, J.H. Sericin for Tissue Engineering. Appl. Sci. **2020**, 10, 8457. [CrossRef]
7. Altman, G.H.; Diaz, F.; Jakuba, C.; Calabro, T.; Horan, R.L.; Chen, J.; Lu, H.; Richmond, J.; Kaplan, D.L. Silk-based biomaterials. Biomaterials **2003**, 24, 401–416. [CrossRef]
8. Thurber, A.E.; Omenetto, F.G.; Kaplan, D.L. In vivo bioresponses to silk proteins. Biomaterials **2015**, 71, 145–157. [CrossRef]
9. Zhang, L.; Zhang, W.; Hu, Y.; Fei, Y.; Liu, H.; Huang, Z.; Wang, C.; Ruan, D.; Heng, B.C.; Chen, W.; et al. Systematic Review of Silk Scaffolds in Musculoskeletal Tissue Engineering Applications in the Recent Decade. ACS Biomater. Sci. Eng. **2021**, 7, 817–840. [CrossRef]
10. Alessandrino, A.; Fregnan, F.; Biagiotti, M.; Muratori, L.; Bassani, G.A.; Ronchi, G.; Vincoli, V.; Pierimarchi, P.; Geuna, S.; Freddi, G. SILKBridge: A novel biomimetic and biocompatible silk-based nerve conduit. Biomater. Sci. **2019**, 7, 4112–4130. [CrossRef] [PubMed]
11. Fregnan, F.; Muratori, L.; Bassani, G.A.; Crosio, A.; Biagiotti, M.; Vincoli, V.T.; Pierimarchi, P.; Geuna, S.; Alessandrino, A.; Freddi, G.; et al. Preclinical validation of SILKBridge for peripheral nerve regeneration. Front. Bioeng. Biotechnol. **2020**, 8, 835. [CrossRef] [PubMed]
12. Biggi, S.; Bassani, G.A.; Vincoli, V.; Peroni, D.; Bonaldo, V.; Biagiotti, M.; Belli, R.; Alessandrino, A.; Biasini, E.; Freddi, G. Characterization of Physical, Mechanical, and Biological Properties of SILKBridge Nerve Conduit after Enzymatic Hydrolysis. ACS Appl. Bio Mater. **2020**, 3, 8361–8374. [CrossRef] [PubMed]

13. Alessandrino, A. Process for the Production of a Hybrid Structure Consisting of Coupled Silk Fibroin Microfibers and Nanofibers, Hybrid Structure thus Obtained and Its Use as Implantable Medical Device. U.S. Patent 10,758,641, 1 September 2020.
14. Fossey, S.A.; Kaplan, D.L. Silk protein. In Polymer Data Handbook; Mark, J.E., Ed.; Oxford University Press, Inc.: Oxford, UK, 1999; pp. 970–974.
15. Bath, N.V.; Nadiger, G.S. Crystallinity in silk fibers: Partial acid hydrolysis and related studies. J. Appl. Polym. Sci. **1980**, 25, 921–932. [CrossRef]
16. Ortiz, R.; Westenberg, R.F.; Langhammer, C.G.; Knaus, W.J.; Chen, N.C.; Eberlin, K.R. Nerve Diameter in the Hand: A Cadaveric Study. Plast. Reconstr. Surg. Glob. Open **2019**, 7, e2155. [CrossRef]
17. Nectow, A.R.; Marra, K.G.; Kaplan, D.L. Biomaterials for the Development of Peripheral Nerve Guidance Conduits. Tissue Eng. Part B **2012**, 18, 40–50. [CrossRef]
18. Chiono, V.; Tonda-Turo, C. Trends in the design of nerve guidance channels in peripheral nerve tissue engineering. Prog. Neurobiol. **2015**, 131, 87–104. [CrossRef]
19. Lauren, E.; Kokai, L.E.; Lin, Y.C.; Oyster, N.M.; Marra, K.G. Diffusion of soluble factors through degradable polymer nerve guides: Controlling manufacturing parameters. Acta Biomater. **2009**, 5, 2540–2550. [CrossRef]
20. Boulet-Audet, M.; Vollrath, F.; Chris Holland, C. Identification and classification of silks using infrared spectroscopy. J. Exp. Biol. 2015, 218, 3138–3149. [CrossRef]
21. Magoshi, J.; Nakamura, S. Studies on Physical Properties and Structure of Silk. Glass Transition and Crystallization of Silk Fibroin. J. Appl. Polym. Sci. **1975**, 19, 1013–1015. [CrossRef]
22. Dumont, C.E.; Born, W. Stimulation of Neurite Outgrowth in a Human Nerve Scaffold Designed for Peripheral Nerve Reconstruc- tion. J. Biomed. Mater. Res. Part B Appl. Biomater. **2005**, 73, 194–202. [CrossRef]
23. Billiar, K.; Murray, J.; Laude, D.; Abraham, G.; Bachrach, N. Effects of carbodiimide crosslinking conditions on the physical properties of laminated intestinal submucosa. J. Biomed. Mater. Res. **2001**, 56, 101–108. [CrossRef]
24. Mine, Y.; Mitsui, H.; Oshima, Y.; Noishiki, Y.; Nakai, M.; Sano, S. Suture Retention Strength of Expanded Polytetrafluoroethylene (ePTFE) Graft. Acta Med. Okayama **2010**, 64, 121–128. [CrossRef]
25. Pensalfini, M.; Meneghello, S.; Lintas, V.; Bircher, K.; Ehret, A.E.; Mazza, E. The suture retention test revisited and revised. J. Mech. Behav. Biomed. Mater. **2018**, 77, 711–717. [CrossRef]

26. Topp, K.S.; Boyd, B.S. Structure and biomechanics of peripheral nerves: Nerve responses to physical stresses and implications for physical therapist practice. Phys. Ther. 2006, 86, 92–109. [CrossRef]

14 Chapter

TRANSCRIPTOMIC ANALYSIS UNCOVERS GENE EXPRESSION ALTERATIONS IN THE SILKWORM (BOMBYX MORI) FOLLOWING HYDROGEN SULFIDE EXPOSURE.

1. INTRODUCTION: AN OVERVIEW

The silkworm, Bombyx mori, serves as a crucial model organism within the Lepidoptera order, a group of holometabolous insects, and holds significant economic value globally. The silkworm is regarded as a model organism due to its suitable life cycle, cultivation, and rapid reproduction. The fat body in insects functions as both a dynamic tissue and a metabolic organ, primarily engaged in synthesis, nutrient storage, and energy metabolism. It has been extensively examined in numerous studies of silkworms to elucidate various biological, physiological, and biochemical processes. The overexpression of BmFoxO in the fat body may inhibit protein translation in transgenic silkworms [4].

The detrimental consequences of hydrogen sulfide (H2S) as an environmental contaminant have been investigated in recent years. Currently, the unusual physiological role of H2S has garnered interest as a signaling molecule. H2S has essential regulatory roles in development, the cardiovascular system, oxidative stress, and inflammation across several organisms [5,6]. Hydrogen sulfide (H2S) has been recognized as the third signaling gas, following nitric oxide (NO) and carbon monoxide (CO), in recent years, having uses in numerous insect species. Miller and Roth (2007) discovered that high concentrations of H2S are harmful to Caenorhabditis worms; nevertheless, exposure to low concentrations of H2S enhanced thermotolerance and lifespan via elevating the activity of SIR-2.1 [7].

Budde and Roth (2010) discovered that HIF-1 is essential when C. elegans are subjected to H2S; consequently, H2S can elevate both the concentration of HIF-1 protein and its nuclear location [8]. H2S serves as an intrinsic regulator of oxidative damage, metabolism, and aging in C. elegans [9]. Wei and Kenyon (2016) discovered that the elimination of germ cells in C. elegans induces an increase in hydrogen sulfide levels in nonreproductive somatic tissues. Hydrogen sulfide exhibited defensive responses that decelerate aging [10]. Treatment with H2S markedly enhanced the survival and longevity of Drosophila melanogaster in arid and food-deprived environments [11,12].

Our prior research indicated that continuous treatment with 7.5 μM of H2S enhanced the developmental (body weight) and economic aspects (cocoon weight, cocoon shell ratio, and cocoon shell weight) of silkworms (Bombyx mori). Following exposure to H2S, the concentrations of inflammation-related metabolites ((6Z,9Z,12Z)-octadecatrienoic acid, hexadecanoic acid, choline phosphate, and malic acid, among others) in the hemolymph of silkworms were markedly elevated in comparison to the control group [13]. Nonetheless, when H2S concentrations above 7.5 μM, it exhibited harmful effects on the silkworm, resulting in reductions in body weight, cocoon weight, cocoon shell weight, and other parameters. RNA-seq technique is extensively employed in silkworm research to examine physiological and biochemical alterations at the transcriptional level under various conditions [14]. Nevertheless, there is an absence of a comprehensive report regarding the impact of H2S on silkworms at the transcriptional level.

This study aims to examine the expression levels of genes in the fat body of silkworms following exposure to H2S. The findings from the transcriptome and bioinformatic analysis revealed markedly distinct responses to H2S and offered valuable insights into the role of H2S in silkworms.

2. MATERIALS AND METHODS

2.1 Silkworm Rearing and H2S Exposure

P50 strains of silkworm were obtained from the Sericulture Research Institute of the Chinese Academy of Agricultural Sciences, Zhenjiang, China. According to the previous study [13], from the 4th instar larvae to the 5th instar larvae, silkworms were divided into two groups (H_2S-treated and control) and reared in desiccators using fumigating treatment at 25 ± 1°C and 12L:12D photoperiod conditions. Silkworms in the H_2S-treated group were treated by NaHS (Shandong West Asia Chemical Company, Jinan, China), a donor of H_2S. The experimental concentration of H_2S was set to 7.5 μM according to the report of Cao et al. (2020) [13]. There were triplicates in each group with 30 silkworms per replicate and the silkworms were fed abundant fresh mulberry leaves twice a day.

2.2 RNA Extraction

The total RNA of the silkworm fat body was extracted on ice from the 1st day (5L1D) to the 5th day (5L5D) at the 5th instar larvae using Trizol reagent (Invitrogen, CA, USA). The concentration and purity of total RNA were analyzed by NanoDrop ND-1000 (Nan- oDrop, Wilmington, DE, USA) and the ratios of A260/A280 and 260/230 were used to measure the quality of total RNA.

2.3 Sequencing, Data Processing, and Assembly

The total RNA of 5L3D was analyzed in the H_2S-treated and control groups by RNA-Sequencing. Total RNA (10 μg) of each sample was subjected to Poly(A) mRNA isolation with poly-T oligo-attached magnetic beads (Invitrogen, CA, USA). The poly(A)RNA was fragmented into small pieces using a Magnesium RNA Fragmentation Module (New Eng- land Biolabs (Beijing) LTD., cat. e6150, Beijing, China) under 94°C for 5–7 min. The cleaved RNA fragments were then reverse-transcribed to create the cDNA by SuperScript™ II Reverse Transcriptase (Invitrogen, CA, USA), which were next used to synthesize U-labeled second-stranded DNAs. An A-base was then added to the blunt ends of each strand, preparing them for ligation to the indexed adapters. Each adapter contained a T-base overhang for ligating the adapter to the A-tailed fragmented DNA. Single- or dual-index adapters were ligated to the fragments, and size selection was performed with AMPureXP beads. After the heat-labile UDG enzyme (New England Biolabs (Beijing) LTD., cat.m0280, Beijing, China) treatment of the U-labeled second-stranded DNAs, the ligated products were amplified with PCR by the following conditions: initial denaturation at 95°C for 3 min; 8 cycles of denaturation at 98°C for 15 s, annealing at 60°C for 15 s, and extension at 72°C for 30 s; and final extension at 72°C for 5 min. The average insert size for the final cDNA library was 300 ± 50 bp. Lastly, we performed the 2 × 150 bp paired-end sequencing (PE150) on an Illumina Novaseq™ 6000 (LC-Bio Technology CO., Ltd., Hangzhou, China) following the vendor's recommended protocol.

The transcriptome was sequenced on basis of the Illumina paired-end RNA-seq approach. The average insert size for the paired-end libraries was 300 bp (± 50 bp). The low-quality reads with sequencing adaptors, sequencing primer, and nucleotide (q quality score (*p* value corrected by BH algorithm, also called FDR value, *q* value, *p* adj value) <20) were removed using Cutadapt software (https://cutadapt.readthedocs.io/en/stable/, accessed on 1 August 2021) [15] before assembly. We used HISAT2 software (https:// daehwankimlab.github.io/hisat2/,version:hisat2-2.0.4, accessed on 1 August 2021) [16] to map reads to the genome. Three important parameters, such as Q20 (the percentage of bases with mass value 20), Q30 (mass value 30), and GC content were verified to evaluate all the reads using the FastQC online tool [17].

The *Bombyx mori* reference genome [18] was used for the alignment of the sample reads using the HISAT package [19]. StringTie (http://ccb.jhu.edu/software/stringtie/, accessed on 1 August 2021) was used for the assembly of the mapped reads of each sample and a comprehensive transcriptome was constructed combining all transcriptomes using Perl scripts [20]. The expression level of mRNAs was used to calculate the fragments per kilobase of exon model per million mapped reads (FPKM) by StringTie (http://www. bioconductor.org/packages/release/bioc/html/ballgown.html, accessed on 1 August 2021) [20]. The FPKM could evaluate the abundance of genes. The DEGs were selected with $\log_2$ (fold change) >1 or $\log_2$ (fold change) < 1 at p-value < 0.05, and q-value > p-value by R package-edgeR (https://bioconductor.org/packages/release/bioc/html/edgeR.html, accessed on 1 August 2021) [21].

2.4 Functional Annotation and Enrichment Analysis of DEGs

The reference sequences for all the silkworm genes were obtained from the NCBI database and BLASTX was used for annotation. All the sequences were aligned against Gene Ontology (GO) and Kyoto Encyclopedia of Genes and Genomes (KEGG) databases using DIAMOND (0.7.12) [22]. GO function or KEGG pathway significant enrichment analysis first mapped all the DEGs to each GO term in the GO database or each KEGG pathway in the KEGG database, and second calculated the number of genes in each GO term or KEGG pathway and finally applied a hypergeometric test to find a GO term or KEGG pathway that were significantly enriched in DEGs compared to the entire genome background.

2.5 Quantitative Real-Time PCR (qRT-PCR)

The qRT-PCR was used to validate the accuracy of DEGs in L5D3 at the transcriptome level and the expression pattern of DEGs in the 5th instar larvae was investigated. Reverse transcription was performed by HiScript II QRT SuperMix (Vazyme, Nanjing, China) as follows: total RNA (1 μg) (without genome contamination) and 4 × gDNA Wiper were mixed with 5 × HiScript II QRT SuperMix and then incubated at 25°C for 10 min, 50°C for 15 min, and 85°C for 2 min [14]. The cDNA was used for real-time PCR detection using the UltraSYBR Mixture (Cwbio, Beijing, China) and qRT-PCR was executed using the Light Cycler® 480 System (Roche, Basel, Switzerland). The primer sequences for target genes were listed in Table S1 using primer 3.0 software (Premier Biosoft International, Palo Alto, CA, USA). The expression level of action A3, a silkworm housekeeping gene, was regarded as an internal reference in standardization [14]. Before Quantitative real-time PCR, the efficiency and specificity of primers were confirmed. The volume of the reaction system was 20 μL, containing 10 μL UltraSYBR Mixture, 1 μL cDNA, 0.5 μL forward and reverse primers and 8 μL ddH_2O. The qRT-PCR program was as follows: initial denaturation at 95°C for 10 min; followed by 40 times of denaturation at 95°C for 10 s, annealing at 60°C for 30 s, and extension at 72°C for 32 s; a melting curve program was set at 95°C for 15 s, 60°C for 1 min, 95°C for 15 s, and a60°C for 15 s with a final cooling step at 4°C for 30 s [23]. The data were analyzed with the LightCycler®96 software (Roche Diagnostics, Indianapolis, IN, USA) using the $2^{-\Delta\Delta Ct}$ method.

2.6 Statistical Analysis

The data were analyzed using one-way ANOVA data analysis. All the data were obtained in triplicate and presented as mean ± standard error (SE) ($n = 3$). Statistical significance at $p < 0.05$ was measured using SPSS 20.0 software (IBM, Endicott, NY, USA).

3. RESULTS

3.1 Transcriptome Sequencing and Assembly

The deep sequencing of RNA from the fat body (5L3D) treated with H_2S using the illumine sequencing and the data processing results were shown in Table 1. The raw data of H_2S-treated and control groups were approximately 7.75 GB and 7.66 GB, respectively. After filtering and removing the low-quality reads, adaptors and poly-N stretches, the valid data and valid reads ratio of each replicate in H_2S-treated group were obtained as follows: 7.55 Gb (99.12%), 7.87 Gb (98.91%), and 7.54 Gb (98.39%) and that of the control group were 7.42 Gb (94.00%), 7.37 Gb (98.99%), and 7.50 Gb (98.12%). The quality scores of Q30 exceeded 98% in each sample. Moreover, the mapped ratio in H_2S-treated and control groups were 97.35%, 97.43%, and 97.72% and 93.19%, 97.35%, and 97.52%, respectively, which highly matched with the silkworm genome (Table S2). The expression interval of genes was mainly focused on the 0.3~3.57 FPKM interval (FI) and the ratio was about 35% gene of the control group and 39% of H_2S-treated group (Table S3). These results revealed a good quality sequencing analysis and the genes were analyzed in the following section.

Table 1. The sequencing data of samples.

Samples	FB_ Control_1	FB_ Control_2	FB_ Control_3	FB_H_2S_1	FB_H_2S_2	FB_H_2S_3
Raw reads	52,598,692 (7.89 Gb)	49,647,104 (7.45 Gb)	50,987,488 (7.65 Gb)	50,804,424 (7.62 Gb)	53,068,042 (7.96 Gb)	51,108,946 (7.67 Gb)
High quality reads	49,443,624 (7.42 Gb)	49,144,228 (7.37 Gb)	50,029,926 (7.50 Gb)	50,357,636 (7.55 Gb)	52,490,302 (7.87 Gb)	50,286,178 (7.54 Gb)
High quality reads ratio (%)	94.00	98.99	98.12	99.12	98.91	98.39
Q30 (%)	98.53	99.08	99.07	98.93	98.98	99.12
GC content (%)	50	48	49	48	48.50	48

Total mapped	46,077,049	47,843,354	48,788,614	49,023,730	51,140,817	49,141,008

3.2 DEG Identification and Analysis

DEGs in H_2S and control groups were calculated according to the FPKM value of the obtained genes and then a large number of genes were differentially expressed after H_2S exposure. The volcano plot of DEGs was shown in Figure 1 and the detailed information of DEGs was listed in Table S4. A total of 1200 DEGs were obtained between the H_2S-treated and control groups. The number of up-regulated and down-regulated DEGs was 977 and 223, and the up-regulated genes were higher than down-regulated genes.

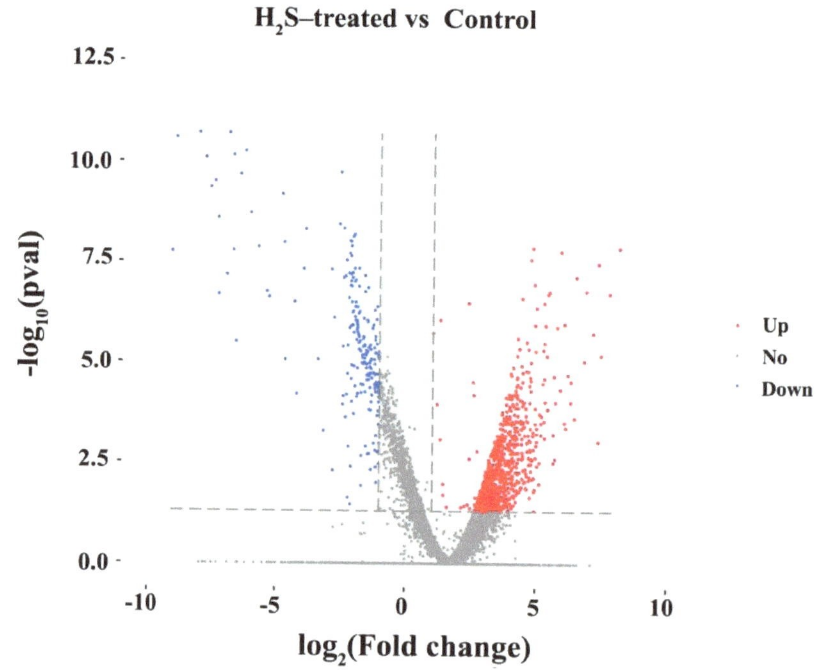

Figure 1. The volcano plot of DEGs. The red points represent the up-regulated genes, the blue points represent the down-regulated genes, and the grey points represent the genes without differentia expression.

3.3 GO Analysis of DEGs

The GO database can provide standard terms for describing the comprehensive properties of genes in organisms as an international classification system. In this study, DEGs were assigned to different GO categories such as biological processes, cellular components, and molecular functions. The major subcategories of DEGs for biological processes were border follicle cell migration, mitotic spindle elongation, myoblast fusion and microtubule-based processes, as well as others. (Figure 2 and Table S5). Through the results of the GO analysis, it was revealed that H_2S could influence these multiple biological processes by regulating the expression of genes in the fat body of silkworms at the molecular level.

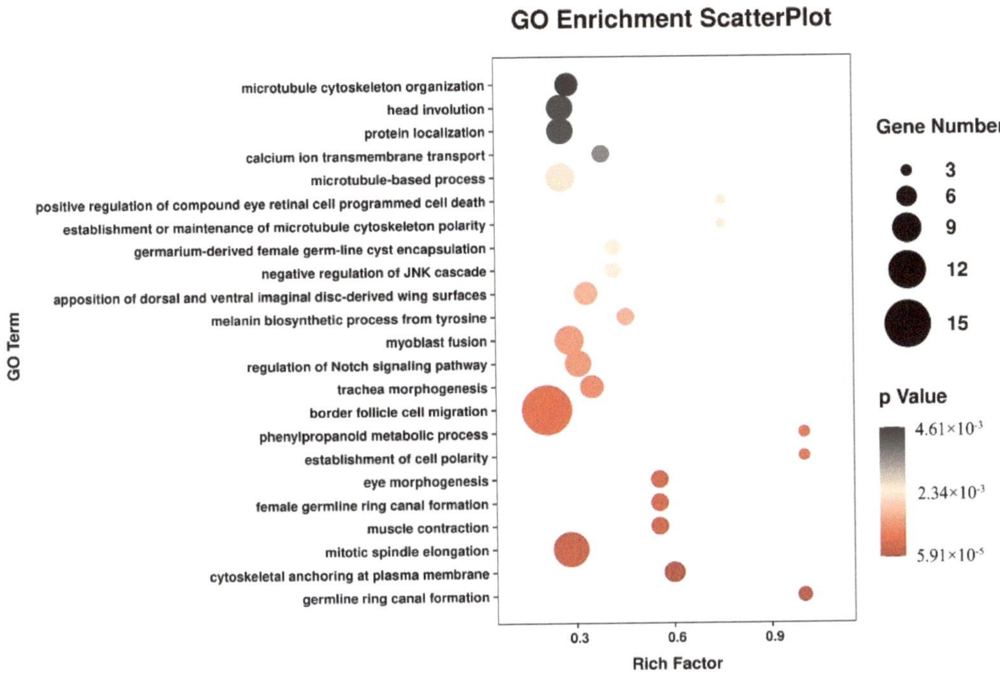

Figure 2. GO classification of DEGs ($p \le 0.01$).

3.4 KEGG Pathway Analysis

To systematically analyze the biological function and metabolic pathways of DEGs, KEGG pathway annotations were obtained. The 561 genes of 1200 DEGs predicted to encode enzymes were mapped into 225 KEGG pathways. These annotated genes were as- signed to six categories including organismal systems (49 pathways), metabolism (74 pathways), human diseases (39 pathways), genetic information processing (19 pathways), environmental information processing (28 pathways), and cellular processes (16 pathways) (Figure 3). In the organismal systems, vascular smooth muscle contraction (7 DEGs) is the most abundant metabolic pathway followed by the insulin signaling pathway (6 DEGs) and toll-like receptor signaling pathway (5 DEGs). The most abundant metabolic cat- egory in metabolism is inositol phosphate metabolism (12 DEGs) followed by glycoly- sis/gluconeogenesis (11 DEGs), glutathione metabolism (10 DEGs), and the citrate cycle (9 DEGs). In human disease, it is mainly related to insulin resistance (17 DEGs). In the category of genetic information processing, it is mainly enriched in ribosomes (33 DEGs) closely related to synthetic genetic material. In environmental information processing, the most abundant pathways are the Hippo signaling pathway (19 DEGs), MAPK signaling pathway (16 DEGs), and phosphatidylinositol signaling system (12 DEGs). In cellular processes, 36 DEGs were mainly enriched in endocytosis followed by focal adhesion (20 DEGs) and regulation of the actin cytoskeleton (15 DEGs).

A significant enrichment analysis was further performed on the metabolic pathways of DEGs, and a total of 26 significantly enriched metabolic pathways were obtained (Table 2). The terms were enriched to multiple molecular pathways such as the transport path-way and cellular community, including endocytosis, focal adhesion, regulation of the actin cytoskeleton, tight junction, and adherens junction; amino acid and carbohydrate metabolisms, such as glycolysis/gluconeogenesis, the citrate cycle (TCA cycle), phenylalanine metabolism, and histidine metabolism; immune-associated signal transduction such VEGF signaling pathway, the TNF signaling pathway, NOD-like receptor signaling, the MAPK signaling pathway, the Hippo signaling pathway-fly, the Toll-like receptor signaling pathway, and the NF-κB signaling pathway (Table S6). The top 10 metabolic pathways affected by H_2S were endocytosis, ribosome, focal adhesion, the Hippo signaling pathway, the MAPK signaling pathway, regulation of the actin cytoskeleton, tight junction, inositol phosphate metabolism, glycolysis/gluconeogenesis, and the TCA cycle (Table 2). These annotations indicated that H_2S could influence the multiple biological pathways in the fat body of silkworms and provide a new perspective to the effects of H_2Son silkworms.

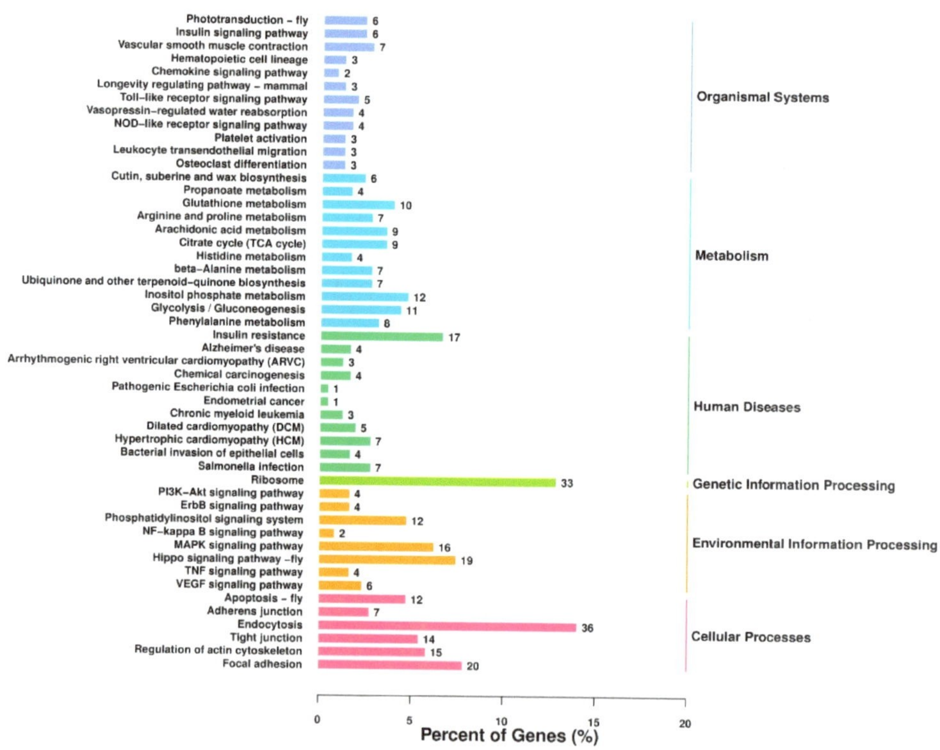

Figure 3. The KEGG classification of DEGs. The x-axis represents the number of genes annotated into the pathway and the proportion of the number of DEGs annotated to the total number of genes. The y-axis represents the name of the enriched KEGG pathways.

Table 2. The main enriched KEGG pathways ($p < 0.05$) in the H_2S-treated and the control groups.

Pathway ID	Pathway Name	Number of DEGs	p-Value
ko03010	Ribosome	33	<0.01
ko04510	Focal adhesion	20	<0.01
ko05132	Salmonella infection	7	<0.01
ko04370	VEGF signaling pathway	6	<0.01
ko04810	Regulation of actin cytoskeleton	15	<0.01
ko00360	Phenylalanine metabolism	7	<0.01
ko04530	Tight junction	13	<0.01
ko04668	TNF signaling pathway	4	0.01
ko00010	Glycolysis/Gluconeogenesis	11	0.01
ko04144	Endocytosis	35	0.01
ko00562	Inositol phosphate metabolism	12	0.01
ko00130	Ubiquinone and other terpenoid-quinone biosynthesis	6	0.01
ko04380	Osteoclast differentiation	3	0.01
ko04670	Leukocyte transendothelial migration	3	0.01
ko00410	beta-Alanine metabolism	7	0.01
ko05100	Bacterial invasion of epithelial cells	4	0.02
ko04611	Platelet activation	3	0.02
ko04520	Adherens junction	7	0.02
ko04621	NOD-like receptor signaling pathway	4	0.02
ko04391	Hippo signaling pathway-fly	19	0.02
ko04010	MAPK signaling pathway	16	0.03
ko00340	Histidine metabolism	4	0.04
ko05410	Hypertrophic cardiomyopathy (HCM)	6	0.04
ko04064	NF-kappa B signaling pathway	2	0.04
ko00020	Citrate cycle (TCA cycle)	9	0.04
ko04962	Vasopressin-regulated water reabsorption	4	0.04

3.5 The Validation of DEGs by qRT-PCR

To validate the accuracy of DEGs obtained from the transcriptome between the H_2S-treated and control groups, 28 candidate genes were selected for qRT-PCR analysis. The selected DEGs were mainly annotated to glycolysis/gluconeogenesis, the TCA cycle, focal adhesion, endocytosis, adherens junction, the NF-κB signaling pathway, the MAPK signaling pathway, and the synthesis of fibroin. Good consistency between the results of qRT-PCR and the transcriptome confirmed the accuracy and reliability of the sequencing data and revealed the significant difference of these genes during H_2S exposure (Figure 4 and Table S3).

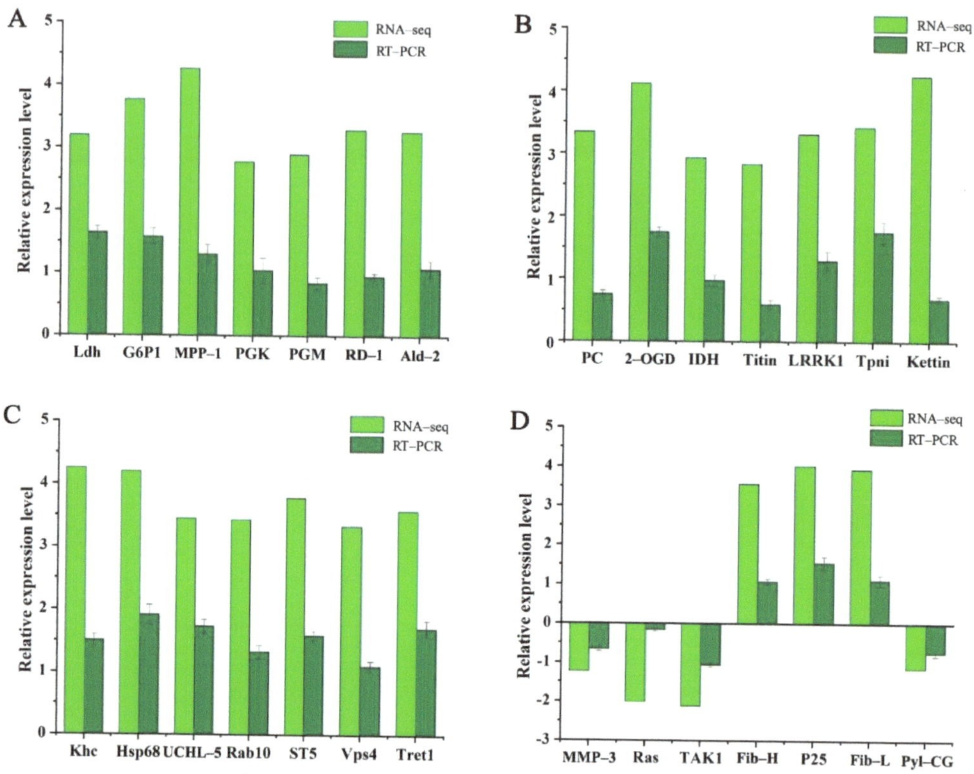

Figure 4. The qRT-PCR validation for DEGs expression after exposure to H_2S. DGEs related to the glycolysis pathway (A), the TCA cycle (B), endocytosis (C), and inflammation and silk protein synthesis (D) were analyzed. The full names of abbreviations of genes were listed in Table S1.

3.6 The Expression Pattern Analysis of DEGs

A total of 11 DEGs was representatively selected to analyze the expression level pat- tern in the 5th instar larvae using qRT-PCR. Genes (relating to the glycolytic pathway and TCA cycle) such as lactate phosphoglycerate kinase (PGK), phosphoglucomutase (PGM), NADP-dependent isocitrate dehydrogenase (IDH), and pyruvate carboxylase (PC) in the H_2S-treated group were up-regulated compared to control group (Figure 4A,B). Mean- while, the levels of these genes were significantly increased at the 5L3D, 5L4D, and 5L5D comparing with 5L1D in the H_2S group (Figure 5A–D). In addition, the expression levels of the members of the Rab family of small GTPases (Rab10) and trehalose transporter 1 (Tret1) in endocytosis were up-regulated after H_2S treatment (Figure 4C). The expression levels of Rab10 and Tret1 were significantly increased in a time-dependent manner (Figure 5E,F). On the contrary, transforming growth factor-β activated kinase1 (Tak1) and matrix metalloproteinase-3 (MMP-3) were significantly decreased in the 5th instar larvae after H_2S exposure (Figure 5G,H). Moreover, the increased expression of fibroin synthesis genes, such as the fibroin heavy chain (Fib-H), fibroin light chain (Fib-L), and glycoprotein P25 (P25) was observed during the 5th larvae stage after H_2S exposure (Figure 4D). At the end of the 5th instar larvae, the expression levels were higher than the early 5th instar larvae (Figure 5I–K).

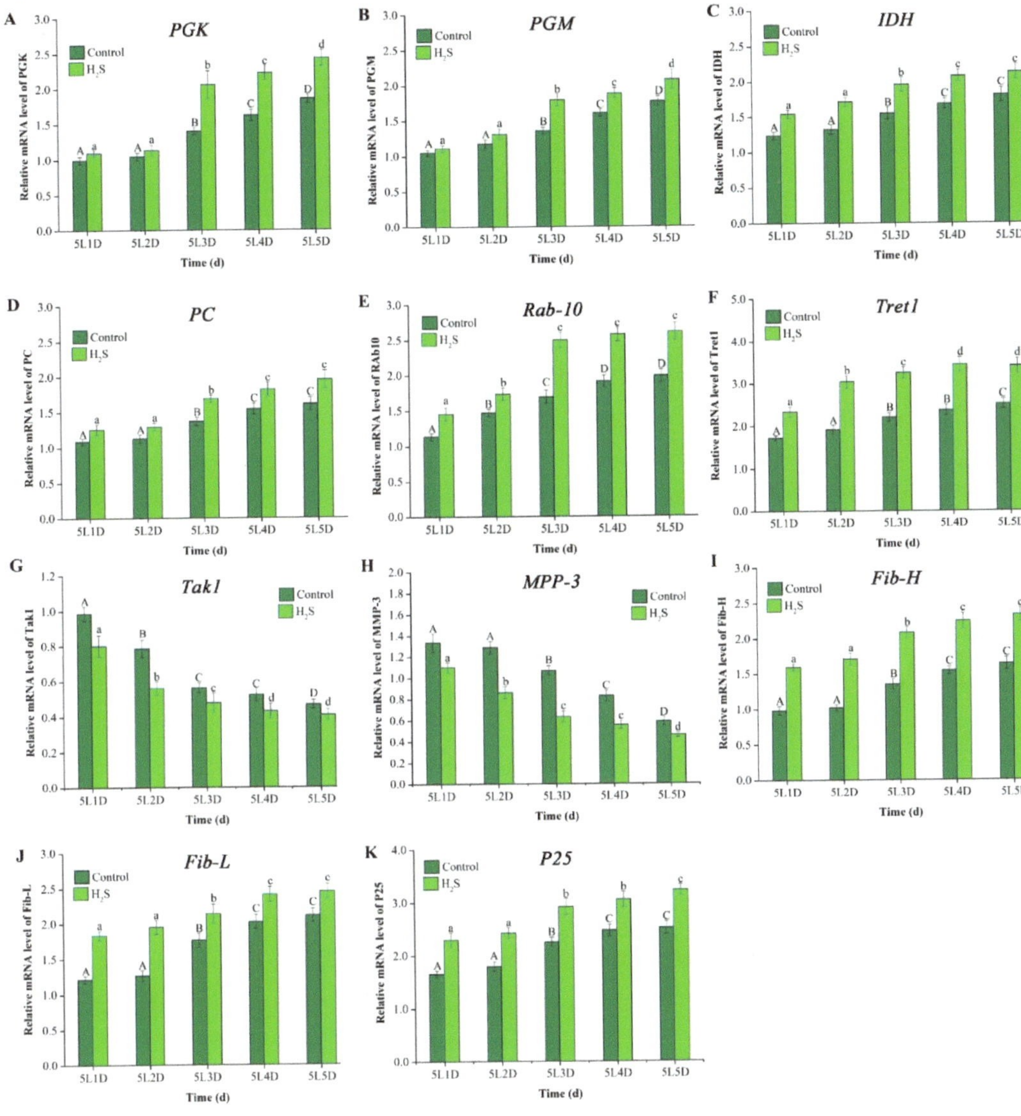

Figure 5. Expression changes of 10 DEGs after H2S exposure at different time points in the 5th instar larvae (5L1D, 5L2D, 5L3D, 5L4D, and 5L5D). (A) PGK; (B) PGM; (C) IDH; (D) PC; (E) Rab-10; (F) Tret1; (G) Tak1; (H) MPP-3; (I) Fib-H; (J) Fib-L; and (K) P25. The lowercase letters (a, b, c, and d) represent the significant difference at $p \leq 0.05$ among the expression levels at different times of the H2S treated group; the uppercase letters (A, B, C, and D) represent the significant difference at $p \leq 0.05$ among the expression levels at different times of the control group. The full names of abbreviations of genes are listed in Table S1.

4. Discussion

H_2S has been accepted as the third transmission gas followed by nitric oxide (NO) and carbon monoxide (CO) in recent years with applications to various insects, such as *C. elegans* [7–10], *D. melanogaster* [11,12], and *B. mori* [13]. RNA-seq technology is widely used in the field of silkworms to investigate the physiological and biochemical changes at the transcriptional level under diverse conditions [4,14,24]. In our previous research, it was found that the development (body weight) and economic characteristics (cocoon weight, cocoon shell ratio, and cocoon shell weight) were increased after continuous treatment with 7.5 μM of H_2S. However, there is no effective report for the effects of H_2S on the silkworm at the transcriptional level [13]. In the present study, the expression levels of genes in the fat body of silkworms after exposure to H_2S was investigated using transcriptomic and bioinformatic analysis. The results showed significantly different responses to H_2S and provided useful insights into the role of H_2S on silkworms.

Using transcriptome, a total of 1200 DEGs were identified in the fat body of silkworms at the 5th instar larvae. After treatment with H_2S, up-regulated DEGs (977) were higher than the down-regulated DEGs (223). These DEGs were mainly involved in glycol- ysis/gluconeogenesis, endocytosis, and the TCA cycle. The carbohydrate metabolism is one of the important energy metabolism pathways in organisms, incorporating glycoly- sis/gluconeogenesis, the TCA cycle, oxidative phosphorylation, and the pentose phosphate pathway [25]. Energy metabolism, especially the metabolic control of carbohydrates, is essential for insect growth and development [26]. Glucose is oxidized to pyruvate in glycolysis/gluconeogenesis [25]. Glycolytic activity is regulated by many enzymes in the glycolytic pathway such as PGK and PGM. PGM can convert glucose 1 phosphate into glucose 6 phosphate, providing the raw materials for the glycolytic pathway [27]. PGK is commonly known as a soluble and membrane-bound protein in different organ- isms [27]. In one of the glycolytic processes, phosphorylation of 1,3-diphosphoglycerate to 3-phosphoglycerate is catalyzed by PGK [28]. Like glycolysis/gluconeogenesis, the TCA cycle also provides the energy and intermediates for diverse biosynthetic pathways [25]. IDH, one of three rate-limiting enzymes in the TCA cycle, catalyzes the TCA cycle reaction wherein isocitrate is oxidized and decarboxylated to produce α-ketoglutarate, NADH, and CO_2 [29]. PC catalyzes the carboxylation of pyruvate into oxaloacetate for gluconeogenesis, the urea cycle, lipid synthesis, and other pathways which are the primary anaplerotic pathways for the TCA cycle where the intermediates are replenished by alanine and lactate [30]. In this study, the results showed that the levels of PGM, PGK, IDH, and PC were up-regulated after H_2S exposure. According to the expression pattern analysis of these genes, gene levels had gradually increased in a time-dependent manner. Especially for 5L4D and 5L5D, a significant difference was observed compared to 5L1D. These results indicated that the expression levels of the genes involved in carbohydrate metabolism were increased with H_2S exposure, suggesting H_2S can accelerate the carbohydrate metabolism of silkworms and provide much energy for silkworm metamorphosis.

The uptake and sorting of eukaryotes and the following recycling or degradation of fluid, membranes, membrane proteins, and macromolecules are regulated by endocytosis [31]. Endocytosis often depends on clathrin and occurs at clathrin-coated pits [32]. Utilizing endocytosis with clathrin-dependent or clathrin-independent uptake mechanisms, cargo proteins and adaptor molecules are delivered to early endosomes and sorted [33]. The Rab family of small GTPases can mediate membrane the trafficking processes of intracellular compartmentation, generation, and maintenance [34]. Rab10, which is closely related to clathrin-independent endocytosis is largely expressed in many model animals directing the trafficking of proteins. In *Caenorhabditis elegans*, Rab10 was required in interneuron postsynaptic glutamate receptor recycling and the transport between early endosomes and recycling endosomes [35]. Rab10 in the H_2S-treated group was up-regulated compared to the control group, which indicated that the H_2S-treated silkworms were more active in supporting endocytosis for the body. Tret1 participated in the transfer of trehalose, the major hemolymph sugar in most of the insect, which is synthesized in the fat body and released into the hemolymph. The transportation of trehalose with the up-regulated expression level of Tret1 after exposure to TiO_2 nanoparticles was accelerated in the silkworm, showing that TiO_2 nanoparticles could facilitate carbohydrate and the nutrition metabolism [36]. In this study, the level of Tret1 was up-regulated with the H_2S supplement to accelerate the transport rate of trehalose. Moreover, the expression levels of Rab10 and Tret1 were increased in a time-dependent manner during the 5th larvae. These results showed that H_2S could accelerate the transmission of mass in the silkworm fat body via up-regulating the expressions of Rab10 and Tret1 to provide material for larva–pupae metamorphosis.

Previous studies reported that H_2S exerted anti-inflammatory effects on T lymphocytes. Rats injected with NaHS inhibited inflammatory processes such as leukocyte infiltration and adherence to the vascular endothelium [37]. In this study, DEGs were annotated to various pathways related to inflammation by KEGG analysis, including the TNF signaling pathway, the MAPK signaling pathway, and the NF-κB signaling pathway.

Tak1 is a serine/threonine kinase and a member of the mitogen-activated protein kinase kinase (MAP3K) family which is activated by receptors such as TGF-β, TNF-α, and IL-1. MMP-3 is a matrix-soluble protein containing the binding site of NF-κB and can participate in the degradation of ECM and activate the precursors of MMPs. The content of MMP-3 was increased in the mouse arthritis model and the administration of baicalin can reduce the gene expression level of MMPs and alleviate inflammatory processes [38]. After H_2S exposure, levels of Tak1 and MMP-3 were down-regulated compared to the control group. Moreover, the expression levels of Tak1 and MMP-3 at the end of 5th instar larvae were significantly decreased compared to the early 5th instar larvae. These results indicated that H_2S might have an anti-inflammatory role in silkworm growth.

In addition, significant changes in the genes related to fibroin synthesis were observed in this study. The domestic silkworm has tremendous economic value and the most important manifestation of its value is silk production [39]. The function of the silk gland is to synthesize silk protein and during the 5th instar larvae it is the biggest organ of *B. mori* [39]. It was reported that Fib-H, Fib-L, and P25 encoding fibroin were expressed in all the larval instars. However, their highest expression was noticed before the last molting period [40]. The expression levels of Fib-L, P25, and Fib-H were up-regulated in response to H_2S compared to the control group. In addition, the expression of Fib-H at the end of the 5th instar larvae (5L4D, 5L5D) was rapidly increased and higher than on the other days. Similarly, the expression levels of P25 and Fib-L were also significantly increased. These results indicated that H_2S could promote the fibroin synthesis by up-regulating the expression levels of Fib-L, P25, and Fib-H, which was beneficial to silk production. These results were similar to the previous study which reported the up-regulated expression of Fib-L, P25, and Fib-H after treatment with titanium dioxide nanoparticles (TiO_2NPs) [41].

5. CONCISE SUMMARY

This article represents the inaugural investigation of the silkworm fat body at the transcriptional level following H2S exposure, utilizing transcriptome analysis during the fifth larval stage. Transcriptome analysis identified multiple genes and signaling pathways, with the majority of differentially expressed genes (DEGs) exhibiting upregulation relative to the control group. The expression of essential genes associated with energy metabolism, transport mechanisms, and fibroin synthesis in silkworms was upregulated following H2S exposure; in contrast, genes implicated in inflammatory processes were downregulated, suggesting that H2S may enhance biological processes related to energy and nutrient metabolism while inhibiting inflammatory responses. These findings offer a new perspective on understanding the molecular mechanisms that govern the effects of H2S in the silkworm fat body.

REFERENCES

1. Goldsmith, M.R.; Shimada, T.; Abe, H. The genetics and genomics of the silkworm, *Bombyx mori*. *Annu. Rev. Entomol.* **2005**, *50*, 71–100. [CrossRef] [PubMed]
2. Liu, L.; Zhang, P.; Gao, Q.; Feng, X.; Han, L.; Zhang, F.; Bai, Y.; Han, M.; Hu, H.; Dai, F.; et al. Comparative transcriptome analysis reveals bmo-miR-6497-3p regulate circadian clock genes during the embryonic diapause induction process in bivoltine silkworm. *Insects* **2021**, *12*, 739. [CrossRef] [PubMed]
3. Zhu, Z.; Tan, Y.; Xiao, S.; Guan, Z.; Zhao, W.; Dai, Z.; Liu, G.; Zhang, Z. Solitary living brings a decreased weight and an increased agility to the domestic silkworm, Bombyx mori. *Insects* **2021**, *12*, 809. [CrossRef] [PubMed]
4. Lu, Z.Y.; Meng, Z.; Wen, M.Y.; Kang, X.; Zhang, Y.; Liu, Q.; Zhao, P.; Xia, Q. Overexpression of BmFoxO inhibited larval growth and promoted glucose synthesis and lipolysis in silkworm. *Mol. Genet. Genomics.* **2019**, *294*, 1375–1383. [CrossRef] [PubMed]
5. Zheng, S.F.; Jin, X.; Chen, M.H.; Shi, Q.; Zhang, H.; Xu, S. Hydrogen sulfide exposure induces jejunum injury via CYP450s/ROS pathway in broilers. *Chemosphere* **2019**, *214*, 25–34. [CrossRef]
6. Magierowski, M.; Magierowska, K.; Hubalewska-Mazgaj, M.; Surmiak, M.; Sliwowski, Z.; Wierdak, M.; Kwiecien, S.; Chmura, A.; Brzozowski, T. Cross-talk between hydrogen sulfide and carbon monoxide in the mechanism of experimental gastric ulcers healing, regulation of gastric blood flow and accompanying inflammation. *Biochem. Pharmacol.* **2018**, *149*, 131–142. [CrossRef] [PubMed]
7. Miller, D.L.; Roth, M.B. Hydrogen sulfide increases thermotolerance and lifespan in *Caenorhabditis elegans*. *Proc. Natl. Acad. Sci. USA* **2007**, *104*, 20618–20622. [CrossRef] [PubMed]
8. Budde, M.W.; Roth, M.B. Hydrogen sulfide increases hypoxia-inducible factor-1 activity independently of von Hippel-Lindau tumor suppressor-1 in *C. elegans*. *Mol. Biol. Cell* **2010**, *21*, 212–217. [CrossRef]
9. Qabazard, B.; Li, L.; Gruber, J.; Peh, M.T.; Ng, L.F.; Kumar, S.D.; Rose, P.; Tan, C.H.; Dymock, B.W.; Wei, F.; et al. Hydrogen sulfide is an endogenous regulator of aging in Caenorhabditis elegans. *Antioxid. Redox. Signal.* **2014**, *20*, 2621–2630. [CrossRef] [PubMed]
10. Wei, Y.; Kenyon, C. Roles for ROS and hydrogen sulfide in the longevity response to germline loss in *Caenorhabditis elegans*. *Proc. Natl. Acad. Sci. USA* **2016**, *113*, E2832–E2841. [CrossRef] [PubMed]

11. Kabil, H.; Kabil, O.; Banerjee, R.; Harshman, L.G.; Pletcher, S.D. Increased transsulfuration mediates longevity and dietary restriction in *Drosophila. Proc. Natl. Acad. Sci. USA* **2011**, *108*, 16831–16836. [CrossRef]
12. Zhong, J.F.; Wang, S.P.; Shi, X.X.; Mu, L.L.; Li, G.Q. Hydrogen sulfide exposure increases desiccation tolerance in *Drosophila melanogaster. J. Insect Physiol.* **2010**, *56*, 1777–1782. [CrossRef]
13. Cao, Y.Y.; Peng, L.L.; Jiang, L.; Thakur, K.; Hu, F.; Tang, S.M.; Wei, Z.J. Evaluation of the metabolic effects of hydrogen sulfide on the development of *Bombyx mori* (Lepidoptera: Bombycidae), using liquid chromatography-mass spectrometry-based metabolomics. *J. Insect Sci.* **2020**, *20*, 4. [CrossRef]
14. Jiang, L.; Peng, L.L.; Cao, Y.Y.; Thakur, K.; Tang, S.M.; Hu, F.; Wei, Z.J. Transcriptome analysis reveals gene expression changes of the fat body of silkworm (*Bombyx mori* L.) in response to selenium treatment. *Chemosphere* **2020**, *245*, 125660. [CrossRef]
15. Martin, M. Cutadapt removes adapter sequences from high-throughput sequencing reads. *EMBnet. J.* **2011**, *17*, 10–12. [CrossRef]
16. Pertea, M.; Kim, D.; Pertea, G.M.; Leek, J.T.; Salzberg, S.L. Transcript-level expression analysis of RNA-seq experiments with HISAT, StringTie and Ballgown. *Nat. Protoc.* **2016**, *11*, 1650–1667. [CrossRef]
17. Beekman, R.; Chapaprieta, V.; Russiñol, N.; Vilarrasa-Blasi, R.; Verdaguer-Dot, N.; Martens, J.H.A.; Duran-Ferrer, M.; Kulis, M.; Serra, F.; Javierre, B.M.; et al. The reference epigenome and regulatory chromatin landscape of chronic lymphocytic leukemia. *Nat. Med.* **2018**, *24*, 868–880. [CrossRef]
18. Xia, Q.; Zhou, Z.; Lu, C.; Cheng, D.; Dai, F.; Li, B.; Zhao, P.; Zha, X.; Cheng, T.; Chai, C.; et al. A draft sequence for the genome of the domesticated silkworm (*Bombyx mori*). *Science* **2004**, *306*, 1937–1940.
19. Kim, D.; Langmead, B.; Salzberg, S.L. HISAT: A fast spliced aligner with low memory requirements. *Nat. Methods* **2015**, *12*, 357–360. [CrossRef]
20. Pertea, M.; Pertea, G.M.; Antonescu, C.M.; Chang, T.C.; Mendell, J.T.; Salzberg, S.L. StringTie enables improved reconstruction of a transcriptome from RNA-seq reads. *Nat. Biotechnol.* **2015**, *33*, 290–295. [CrossRef]
21. Robinson, M.D.; McCarthy, D.J.; Smyth, G.K. edgeR: A Bioconductor package for differential expression analysis of digital gene expression data. *Bioinformatics.* **2010**, *26*, 139–140. [CrossRef]]
22. Kanehisa, M.; Araki, M.; Goto, S.; Hattori, M.; Hirakawa, M.; Itoh, M.; Katayama, T.; Kawashima, S.; Okuda, S.; Tokimatsu, T.; et al. KEGG for linking genomes to life and the environment. *Nucleic Acids Res.* **2008**, *36*, D480–D484. [CrossRef]

23. Zhang, F.; Zhang, Y.Y.; Ma, R.H.; Thakur, K.; Han, J.; Hu, F.; Zhang, J.G.; Wei, Z.J. Multi-omics reveals the anticancer mechanism of asparagus saponin-asparanin A on endometrial cancer Ishikawa cells. *Food Funct.* **2021**, *12*, 614–632. [CrossRef]

24. Wang, J.J.; Bai, W.W.; Zhou, W.; Liu, J.; Chen, J.; Liu, X.Y.; Xiang, T.T.; Liu, R.H.; Wang, W.H.; Zhang, B.L.; et al. Transcriptomic analysis of two Beauveria bassiana strains grown on cuticle extracts of the silkworm uncovers their different metabolic response at early infection stage. *J. Invertebr. Pathol.* **2017**, *145*, 45–54. [CrossRef]

25. Adam, G.M.; Naama, K. NADH ties One-Carbon metabolism to cellular respiration. *Cell Press.* **2020**, *31*, 660–662.

26. Chen, Q.M.; Ma, Z.G.; Wang, X.; Li, Z.; Zhang, Y.; Ma, S.; Zhao, P.; Xia, Q. Comparative proteomic analysis of silkworm fat body after knocking out fibroin heavy chain gene: A novel insight into cross-talk between tissues. *Funct. Integr. Genomics.* **2015**, *15*, 611–637. [CrossRef]

27. Long, W.; Wu, J.; Shen, G.; Zhang, H.; Liu, H.; Xu, Y.; Gu, J.; Jia, L.; Lin, Y.; Xia, Q. Estrogen-related receptor participates in regulating glycolysis and influences embryonic development in silkworm Bombyx mori. *Inesct Mol. Biol.* **2020**, *29*, 160–169. [CrossRef]

28. Efe, S.; Duvernell, D.D.; Matzkin, M.L.; Duan, Y.H.; Zhu, C.T.; Verrelli, B.C.; Eanes, W.F. Single-Locus latitudinal clines and their relationship to temperate adaptation in metabolic genes and derived alleles in *Drosophila melanogaster*. *Genet. Soc. Am.* **2004**, *168*, 923–931.

29. Long, L.; Wang, C.C.; Zhao, X.Y.; Guan, J.X.; Lei, C.L.; Huang, Q.Y. Isocitrate dehydrogenase-mediated metabolic disorders disrupt active immunization against fungal pathogens in eusocial termites. *J. Pest. Sci.* **2020**, *93*, 291–301.

30. Meyer, S.; Gessner, D.K.; Wen, C.P.; Most, E.; Liebisch, G.; Zorn, H.; Ringseis, R.; Eder, K. The Antisteatotic and Hypolipidemic Effect of Insect Meal in Obese Zucker Rats is Accompanied by Profound Changes in Hepatic Phospholipid and 1-Carbon Metabolism. *Mol. Nutr. Food Res.* **2019**, *63*, e1801305. [CrossRef]

31. Shi, A.B.; Chen, C.C.; Banerjee, R.; Glodowski, D.; Audhya, A.; Rongo, C.; Grant, B.D. EHBP-1 functions with RAB-10 during endocytic recycling in Caenorhabditis elegans. *Mol. Biol. Cell.* **2010**, *21*, 2930–2943. [CrossRef]

32. Le Roy, C.; Wrana, J.L. Clathrin- and non-clathrin-mediated endocytic regulation of cell signaling. *Nat. Rev. Mol. Cell Biol.* **2005**, *6*, 112–126.

[CrossRef]

33. Nichols, B. Caveosomes and endocytosis of lipid rafts. *J. Cell Sci.* **2003**, *116*, 4707–4714. [CrossRef]
34. Stenmark, H. Rab GTPases as coordinators of vesicle traffic. *Nat. Rev. Mol. Cell Biol.* **2009**, *10*, 513–525. [CrossRef]
35. Glodowski, D.R.; Chen, C.C.; Schaefer, H.; Grant, B.D.; Rongo, C. RAB-10 regulates glutamate receptor recycling in a cholesterol- dependent endocytosis pathway. *Mol. Biol. Cell.* **2007**, *18*, 4387–4396. [CrossRef]
36. Tian, J.H.; Xue, B.; Hu, J.H.; Li, J.X.; Cheng, X.Y.; Hu, J.S.; Li, F.C.; Chen, Y.H.; Li, B. Exogenous substances regulate silkworm fat body protein synthesis through MAPK and PI3K/Akt signaling pathways. *Chemosphere* **2017**, *171*, 202–207. [CrossRef]
37. Zanardo, R.C.; Brancaleone, V.; Distrutti, E.; Fiorucci, S.; Cirino, G.; Wallace, J.L. Hydrogen sulfide is an endogenous modulator of leukocyte-mediated inflammation. *FASEB J.* **2006**, *20*, 2118–2120. [CrossRef]
38. Jia, Q.; Wang, T.T.; Wang, X.Y.; Xu, H.; Liu, Y.; Wang, Y.; Shi, Q.; Liang, Q. Astragalin suppresses inflammatory responses and bone destruction in mice with collagen-induced arthritis and in human fibroblast-likesynoviocytes. *Front. Pharmacol.* **2019**, *10*, 94. [CrossRef]
39. Hu, W.; Wang, X.; Ma, S.; Peng, Z.; Cao, Y.; Xia, Q. CRISPR-Mediated endogenous activation of fibroin heavy chain gene triggers cellular stress responses in *Bombyx mori* Embryonic Cells. *Insects* **2021**, *12*, 552. [CrossRef]
40. Kunz, R.I.; Brancalhao, R.M.; Ribeiro, L.F.; Natali, M.R. Silkworm sericin: Properties and biomedical applications. *Biomed. Res. Int.* **2016**, *2016*, 8175701. [CrossRef]
41. Ni, M.; Zhang, H.; Li, F.C.; Wang, B.B.; Xu, K.Z.; Shen, W.D.; Li, B. Nanoparticulate anatase TiO_2 (TiO_2NPs) upregulates the expression of silkworm (*Bombyx mori*) neuropeptide receptor and promotes silkworm feeding, growth, and silking. *Peptides* **2015**, 68, 64–71. [CrossRef]

15 Chapter

PATHOGENICITY ASSESSMENT AND GENOMIC ANALYSIS OF TWO DISTINCT GEOGRAPHIC STRAINS OF BMNPV.

1. INTRODUCTION: AN OVERVIEW

Bombyx mori nuclear polyhedrosis virus (BmNPV), a circular double-stranded DNA virus, is the inaugural insect baculovirus identified [1,2]. B. mori hemolytic septic illness, induced by BmNPV, represents the most serious viral affliction in silkworms, adversely impacting the sericulture sector [3]. It is a subacute illness that manifests over 3–5 days, with a more rapid onset during summer and autumn. Infected silkworms exhibit distinct characteristics such as erratic movement, a whitish and glossy body with intersegmental enlargement, skin that readily ruptures and exudes a milky white fluid, culminating in mortality. The opalescent bodily fluid harbors BmNPV virus particles that infect the mulberry leaves inhabited by silkworms. Other silkworms consume the contaminated leaf, resulting in oral illnesses or infections through wounds after interacting with other injured silkworms. The disease epidemic in rural production regions often transpires around the mid to late phases of the fifth instar of B. mori. It is challenging to avert and results in a substantial decrease in crop volume, occasionally leading to a complete absence of harvest. Developing silkworm strains with enhanced resistance to BmNPV is the most cost-effective and efficient method to mitigate losses in sericulture production. In the past decade, multiple institutions in China have developed several silkworm varieties that are resistant to BmNPV; for instance, "Huakang No.2" demonstrates over 100 times enhanced resistance to B. mori hemolynamic septic disease compared to the typical summer and autumn species "Qiufeng × Baiyu". The silk yield of these cultivars also demonstrates their exceptional disease resistance characteristics [8–11].

Similar to other viruses, genetic variation has been seen in BmNPV and distinguishes the various geographical strains. Moreover, many regional strains of BmNPV exhibit disparities in their appearance and pathogenicity [17,18]. The pathogenicity of local strains of BmNPV from the same province varies in relation to B. mori. Bai et al. [19] and Tang et al. [12] discovered that the pathogenicity of eight BmNPV strains from several areas in Yunnan varied, demonstrating differing infectivity rates in B. mori. Wang et al. [20] indicated that the pathogenicity of BmNPV from various origins in Guangxi Province varied across different silkworm kinds, presenting a novel challenge for silkworm breeding against BmNPV. Consequently, it is essential to investigate the pathogenic processes of various mutant strains of BmNPV concerning silkworms. We identified two BmNPVs from Zhenjiang, Jiangsu Province, and Luliang, Yunnan Province, exhibiting differing pathogenicities towards the same silkworm kinds. This study utilized Illumina second-generation sequencing and PacBio third-generation sequencing technologies to analyze the genomes and investigate the mechanisms underlying the pathogenicity variations among distinct BmNPV strains..

2. MATERIALS AND METHODS

2.1 Materials

BmNPV strains were obtained from different geographic regions: the BmNPV Zhen- jiang strain (BmNPV ZJ) was obtained from Zhenjiang, Jiangsu province, and the BmNPV Yunnan strain (BmNPV YN) was obtained from Luliang, Yunnan province. These two strains were isolated, purified, and preserved by our teams.

The silkworm variety "Baiyu", bred by the Institute of Sericulture of the Chinese Academy of Agricultural Sciences, was the parent of the control variety of silkworm varieties approved by the state for use in summer and autumn and was maintained in our laboratory.

2.2 Methods

2.2.1 Virus Collection and Purification

The budded virus (BV) of BmNPV ZJ was preserved and provided by the Pathology Laboratory of the Institute of Sericulture of the Chinese Academy of Agricultural Sciences. The BV of BmNPV YN was preserved and provided by the Institute of Sericulture of the Yunnan Academy of Agricultural Sciences. In the autumn of 2019, our team used the BV to puncture the fifth instar of silkworms. After the onset of the disease, the blood of the infected silkworm was collected, filtered, and centrifuged three to four times to obtain the purified virus solution. Then, it was diluted with an appropriate amount of double-distilled water and counted using a hemocytometer. The concentrations of BmNPV ZJ and BmNPV YN were 2.45×10^9 and 3.66×10^9, respectively; they were stored at 4°C for further use.

2.2.2 Determination of Pathogenicity of Different BmNPV to B. mori

The original viral solutions of the two BmNPV strains were diluted with sterile water to a total of five concentrations, from 1 × 10^4–1 × 10^8, and smeared on mulberry leaves to feed the Baiyu larvae. The mulberry leaves with smooth surfaces were cut into small pieces (2.7 cm × 2.7 cm) using a punch, and 50 μL of different concentrations of polyhedral suspension was added dropwise on the upper surface of each leaf piece. After smearing evenly, these were fed to the second instar larvae of the silkworm, and leaves with the same volume of sterile water were used as blank controls. There were a total of 90 silkworm larvae per treatment, with three replicates (30 silkworms each), for each BmNPV concentration [20]. Four mulberry leaves carrying the virus were fed to each silkworm in each area at one time. After 24 h, the mulberry leaves carrying the virus were replaced with plain mulberry leaves with no virus. The growth and development of silkworms were observed every day, and the diseased silkworms were removed in time to avoid cross-infection. The number of dead silkworms that were infected with BmNPV was recorded after their third instar stage. The lethal concentration 50 (LC_{50}) of Baiyu was calculated using the Statistical Product and Service Solutions (SPSS, https://www.ibm.com/cn-zh/analytics/spss-statistics-software accessed on 26 September 2021), and the pathogenicities of BmNPV YN and BmNPV ZJ were compared.

2.2.3 Negative-Staining Electron Microscopy to Observe BmNPV Particles

The purified BmNPV ZJ and BmNPV YN were suspended on a copper mesh supported by a polyvinyl alcohol formaldehyde membrane (formvar membrane). The samples were stained with 2% phosphotungstic acid (pH 7.2) for 20 s. After washing and drying, the mor- phologies of the virus particles were observed with negative-staining electron microscopy.

2.2.4 Sequencing of BmNPV ZJ and BmNPV YN Genomes

DNA was extracted from BmNPV YN and BmNPV ZJ according to the alkaline lysis method [12]. The concentrations of the extracted viral DNA were measured using the Qubit quantitative detector. The DNA quality was assessed using 1% agarose gel electrophoresis. The high-quality viral DNA was used for genome sequencing using Illumina second- generation and PacBio third-generation sequencing technologies (Sangon Biotech Co., Ltd., Shanghai, China).

2.2.5 Analysis of BmNPV ZJ and BmNPV YN Genome Sequences

Using the genome of BmNPV T3 (L33180.1, 128,413 bp) as the reference, the sequenc- ing data were assembled and corrected by SPAdes (https://cab.spbu.ru/software/spades/ accessed on 26 September 2021) and Primer-initiated Sequence Synthesis for Genomes (PrlnSeS-G, https://updeplasrv1.epfl.ch/prinses/ accessed on 26 September 2021) [21,22]. The Gene Ontology, Kyoto Encyclopedia of Genes and Genomes, Clusters of Orthologous Groups (COGs) of proteins, non-redundant proteins (NR), curated protein families (PFAM), Swiss-Prot, and TrEMBL databases were used for functional annotation of the virus genes. The genome sequences of these two BmNPVs were compared with other published Bm- NPV genome sequences. For multiple sequence alignment, the ClustalW parameters of Molecular Evolutionary Genetics Analysis (MEGA 7.1, https://megasoftware.net/ accessed on 26 September 2021) were used. The baculovirus repeated orfs gene (*bro)* of the BmNPV exists in diverse copies and nucleotide sequences in different ecological environments; therefore, this gene was also used for molecular identification of different strains of BmNPV [15,16,23]. The bootstrap statistical method was used for the calculation of 1000 replicates. The phylogenetic tree of BmNPV was constructed using the *bro-d* gene sequences of different BmNPV strains.

2.2.6 Detection of Gene-Relative Expression of BmNPV by qPCR

The two strains of BmNPV were diluted to 1 × 108 and fed to the fifth instar of the Baiyu silkworms, feeding 7 μL to each silkworm; there were five silkworms per treatment. The disease incidence rate among the silkworms was 100% at this concentration. At 12, 24, 48, 72, and 96 h, the silkworms were dissected and RNA was extracted from their midgut tissues. The very early gene *ie-1*, early gene *p143*, late gene *vp39*, and very late gene *polh* of BmNPV were chosen for quantitative polymerase chain reaction (qPCR) analysis. The primers for viral genes were designed in Primer 6.0 and their details are listed in Table 1. Reaction conditions for qPCR were: 95°C, 30 s; 95°C, 10 s; and 55°C, 30 s, for a total of 40 cycles. Using *actin-3* (U49854) as the reference gene, the relative expression levels of the genes in the two BmNPV strains were calculated with the 2-ΔΔCT method.

Table 1. Primers used in qPCR.

Gene	Primer Sequences	Length/(bp)
actin-3	F: 5′ -CGGCTACTCGTTCACTACC-3′ R: 5′ -CCGTCGGGAAGTTCGTAAG-3′	147
ie-1	F: 5′-GGACGAATACTTGGACGAT-3′ R: 5′-GAGAACCTGTTGGAATTGTAG-3′	237
p143	F: 5′, -GCACGGCAATACTTATCATC-3′ R: 5′, -TGAGCACCAACAATAGTCC-3′	120
vp39	F: 5′-ACACGGAGGAATTGAGATT-3′ R: 5′ -GATGTCACTGCTTCTTATCG-3′	116
polh	F: 5′-CTACAAGTTCCTCGCTCAA-3′ R: 5′-CTCGCTGTGGATGTTCAT-3′	163

2.2.7. BmNPV YN and BmNPV ZJ Differential Genes Analysis

Referring to the design process proposed by the Genome Analysis Toolkit (GATK) [24], the effective data for the BmNPV YN and BmNPV ZJ genomes were compared using the Burrows–Wheeler Aligner (BWA, http://bio-bwa.sourceforge.net/ accessed on 26 September 2021). To convert and sort the results, sequence alignment/ map Tools (SAMtools, http://www.htslib.org/doc/samtools.html accessed on 26 September 2021) was used and the comparison results were statistically analyzed. The genotype differences between Bm- NPV YN and BmNPV ZJ were assessed using the HaplotypeCaller software, and the single nucleotide polymorphism (SNP) and insertion and deletion (InDel) information of the two strains were obtained. After quality control (Table 2), the SNP and InDel information were annotated using SNP effect software (SnpEff, https://pcingola.github.io/SnpEff/ accessed on 26 September 2021) [25]. According to the annotation results and BLAST analysis of the differential genes between BmNPV YN and BmNPV ZJ, candidate genes associated with the difference in pathogenicity of the BmNPV YN and BmNPV ZJ were screened.

Table 2. SNP and InDel quality control standards.

Quality Control Items	SNP	InDel
QualByDepth	≥2.0	≥2.0
RMSMappingQuality	≥40.0	-
FisherStrand	≤60.0	≤200.0
StrandOddsRatio	≤3.0	≤10.0
MappingQualityRankSumTest	≥-12.5	≥-12.5
ReadPosRankSumTest	≥-8.0	≥-8.0

3. RESULTS

3.1 The Pathogenicity of BmNPV ZJ and BmNPV YN to B. mori

The mortality rate of Baiyu larvae increased with the increase in the concentration of BmNPV ZJ and BmNPV YN (Figure 1). At the same concentration, BmNPV YN exhibited stronger pathogenicity than BmNPV ZJ. In addition, the mortality rates at 1×10^7 and 1×10^8 of both BmNPVs differed significantly.

The median lethal dose of the two viruses in the Baiyu larvae was calculated by SPSS v21 (https://www.ibm.com/cn-zh/analytics/spss-statistics-software accessed on 26 September 2021) software. The LC_{50} concentration of BmNPV YN in Baiyu larvae was about 10 times less than that of BmNPV ZJ (Table 3), indicating that BmNPV YN could even cause death in half of these larvae at a low concentration.

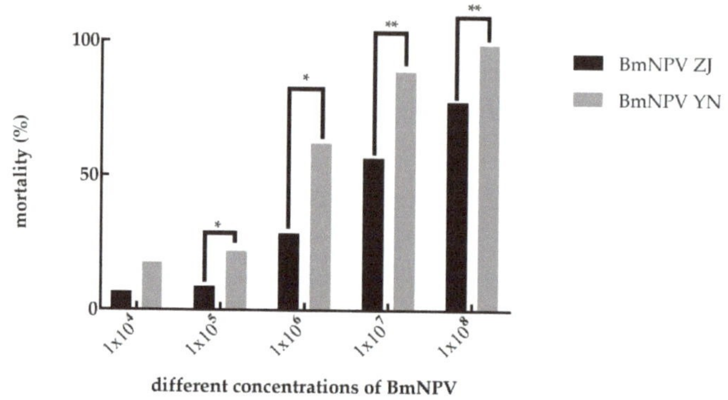

Figure 1. A comparison of the pathogenicities of different concentrations of the two BmNPV strains against Baiyu larvae. Note: "*" represents p < 0.05; "" represents p < 0.01.**

Table 3. Comparison of virulence of BmNPV ZJ and BmNPV YN in Baiyu larvae.

Variety	Strain	Regression Equation	LC50	95% Confidence Value	Slope of Regression Line/SE
Baiyu	BmNPV ZJ	y = −4.267 + 0.627x	6.45×10^6	3.88×10^6–1.14×10^7	0.056
Baiyu	BmNPV YN	y = −4.426 + 0.796x	3.62×10^5	1.98×10^5–6.45 × 105 0.063	0.063

3.2 The Morphology and Size of BmNPV ZJ and BmNPV YN

The purified BmNPV ZJ and BmNPV YN virus particle suspensions were observed by electron microscopy (Figure 2). The sizes of BmNPV ZJ and BmNPV YN appeared similar, with diameters of about 2.2–4.0 μm. The BmNPV polyhedra appeared mostly hexagonal, while a small number of particles appeared quadrilateral and irregular. The edges of BmNPV YN particles appeared sharper compared with those of BmNPV ZJ.

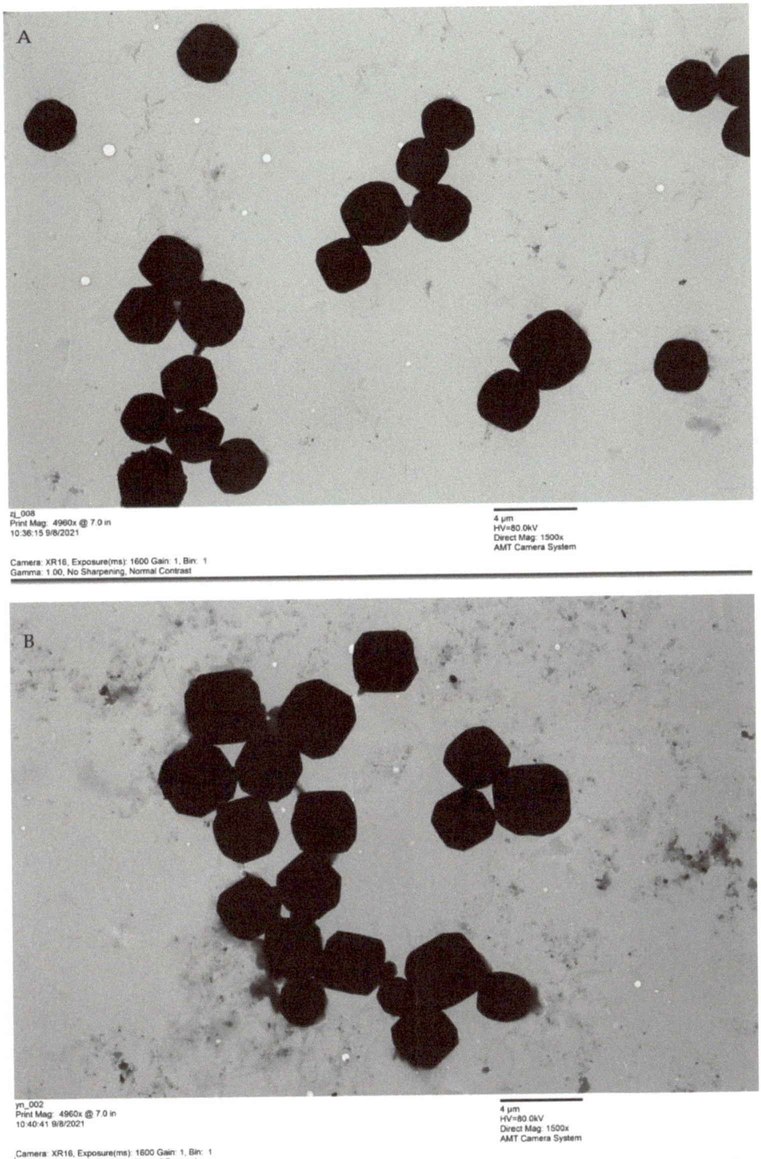

Figure 2. Morphological observation of BmNPV polyhedra by electron microscopy (×1500): (A) BmNPV ZJ and (B) Bm- NPV YN.

3.3 The Relative Expression of Genes of BmNPV ZJ and BmNPV YN in the Midgut of Baiyu

The qPCR analysis of the genes of BmNPV ZJ and BmNPV YN in the midgut of Baiyu larvae (Figure 3A) revealed that the relative expression of *ie-1* in BmNPV ZJ was higher than that in BmNPV YN at 12–96 h, with the highest level at 96 h. The expressions of *p143*, *vp39*, and *polh* in BmNPV ZJ were also higher than BmNPV YN before 96 h (Figure 3B–D).

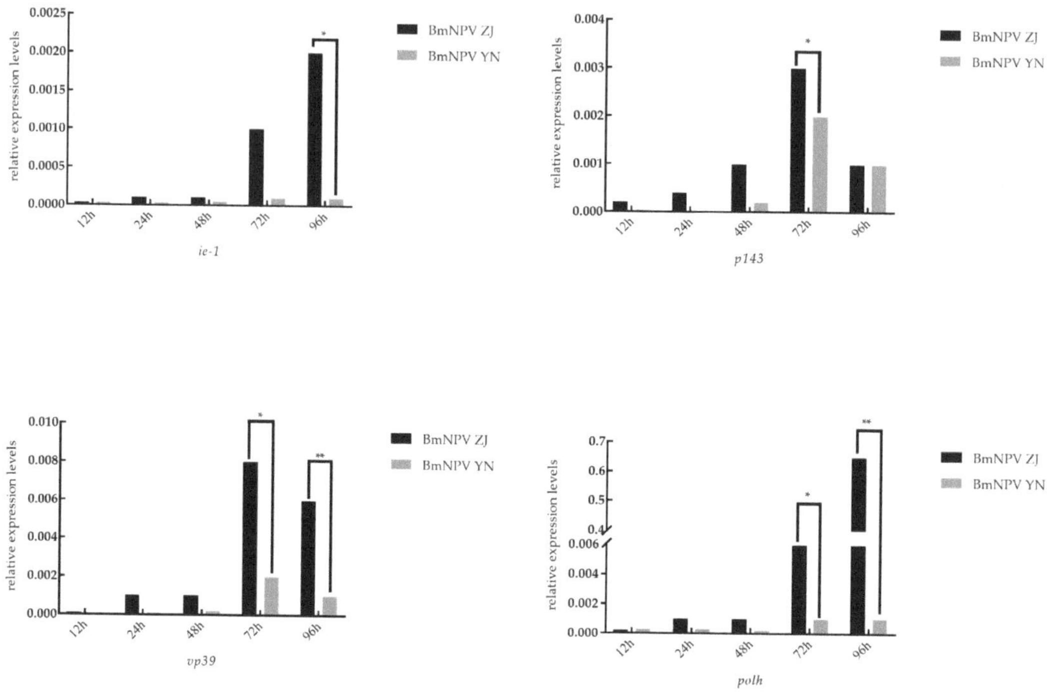

Figure 3. The relative expression of genes of the two BmNPV strains in the midgut of Baiyu larvae: **(A)** *ie-1*; **(B)** *p143*; *vp39*; **(D)** *polh*. Note: "*" represents $p < 0.05$; "**" represents $p < 0.01$.

3.4 Structural Characteristics of Virus Genomes

3.4.1 Genome Characteristics of BmNPV ZJ

The genome of BmNPV ZJ was estimated to be 135,895 bp long, with 40.39% GC content. It comprised 141 predicted protein-coding sequences (CDSs), accounting for 83.86% of the entire genome sequence, of which 70 CDSs were on the positive strand and 71 on the negative strand. The average length of this genome was estimated to be 808 bp. Further analysis indicated 91 putative genes of 500 to 1000 bp lengths and 40 genes with ≥ 1000 bp lengths. Only 10 genes were less than 500 bp in length. In total, 139 genes were annotated. The annotated genes fibroblast growth factor (*fgf*), chi-tinase gene (*chi-a*), and alkaline exonuclease gene (*alk-exo*) were found in the BmNPV ZJ genome, but their homologous reading frames were absent in BmNPV YN. Further, late expression factor 12 (*lef-12*) and *AcMNPV 94K* were not annotated in the BmNPV T3 reference strain (Figure S1). Compared with the genome of BmNPV T3, some genes in these two viral genomes had SNP mutations, resulting in premature stop codons, and truncated or abnormal proteins. One putative gene with an unknown function was also found.

3.4.2 Genome Characteristics of BmNPV YN

The genome of BmNPV YN was estimated to be 143,180 bp long, with 39.71% GC content. It comprised 146 CDSs, accounting for 89.37% of the whole genome sequence, of which 76 CDSs were on the positive strand and 69 on the negative strand. The average length of this genome was estimated to be 882.52 bp, with 96 genes of 500 to 1000 bp lengths and 43 genes more than 1000 bp in length. Six genes were found to be of less than 500 bp in length and 137 genes were annotated. Further analysis revealed *Orf20*, late expression factor 7 (*lef-7*), and *p26* genes in the BmNPV YN genome, but the homologous genes were absent in the BmNPV ZJ genome. *Lef-12* was not annotated in the BmNPV T3 reference strain (Figure S2). In addition, seven putative genes with unknown functions were found.

3.4.3 Genome Comparison of BmNPV

The genome sequences of the two BmNPVs were compared with that of the BmNPV T3 strain as a reference, as shown in Table 4.

Table 4. General features of BmNPV genomes.

Items	BmNPV ZJ	BmNPV YN	BmNPV T3
Size (base)	135,895	143,180	128,413
G + C content (%)	40.39	39.71	40
Protein coding genes	141	145	136
Min length (base)	111	111	61
Max length (base)	2964	4050	1222
Average length (base)	808.23	882.52	852.154
Total coding gene (base)	113,961	127,965	115,893
Coding ratio (%)	83.86	89.37	90.25

The nucleotide sequences of the BmNPV ZJ and BmNPV YN genomes were analyzed by DNAMAN software (https://www.lynnon.com/dnaman.html accessed on 26 September 2021). The genes of BmNPV ZJ and BmNPV YN were 92.6–100.0% homologous. A comparison with the genome sequences of the other six BmNPV reference strains showed homologies of BmNPV ZJ and BmNPV YN in the range of 92.0–100.0% and 93.3–100.0%, respectively (Table S1). The *bro-d* gene sequence of BmNPV ZJ and BmNPV YN showed 95.9% homology, and the homologies with the other BmNPV reference strains were 92.1–99.6% and 93.3–97.3%, respectively (Table 5). The amino acid sequence homology of the BRO-D (encoded by the *bro-d*) of BmNPV ZJ and BmNPV YN was 96.6%, and the homologies with other BmNPVs were 92.0–99.7% and 92.2–97.1%, respectively (Table 5). The amino acid of the BRO-D protein of the two strains was mainly mutated at the N-terminal (Figure 4), which was consistent with the observation of Zhou et al. [26].

Table 5. Nucleotide and amino acid homology of *bro-d* gene sequences.

Reference Strain	Nucleotide Homology/(%)		Reference Strain	Amino Acid Sequence/(%)	
	BmNPV ZJ (N)	BmNPV YN (N)		BmNPV ZJ (AA)	BmNPV YN (AA)
Brazilian	94.3	94.2	Brazilian	93.7	92.5
Cubic	99.6	95.7	Cubic	99.7	96.3
Guangxi	92.1	93.3	Guangxi	91.7	92.2
India	96.8	97.3	India	96.6	97.1
T3	95.9	94.1	T3	92.0	92.5
Zhejiang	95.5	94.3	Zhejiang	94.0	92.8
Zhenjiang	-	95.9	Zhenjiang	-	95.9
Yunnan	96.9	-	Yunnan	96.6	-

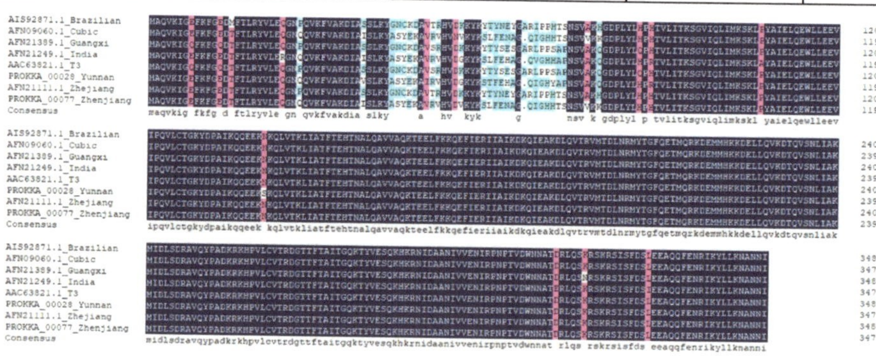

Figure 4. Alignment of amino acid sequences of BRO-D protein from different BmNPV strains.

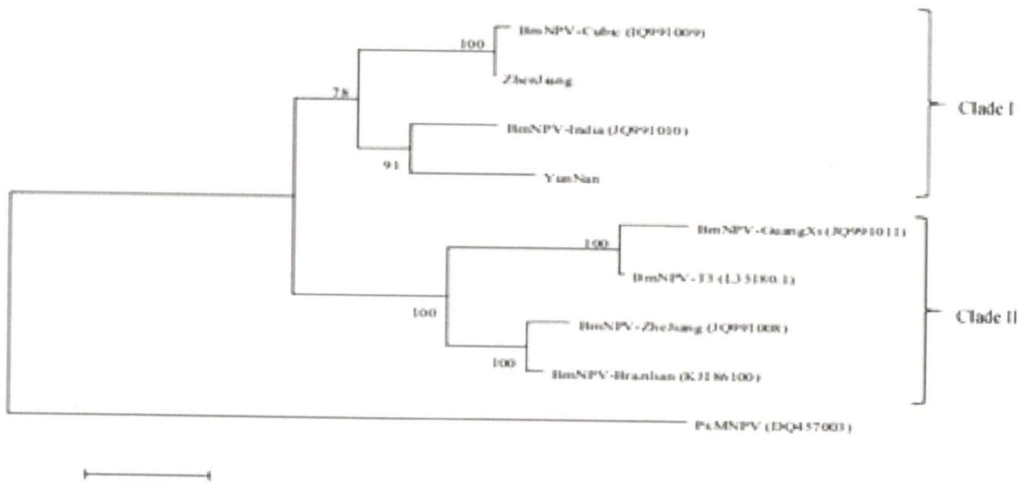

Figure 5. Phylogenetic tree constructed based on bro-d gene sequences of BmNPV isolates.

The *bro-d* gene was chosen for the evolutionary analysis of the two BmNPVs. Using *Plutella xylostella* multiple nucleopolyhedrovirus (PxMNPV) as the exogenous reference, the phylogenetic tree was plotted using the MEGA 7.1 software (https://megasoftware.net/ accessed on 26 September 2021). The results are shown in Figure 5. The eight BmNPV strains were divided into two clades. Both BmNPV ZJ (Zhenjiang) and BmNPV YN (Yunnan) were grouped in clade I, and BmNPV ZJ was closer to the Cubic strain (IQ991009) while BmNPV YN was closer to the India strain (JQ991010). These finding are similar to those reported by Tang et al. [12].

3.5 Detection and Analysis of Single Nucleotide Polymorphism and InDels in the Two Viral Genomes

Using BmNPV ZJ as the reference genome, a total of 893 SNP loci and 132 InDel variants were identified in BmNPV YN. The number of SNPs of transitions and transversions were 693 and 195, respectively. The SNP loci of the BmNPV YN strain were mainly enriched in the coding region, accounting for about 82.4% of the total number of SNPs. Among them, there were 487 synonymous mutations, 247 missense mutations, and 2 nonsense mutations, resulting in 102 different gene sequences (Table 6).

Table 6. Summary statistics of SNP base changes of BmNPV YN.

Type	SNP Number	Region
Transitions	693	
Transversions	195	
Mutations in coding region	736	
Intergenic mutation	151	
Synonymous mutation	487	CDS
Missense mutation	247	CDS
Nonsense mutation	2	CDS
Other mutations that could not be accurately judged	6	

Note: The reference genome was BmNPV ZJ.

Analysis of InDel mutations by SnpEff software showed that the numbers of InDel mutations were similar between coding and non-coding regions (Table 7). Further analysis revealed 38 in-frame mutations and 25 frameshift mutations, resulting in 30 differential gene sequences.

Table 7. InDel annotations of BmNPV YN.0

Type	InDel Number	Region
Mutations in coding region	66	
Intergenic mutation	64	
Codon mutation Code insertion Code deletion	19	CDS
	19	CDS
Frameshift mutation	25	CDS
Other mutations that could not be accurately judged	6	

Note: The reference genome was BmNPV ZJ.

3.6. Differential Genes Analysis

The BmNPV ZJ and BmNPV YN strains were observed to have 106 different annotated genes sequences, which included non-synonymous mutations, synonymous mutations, and frameshift mutations. Among these, 76 differential nucleotide sequences led to differences in amino acid sequences in the two viruses (Table 8). The core genes [27] (shared genes of the baculovirus) accounted for 28.94% and non-core genes accounted for 47.37% of all differential genes, while some genes had not been reported earlier. Thirty annotated genes were associated with the viral genome replication and transcription, and five genes were related to the oral infection of viruses. These differences in genes may have been the cause of the difference in pathogenicity between the two virus strains.

Table 8. Viral functional genes with non-synonymous mutations and frameshift

Type	BmNPV ZJ	BmNPV YN	Gene Name	Mutation Type	Biological Function	Description
Core	PROKKA 00060	PROKKA 00011	*pif-2*	Transversions	Viral infection	The composition of the membrane, necessary for
	PROKKA 00082	PROKKA 00117	*p74 (pif-0)*	Transversions	Viral infection	It is related to the life cycle of the virus and participation in the oral infection of the virus
	PROKKA 00108	PROKKA 00135	*pif-1*	Transitions and transversions	Viral infection	The composition of the membrane, related to oral infections
	PROKKA 00113	PROKKA 00108	*pif-5 (odv-e56)*	PROKKA 00108	Viral infection	Determines the virus host range, related to oral infections
	PROKKA 00009	PROKKA 00056	*Vlf-1*	Transversions and InDel	Replication, transcription	Late gene expression
	PROKKA 00020	PROKKA 00067	*dna pol*	Transitions and	Replication, transcription	The catalytic activity, replication of the viral genome, and host DNA polymerase cannot replace viral enzymes in this process
	PROKKA 00035	PROKKA 00082	*lef-8*	Transitions	Replication, transcription	Late gene expression
	PROKKA 00043	PROKKA 00090	*p47*	Transitions	Replication, transcription	Regulation of viral transcription
	PROKKA 00067	PROKKA 00018	*lef-1*	InDel and transversions	Replication, transcription	Encoding DNA promoter and interacting with LEF-2
	PROKKA 00073	PROKKA 00024	lef-2	Transitions	Replication, transcription	Virus replication and late gene expression
	PROKKA 00092	PROKKA 00036	lef-5	Transitions	Replication, transcription	Regulation of viral transcription
	PROKKA 00095	PROKKA 00033	AcMNPV orf103	Transitions	Replication, transcription	Virus replication, influences virus particle formation

	PROKKA 00114	PROKKA 00109	*ie-1*	Transitions	Replication, transcription	The essential transactivated protein that initiates viral DNA replication and bro promoter transcription
	PROKKA 00123	PROKKA 00045	*lef-4*	Transitions and	Replication, transcription	Regulation of viral transcription
	PROKKA 00128	PROKKA 00040	*dna hel/p143*	Transversions	Replication, transcription	DNA helicase activity, host domain determinant
	PROKKA 00132	PROKKA 00126	*p24*	Transitions	Replication, transcription	Regulation of viral transcription
	PROKKA 00003	PROKKA 00050	*p95*	Transitions	Structural protein	Composition of BV and ODV
	PROKKA 00006	PROKKA 00053	*gp41*	Transitions and transversions	Structural protein	Exists only in ODV, determining the manner and the ability of the virus to invade the host
	PROKKA 00008	PROKKA 00055	*AcMNPV orf78*	Transitions	Structural protein	Related to BV production and M-ODV formation
	PROKKA 00093	PROKKA 00035	*p40*	Transitions, transversions, and InDe	Structural protein	Includes body virus envelope components related to specific infection of host cells
	PROKKA 00100	PROKKA 00143	*AcMNPV orf109*	Transversions	Structural protein	Participates in viral nucleocapsid assembly
	PROKKA 00122	PROKKA 00046	*vp39*	Transitions and InDel	Structural protein	Related to virus infection
Non-core	PROKKA 00101	PROKKA 00142	*AcMNPV orf110*	Transitions	Viral infection	Related to oral infections
	PROKKA 00024	PROKKA 00071	*fp25K (ac61)*	InDel	Auxiliary function	Involved in BV and ODV formation, implicated in host degradation after death
	PROKKA 00058	PROKKA 00009	*pkip*	Transitions	Auxiliary function	Related to the BV nucleocapsid component

Non-core	PROKKA 00066	PROKKA 00017	*ecdysteroid UDP-glucosyl transferase (egt)*	Transversions	Auxiliary function	Hinders larvae molting and pupation
	PROKKA 00070	PROKKA 00021	*pk1*	Transversions and InDel	Auxiliary function	Regulation of polyhedrin gene promoter activity
	PROKKA 00078	PROKKA 00029	*ptp*	Transitions	Auxiliary function	BV components, essential components for effective infection of larvae brain tissue
	PROKKA 00086	PROKKA 00129	*chitinase A*	Transitions	Auxiliary function	Related to virus transmission
	PROKKA 00140	PROKKA 00128	*viral cathepsin-like protein (v-cath)*	Transversions	Auxiliary function	Related to host liquefaction and degradation
	PROKKA 00016	PROKKA 00063	*AcMNPV orf69*	Transitions	Replication, transcription	Late gene expression
	PROKKA 00018	PROKKA 00065	*lef-3*	Transitions	Replication, transcription	SS-DNA binding and destruction of helical
	stability PROKKA 00031	PROKKA 00078	*lef-10*	Transitions	Replication, transcription	Regulation of viral transcription
	PROKKA 00042	PROKKA 00089	*lef-12*	Transitions	Replication, transcription	Late gene transcirition
	PROKKA 00052	PROKKA 00002	*bro-a*	Transitions and transversions	Replication, transcription	DNA binding protein, complementary to BRO-C
	PROKKA 00054	PROKKA 00005	*lef-6*	Transitions and transversions	Replication, transcription	Virus replication and late gene expression, affecting host cell apoptosis
	PROKKA 00057	PROKKA 00008	*dbp (DNA binding protein)*	InDel	Replication, transcription	SS-DNA binding protein co-localized with ie-1 and lef-3 in viral replication mechanism

	PROKKA 00069	PROKKA 00020	*bm (br) orf-4*	Transitions and transversions	Replication, transcription	Early gene expression of virus
	PROKKA 00071	PROKKA 00022	*orf1629*	Transitions and transversions	virus Replication, transcription	Virus replication
	PROKKA 00076	PROKKA 00027	*Bro-b*	Transitions, transversions, and InDel	Replication, transcription	DNA binding protein
	PROKKA 00077	PROKKA 00028	*Bro-b*	Transitions, transversions, and InDel	Replication, transcription	Virus replication and gene expression regulation
	PROKKA 00080	PROKKA 00105	pe38	Transitions, transversions, and InDel	Replication, transcription	Virus replication and gene expression regulation
	PROKKA 00090	PROKKA 00038	Bro-c	Transitions, transversions, and InDel	Replication, transcription	DNA binding protein, complementary to BRO-A
	PROKKA 00091	PROKKA 00037	39k	InDel	Replication, transcription	Virus replication and gene expression regulation
	PROKKA 00097	PROKKA 00030	he65	Transitions and transversions	Replication, transcription	Virus replication
	PROKKA 00111	PROKKA 00106	ie-2	InDel	Replication, transcription	Virus replication and gene expression regulation
	PROKKA 00120	PROKKA 00115	*ie-0*	Transitions	Replication, transcription	Regulationofviraltranscription
	PROKKA 00121	PROKKA 00116	*me53*	Transitionsand transversions	Replication, transcription	RelatedtoBVandODVproduction
	PROKKA 00021	PROKKA 00068	*gp37*	Transitions	Structural protein	The formation of auxiliary components of polyhedraisinvolvedinthetransportof virus particlesinthecell

	PROKKA 00037	PROKKA 00084	odv-e66	Transitions and transversions	Structural protein	Participation in BV and ODV morphogenesis and oral infection
	PROKKA 00038	PROKKA 00085	bmnpvcubigcp037	InDel	Structural protein	Replication of the virus, regulation gof the transport of virus particles
	PROKKA 00065	PROKKA 00016	Bv/odv-e26	Transitions	Structural protein	Related to BV and ODV envelopes
	PROKKA 00096	PROKKA 00032	vp80	Transitions	Structural protein	Required for virus replication, BV production, and nucleocapsid maturation
	PROKKA 00135	PROKKA 00123	AcMNPV orf132	Transitions and transversions	Structural protein	Involved in BV and ODV formation
	PROKKA 00139	PROKKA 00118	p10	Transitions	Structural protein	Partcipates in the morphogenesis of viral polyhedra and promotes the release of polyhedra from infected insect cells
	PROKKA 00015	PROKKA 00062	iap2	Transitions	Apoptosis	Cell apoptosis inhibiting factor
	PROKKA 00055	PROKKA 00006	iap1	Transitions, transversions, and InDel	Apoptosis	Induction of apoptosis

4. DISCUSSION

BmNPV is a serious threat to silkworms and causes huge economic losses to the sericul- ture industry. Cultivating silkworm varieties resistant to a hemolymph-type septic diseases is the most effective and economical measure to minimize and mitigate these losses [28]. The pathogenicity of BmNPV varies in different geographic regions [18,19]. Furthermore, the underlying molecular mechanism that causes the variations in the pathogenicity of different strains of BmNPV remains unknown and poses a great challenge to the effective prevention of BmNPV. We compared the genomes of two BmNPV strains with different pathogenicities and found differences in some genes related to viral replication and infection.

In this study, the pathogenicities of BmNPV ZJ and BmNPV YN was determined and compared. The LC_{50} of BmNPV YN and BmNPV ZJ against Baiyu larvae were 3.62×10^5 and 6.45×10^6, respectively, indicating the semi-lethality of BmNPV YN in silkworms at a low concentration. Subsequently, the electron microscopic observation morphologies of virus particles revealed their hexagonal shapes with similar sizes (about 2.2–4.0 μm). These were typical silkworm nuclear polyhedrosis viruses, but the edges of BmNPV YN were clearer than those of BmNPV ZJ. The relative expressions of the *ie-1*, *p143*, *vp39*, and *polh* genes, expressed, respectively, at the very early, early, late, and very late stages of BmNPV ZJ, in the midgut of Baiyu larvae were higher than those of BmNPV YN within 12–96 h. This further indicates that the lethal concentration of BmNPV ZJ was higher than that of BmNPV YN. After a host is infected by a virus, in general, viruses with a high rate of proliferation tend to exhibit more robust pathogenicity. However, some more lethal virus strains may kill the host even with a lower rate of proliferation [29–31]. BmNPV YN might be such a type of lethal strain which, even in lower numbers, can kill the host. This inference is consistent with the finding of a tenfold lower LC_{50} for BmNPV YN than for BmNPV ZJ in the second instar of the Baiyu silkworm larvae. Therefore, when the virus was fed at a high concentration (10^8) to silkworms, the expression of the virus gene of BmNPV YN was lower, but it was enough to kill the host. Likewise, Ma [32] found that the mutation and recombination in genomes of different strains of *Helicoverpa armigera* nucleopolyhedrovirus might be the cause of the difference in their pathogenicity even when they are distributed in the same geographic location. Hence, the difference in pathogenicity of our strains might have been because of the variation in their genomes due to the differences in the climates of Zhenjiang, Jiangsu Province, and Luliang, Yunnan Province, which are located, respectively, along the southeast coast and the inland southwest of China.

To further explore the reasons for the difference in pathogenicity between BmNPV YN and BmNPV ZJ to *B. mori*, the whole genomes of the virus strains were sequenced and analyzed. The genome size of BmNPV ZJ was estimated to be 135,895 bp, with

40.39% GC content and encoding 141 genes, while the genome size of BmNPV YN was estimated to be 143,180 bp, with 39.71% GC content and encoding 145 genes. These data demonstrated that the genome sizes of the two viruses and the numbers of putative genes changed. For further analysis, we used the *bro-d* gene, which is highly conserved in different strains of BmNPV from varying geographical regions. BmNPV ZJ exhibited the highest homology with the Cubic strains published earlier in the GenBank, which might be due to the closer origin region of BmNPV. These findings suggest that BmNPV from different regions must be used for screening the BmNPV-resistant silkworm varieties for their wider adaptability.

A comparative analysis of BmNPV YN and BmNPV ZJ genomic sequences revealed a large number of SNP and InDel mutations in the two genomes. The nucleotide sequence ho- mology of the two virus genomes was 92.6–100.0%, and a further BLAST analysis revealed 76 different genes between BmNPV YN and BmNPV ZJ. These genes included 22 core genes and 36 non-core genes, mainly the oral infection factor genes, viral replication genes, host molting, and pupation genes. The alterations of genes sequences affect the function of the encoded proteins. Therefore, we speculated that the difference in the genomes of the two viruses might have been the cause of the difference in their pathogenicity in silkworms.

Thus, the roles of viral genes and the mechanism of action of important proteins encoded by these genes need to be studied in the future.

5. CONCISE SUMMARY

In summary, the BmNPV YN strain has a comparatively lower proliferation rate however possesses greater pathogenicity than the BmNPV ZJ strain. The two viruses vary in genomic size and the quantity of coding genes. Of these, 76 genes are distinct, including those linked to BmNPV viral replication and infection. The data demonstrate that varying geographical conditions cause mutations and rearrangements in the genomes of distinct BmNPV strains, ultimately resulting in divergent pathogenicity. The findings of this study may serve as a reference for further investigation into the molecular mechanisms underlying the variations in pathogenicity among virus strains and may also offer assistance in the development of novel pesticides..

REFERENCES

1. Yuan, M.J.; Wu, W.B.; Liu, C.; Wang, Y.J.; Hu, Z.Y.; Yang, K.; Pang, Y. A highly conserved baculovirus gene *p48* (*ac103*) is essential for BV production and ODV envelopment. *Virology* **2008**, *379*, 87–96. [CrossRef] [PubMed]
2. Maeda, S.; Majima, K. Molecular cloning and physical mapping of the genome of *Bombyx mori* nuclear polyhedrosis virus. *J. Gen. Virol.* **1990**, *71*, 1851. [CrossRef] [PubMed]

3. Li, J.Q. Summary table of nationalsericulture production in 2019. *Chinses Seric.* **2020**, *41*, 70–72.

4. Xu, A.Y.; Qian, H.Y.; Sun, P.J.; Liu, M.Z.; Lin, C.Q.; Li, G.; Li, L.; Zhang, Y.H.; Zhao, G.D. Breeding of a new silkworm variety "Huakang 3" with resistance to *Bombyx mori* Nucleopolyhedrovirus disease. *Sci. Seric.* **2019**, *45*, 201–211.

5. Xu, A.Y.; Lin, C.Q.; Qian, H.Y.; Sun, P.J.; Zhang, Y.H.; Liu, M.Z.; Li, L. Breeding of a new silkworm variety "Huakang 2" with tolerance to *Bombyx mori* nucleopolyhedrovirus disease. *Sci Seric.* **2013**, *39*, 275–282.

6. Shi, M.N.; Bi, L.H.; Gu, J.D.; Fei, M.H.; Qi, G.J.; Wei, B.Y.; Huang, J.T.; Huang, L.L.; Su, H.M.; Meng, Y.Y. Breeding of a new highly resistant nuclear polyhedrosis disease silkworm variety Guican N2. *Guangxi Seric.* **2012**, *49*, 1–12.

7. Xu, A.Y.; Lin, C.Q.; Qian, H.Y.; Sun, P.J.; Zhang, Y.H.; Liu, M.Z.; Li, L. Brief introduction of new varieties of silkworm resistant to BmNPV. In Proceedings of the Tenth Domestic (Tussah) Silkworm Genetics and Breeding and Seed Breeding, Jilin, China, 19 July 2013; Chinese Silkworm Society: Jilin, China, 2014; pp. 184–185.

8. Zang, Y.L.; Hunang, L.; Wang, Y.W.; Wang, H.L.; Gao, H.J. A preliminary report on the feeding of silkworm varieties of "Huakang 1" and "Huakang 2" with tolerance to BmNPV disease in Taian. *Hebei J. For. Orchard. Res.* **2014**, *29*, 435–438.

9. Hu, S.Y.; Luo, C.B.; Sun, Y.P.; Yang, S.T.; Jiang, H.; Yang, B.; Yu, W.Z.; Huang, G.H. A brief report on feeding test of new silkworm variety Huakang 2 in Qingzhen city. *Chinses Seric.* **2016**, 24–27. [CrossRef]

10. Luo, P. Discussion on the prevention and control methods of silkworm nuclear polyhedrosis virus disease in subtropical regions. *Chinses Seric.* **2017**, *38*, 59–62,67.

11. Shi, M.N.; Tang, L.; Huang, H.Y.; Wei, T.X.; Pan, Z.X.; Tang, T.X.; Pan, Z.X.; Qi, G.J.; Pu, Y.X.; Mo, Y.X. Application of new silkworm variety 'Guican N2'. *Guangxi Seric.* **2015**, *52*, 27–31.

12. Tang, F.F.; Zhang, Y.H.; Shao, Y.L.; Zhu, F.; Bai, X.R. Virulence and phylogenetic analysis of *Bombyx mori* nucleopolyhedrovirus isolates from Yunnan, southwestern China. *Acta Entomol. Sin.* **2018**, *61*, 42–51.

13. Fu, J.Y.; Xi, Y.; Tang, M.J.; Yin, K.S. Study on the relationship between virulence and genetic structure of four wild isolates of *Euproctis pseudoconspersa* nuclear polyhedrosis virus. *J. Tea Sci.* **2011**, *31*, 289–294.

14. Fuxa, J.R. Ecology of insect nucleopolyhedroviruses. *Agric. Ecosyst. Environ.*

2004, *103*, 27–43. [CrossRef]

15. Tang, F.F.; Shao, X.L.; Zhong, J.; Zhang, Y.H.; Huang, P.; Dong, Z.P.; Liao, P.F.; Bai, X.R. A preliminary study on molecular identification of *Bombyx mori* nucleopolyhedrovirus strains. *Sci. Seric.* **2014**, 1030–1035. [CrossRef]
16. Zhou, J.B. Identification of Thai Strain of *Bombyx mori* Nuclear Polyhedrosis Virus and Analysis of *bro* Gene Family. Master's Thesis, Anhui Agricultural University, Hefei, China, 2012.
17. Xu, Y.P.; Cheng, R.L.; Xi, Y.; Zhang, C.X. Genomic diversity of Bombyx mori nucleopolyhedrovirus strains. *Genomics* **2013**, *102*, 63–71. [CrossRef] [PubMed]
18. Sun, J.C.; Jin, F.L.; Xu, X.Y.; Tan, P.C. Comparative study on median lethal dose of BmNPV to *B. mori* in different regions. *Guangdong Seric.* **2001**, *035*, 35–37.
19. Bai, X.R.; Ran, R.F.; Dong, Z.P.; Dong, J.H.; Huang, P. Study on virulence of BmNPV to *Bombyx mori* in different areas of Yunnan. *Southwest China J. Agric. Sci.* **2010**, *23*, 2098–2101.
20. Wang, X.; Huang, X.H.; Jiang, M.G.; Tang, L.; Dong, G.Q.; Huang, S.H.; Shi, M.N.; Pan, Z.X. Epidemic factors of *Bombyx mori* hemolymph-type septic disease in Guangxi and its molecular phylogenetic analysis. *J. South. Agric.* **2020**, *051*, 669–676.
21. Bankevich, A.; Nurk, S.; Antipov, D.; Gurevich, A.A.; Dvorkin, M.; Kulikov, A.S.; Lesin, V.M.; Nikolenko, S.I.; Pham, S.; Prjibelski, A.D. SPAdes: A new genome assembly algorithm and its applications to single-cell sequencing. *J. Bioinf. Comput. Biol.* **2012**, *19*, 455–477. [CrossRef] [PubMed]
22. Massouras, A.; Hens, K.; Gubelmann, C.; Uplekar, S.; Decouttere, F.; Rougemont, J.; Cole, S.T.; Deplancke, B. Primer-initiated sequence synthesis to detect and assemble structural variants. *Nat. Methods* **2010**, *7*, 485–486. [CrossRef] [PubMed]
23. Kang, W.; Suzuki, M.; Zemskov, E.; Okano, K.; Maeda, S. Characterization of Baculovirus Repeated Open Reading Frames (bro) in Bombyx moriNucleopolyhedrovirus. *J. Virol.* **1999**, *73*, 10339–10345. [CrossRef]
24. Mckenna, A.; Hanna, M.; Banks, E.; Sivachenko, A.; Cibulskis, K.; Kernytsky, A.; Garimella, K.; Altshuler, D.; Gabriel, S.; Daly, M. The Genome Analysis Toolkit: A MapReduce framework for analyzing next-generation DNA sequencing data. *Genome Res.* **2010**, *20*, 1297–1303. [CrossRef] [PubMed]
25. Cingolani, P. A program for annotating and predicting the effects of single nucleotide polymorphisms, SnpEff: SNPs in the genome of Drosophila melanogaster strain w1118; iso-2; iso-3. *Fly* **2012**, *6*, 1–13. [CrossRef]

26. Zhou, J.B.; Li, X.Q.; De-Eknamkul, W.; Suraporn, S.; Xu, J.P. Identification of a new *Bombyx mori* nucleopolyhedrovirus and analysis of its *bro* gene family. *Virus Genes* **2012**, *44*, 539–547. [CrossRef]
27. Herniou, W.A.; Olszewski, J.A.; Cory, J.S.; O'Reilly, D.R. The genome sequence and evolution of baculoviruses. *Annu. Rev. Entomol.* **2003**, *48*, 211–234. [CrossRef] [PubMed]
28. Qian, H.Y.; Li, G.; Zhao, G.D.; Liu, M.Z.; Sun, P.J.; Xu, A.Y. Identification of new silkworm vrieties resistant to nuclear polyhedrosis Virus to BmNPV. *Chinses Seric.* **2017**, *38*, 15–22.
29. Huang, C.; Fan, X.H.; Jiang, Y.H.; Song, D.Z.; Gao, L.Q.; Huang, Q.G.; Lai, Z.P. Anti-tumor effect of Newcastle disease virus strain D817 against nude mouse xenografts of human colon carcinoma. *Zhonghua Zhong Liu Za Zhi Chin. J. Oncol.* **2009**, *31*, 490.
30. Wang, Y.Y.; Lu, J.; Liu, H.; Ye, R. Comparison of the infectivity of three virus strains of the genus A-coronavirus to cells derived from different hosts. *China Sci.* 2013. Available online: http://www.paper.edu.cn/releasepaper/content/201301-1114 (accessed on 26 September 2021).
31. Lu, W.Q.; Wei, P.W. Changes and Comparison of virus concentrations of six SMV strains in different soybean varieties. *Soybean Sci.* **1991**, *6*, 27–34.
32. Ma, X.C. Sequence Analysis of Genomes from Three Nucleopolyhedroviruses. Ph.D. Thesis, Zhejiang University, Hangzhou, China, 2006.

Index

V

W

X

Y

www.ingramcontent.com/pod-product-compliance
Lightning Source LLC
LaVergne TN
LVHW071246210926
848486LV00008B/357